The Crest of the Wave

The Crest of the Wave

Adventures in Oceanography

Willard Bascom

1817

HARPER & ROW, PUBLISHERS, New York

Cambridge, Philadelphia, San Francisco, Washington

London, Mexico City, São Paulo, Singapore, Sydney

FIRST EDITION

Designer: Kim Llewellyn

Copy editor: Ann Finlayson

Library of Congress Cataloging-in-Publication Data
Bascom, Willard.
The crest of the wave.
I. Oceanography—History.   I. Title.
GC29.B37   1988   551.46'009'04   87-46114
ISBN 0-06-015927-8

---

88 89 90 91 92 RRD 10 9 8 7 6 5 4 3 2 1

Dedicated to
four old friends who were the leaders of oceanography during much
of the period covered by this book:

Dr. Columbus Iselin,
Director, Woods Hole Oceanographic Institution

Dr. Maurice Ewing,
Director, Lamont Geological Observatory

Dr. Roger Revelle,
Director, Scripps Institution of Oceanography

Dean Morrough P. O'Brien,
Dean of Engineering, University of California at Berkeley

Contents

Illustrations appear following pages 114 and 306.

$\mathcal{A}$ppreciation

In the year and a half it took to write this book, many persons were contacted for help in getting the stories straight, the dates correct, and the quotes confirmed. Some furnished references or reminded me of stories; others willingly read early drafts and made suggestions for additions or deletions.

Some of these friends are mentioned as participants in the stories; others were onlookers or librarians, or contributed in other ways. By far the most helpful was my wife Rhoda, who not only insisted that I set these stories down but who witnessed some of the events and reminded me of details, but typed many sets of revisions into her word processor. She is as brave, kind, and happy a partner as any explorer ever had.

Thanks also to the next generation: Rodney and Sarah, Mercedes and Otto, who were no more disturbing than a herd of sea lions while I was concentrating.

At the Scripps Institution of Oceanography many helped by reading early drafts and commenting, including Drs. Roger Revelle, Walter Munk, Gustaf Arrhenius, Bob Livingston, Bill Van Dorn, and Dru Binney. The University of California at Berkeley engineering group, Dean M. P. O'Brien and Professors Bob Wiegel and Joe Johnson, were most generous with their recollections and suggestions. The Woods Hole Oceanographic staff, especially Allyn Vine, Carolyn Wynn, Bill Dunkle, and Charles Innes, contributed suggestions or references. John Coleman and Dr. Douglas Cornell, formerly of the National Academy of Sciences staff, reviewed some sections and straightened out my recollections of the events of long ago.

Some of my old associates at Ocean Science and Engineering, Inc., who commented on early stages of the manuscript and contributed anecdotes.

These included my brother, Bob Bascom, Ed Horton, François Lampietti, Skip Johns, Jack Mardesich, Peter Johnson, and Jeff Savage.

Captain Doug Fane, Cresson Kearny, and Roberto Frassetto read sections and made both factual and literary corrections relating to underwater swimmers. So did Dr. Jerry Cohen, and Prof. T. D. Luckey, who commented on weapons effects and metals toxicity. Al Loomis of the Jet Propulsion Laboratory provided the photo of an SAR scan and Robert A. Carlyle of the Navy those of the nuclear weapons tests. Brian Bielmann of *Surfing* magazine took the cover photo.

Dr. Alan Mearns and Bill Garber reviewed and touched up the chapter on ocean pollution. Harold Goodwin, author of dozens of popular science books, contributed ideas and enthusiasm.

Finally, my apologies to any who contributed but have been inadvertently omitted from this list.

The comments from all the above could be characterized as encouraging as well as helpful. Everyone seemed to be glad that many of these previously untold stories would soon come to light. For that I thank them particularly. However, I alone am responsible for any errors that may remain, unnoticed, in the text.

—W. B.

Preface

Oceanography is not so much a science as a collection of scientists who find common cause in trying to understand the complex nature of the ocean. In the vast salty seas that encompass the earth, there is plenty of room for persons trained in physics, chemistry, biology, geology, and engineering to practice their specialties. Thus, an oceanographer is any scientifically trained person who spends much of his career on ocean problems.

Until World War II, American oceanography consisted mainly of a few marine biologists based at the Scripps Institution of Oceanography in La Jolla, California, and the Woods Hole Oceanographic Institution in Massachusetts at the southern end of Cape Cod. These fellows went to sea on old sailing yachts to collect plankton samples, measure temperature and salinity and tow trawls or dredges along the bottom. The Navy, except for a few farsighted men, still showed little interest in the general scientific study of the ocean. Few members of the public had heard the word "oceanography."

Then, quietly, during the 1940s and 50s scientists of all breeds began to feel the rising swell of interest and to appreciate that new opportunities in the ocean sciences were opening. This new generation soon demonstrated that it was possible to probe the ocean and the rocks beneath it with sound, to link the movements of ocean and the atmosphere, to take samples of the bottom that went deeper and deeper into geologic history, to make better measurements of the motions of the ocean, and to build improved ships and equipment.

Oceanography gathered momentum because it looked like it a fun way to spend one's life. An eagerness to become involved in this new and exciting occupation swept like a tidal wave across academia, industry, and government. Recruits poured in, and new niches were created that suited each one's special talents.

Now there are tens of thousands of ocean scientists in dozens of modern laboratories. They are equipped with modern vessels, computers, instruments, aircraft, satellite links, deep-diving submarines, and drilling ships. Available funds increased from a few hundred thousand to many hundreds of millions of dollars a year. The Navy's involvement in oceanography, stimulated by its need to operate (and catch) new types of fast, silent submarines, increased by two orders of magnitude. The world fishing industry tripled its catch, and the oil industry came to depend heavily on offshore production. The general public became interested in skin diving, sea animals, underwater archaeology, and coastal conservation.

This book is about the adventures of some of the men who first explored the new opportunities in ocean sciences, including military sciences, and caused the onrushing wave of oceanography to rise to a higher peak than anyone expected. When that happened, all those involved were swept along on its crest.

I

Surveying the Surf

BECOMING AN OCEANOGRAPHER

The door to oceanography opened unexpectedly for me in Spenger's Seafood Restaurant, Berkeley, California. It was October 1945 and I, a young mining engineer just arrived from the mountains of Colorado, was dining informally with some new friends. Six feet tall, slim, hard and dark-haired, I felt freed by the recent end of World War II to try something new in some different part of the world. Across the dinner table was John Isaacs, a young civil engineer who was employed by the University of California to study waves under a Navy contract. His wife, sister, and infant daughter Cathy filled the other places.

Isaacs was a big, good-natured fellow with pale skin, blue eyes, and a great shock of blond hair, who was a couple of years older than me. He was interested in practically everything on earth and liked to talk about it, but his first love was the ocean. From the first he fascinated me with stories about his life as a commercial fisherman along the Oregon coast. John's telling may have made the events seem a little larger than life. His waves were higher, his escapes narrower, and his encounters funnier than other witnesses might have confirmed. No matter; I was charmed by tales of a life-style quite different from anything I had experienced.

In return he listened to my stories about working in the mines and tunnels of the Rocky Mountains. After an hour of rapid crosstalk, we discovered similarities in our tastes for music, books, adventure, and lifetime goals. Best of all, we had a similar sense of humor.

1

A waiter approached our table, balancing a tray holding a new bottle of wine and some stemmed glasses. Two-year-old Cathy, flailing about in a high-chair next to me, knocked one of the glasses off the tray. In an unthinking reaction I reached out and caught the glass in midair. Almost as quickly John stabbed his finger at me and said, "How would you like to work for the University of California?"

"Doing what?"

"The beach surveying party is headed for the north coast next week, and our engineer just quit. We need a replacement."

That was not much of an explanation of a career opportunity, and it was not immediately apparent to me why anyone would want to survey beaches. I had never seen the Pacific Ocean and knew nothing of waves or beaches or university research, but John's stories made the expedition seem adventurous, and this was a chance to trade the wet dark life underground for the bright outdoors.

"Sure. I'll give it a try."

The following morning a clean-cut young professor of hydrodynamics named Dick Folsom looked over the form I had filled out and asked a lot of questions: "Why don't you have a college degree?" The answer was, that despite good grades, a difference of opinion with the president of the Colorado School of Mines had caused me to leave a few months before graduation. "Why is it you worked for short periods in so many mines?" That was the best way to learn the mining business. "Why weren't you in the armed services?" They wouldn't take me because my jaw had been broken in a mine accident, and the sides moved independently.

Finally Professor Folsom tried me on technical matters. "What's the hydraulic radius of a pipe?" I replied that I didn't have the faintest idea. He could see at once that that was correct, and he relaxed, secure in the knowledge that the new man would not be much threat to the faculty of the Fluid Mechanics Laboratory. Finally he sighed, thought about the immediate need for an engineer, and with some reluctance hired me as a staff member of the University of California's engineering department for $250 a month. He repeatedly emphasized that the generosity of the offer was because of the temporary nature of the job. I agreed wholeheartedly with the temporary part, thinking that, after a look at the Pacific coast, I might go back to the mining business.

We grinned and shook hands. "Welcome," he said, "to the Waves Project." Then he sent me to look at the equipment we would use, stored at the old swimming pool. Behind a high wooden fence the pool area contained a fascinating collection of junk, including the wreckage of the tiny airplane in which Admiral Byrd had flown over the South Pole and the residue of many best-forgotten experiments.

A large white, generally oval, ring of plaster whose uneven surface rose above the water here and there, occupied much of the pool. Hanging over its center on a piece of rope was a circular iron plate about an inch thick and 2 feet in diameter. Occasionally one of the fellows in white coats working around the pool would holler: "Start all the recorders." Then he would reach out with a blowtorch on the end of a long pole and burn through the rope. The plate would fall, making a loud smack and a wide splash; waves would radiate outward in all directions and wash over the plaster humps. Then the white coats would fish out the plate and drop it from a higher point to get a bigger splash and higher waves. Puzzled, I asked the head man, "What the heck are you guys doing?"

He tensed up. "Are you cleared?" I looked blank. He glanced suspiciously about to make sure no one could overhear. "Don't tell *anyone* what you've seen. This is secret stuff."

That was my first contact with military security. I was nobody's fool; why would I mention that some guys in white coats spent their days splashing water in a junkyard? Some other guys in white coats might come to take me to the funny farm. It takes a while to become accustomed to the curious ways of university research and the workings of the military security system, but in time such odd carryings-on seemed normal. (Later a "confidential" clearance arrived, and I was in it too—whatever "it" was. But months went by before I figured out what was going on at the old pool.)

Presently John Isaacs showed up to tell me about the principal possessions of the beach party: two dukws. A dukw is an amphibious truck. It was invented in 1941 by Rod Stevens, Dennis Puleston, and Palmer Putnam, three engineers who wrapped a 30-foot-long sheet-metal hull around a General Motors six-wheel truck and added a propeller. Dukws were used during WWII to move military cargoes ashore in amphibious operations. John explained that dukws make wonderful surf boats, because their wheels hang down and give them extra stability in big waves so they won't flip over in a breaker. Unfortunately, he added, they can also swamp and sink rather easily. Dukws can also slog their way over soft sand and easily make the transition to rolling on hard highways because the tire pressure can be controlled by valves at the driver's seat. In soft sand the driver lets the air out of the tires to give them more bearing and traction; a built-in compressor reinflates them for hard ground. With power in all six wheels and a lot of low gears, a dukw can go nearly anywhere.

The beach party departed Berkeley a few days later, and since it was temporarily without Army-furnished dukw drivers, John and I each drove a dukw north along the coast to Humboldt Bay, where we would start work in the winter waves. The others, including Marine Lieutenant Dickey, Navy photographer Wilson, and some survey helpers tagged along in cars. Our party's arrival there was timed to coincide with that of the biggest waves the North Pacific could muster.

Dukws are impressive vehicles on the highway; 8 feet wide and high and 30 feet long, they tower formidably over an ordinary car. With their wartime olive drab paint and wedge-shaped bow, the thin-skinned dukws were perceived by innocent motorists to be a heavily armored vehicle and given as wide a berth as possible. Sometimes cars would pull off the road entirely to get out of the way or to gape at us disappearing in the distance at 50 miles an hour. Bluffing cars was easy; the real challenge came from the big log trucks that dominated the coast highway. There is a lot more mass in a load of logs than in a dukw, but the log-truck drivers didn't know that as we approached each other at 100 miles an hour on the often slick and narrow pavement. There were places on the old highway through the redwoods where the road was squeezed between two big trees, and on one occasion, while passing a log truck in such a place, my dukw ticked bark on both sides at the same instant.

THE SURF AT TABLE BLUFF

Humboldt Bay, 300 miles north of San Francisco, is shallow and muddy, about a quarter mile wide and three or four miles long. A ridge of sand perhaps 200 yards wide separates the calm bay from the ocean. Near the middle of the bay this sandy barrier is dissected by a pair of rock jetties that contain the entrance channel through which tidal currents rush in and out twice a day. At the south end of the bay, a green meadow slopes upward toward the ocean and terminates against the beach in a cliff more than 100 feet high, known as Table Bluff. Atop the bluff there was a light and two tall radio towers operated by the Coast Guard. The surf below the lighthouse was the place selected for our party to make definitive measurements of the encounter between the great North Pacific swell and one of the roughest beaches known.

The beach party left for the northern coast encouraged by a pronouncement from the big boss, Mike O'Brien, dean of engineering at Berkeley: "If you can work in the winter surf north of Cape Mendocino, you can work anywhere." We accepted that statement as a fact without thinking through its implications. He did not mention that we might drown while finding out.

Late on a rainy November afternoon, our expedition arrived at a motel on the inner side of the bay. John stood atop the dukws and pointed across the bay and the sand spit to exploding masses of white water. "Out there," he said, "is where we will work. Those are *big* breakers—maybe thirty feet high." It was the first time I had seen the Pacific Ocean and, like the others, was not sufficiently experienced to make a suitable comment. What did we know? Thirty feet didn't sound big for an ocean the size of the Pacific, but that depends on where you are, relative to the wave.

The following day Isaacs, Lieutenant Dickey, our Marine Corps liaison officer, and I visited the beach for a closer look. Dickey had left one arm on the beach at Iwo Jima and won a medal for having unloaded badly needed trucks from a ship into landing craft amid high waves. He was not especially proud of this accomplishment, pointing out that "it cost six trucks and two landing craft to get one truck ashore." Nor did he claim any knowledge of waves: "I was too busy with the war to think much about 'em."

The sky was dark, and it was raining and blowing as we picked our way over the crest of a driftwood-strewn sand ridge to confront the Pacific. To both north and south, a rough spume-crossed beach stretched away to foggy vanishing points; we could see about one quarter of the fifteen miles of desolate, unbroken beach between the Eel River and the Humboldt Bay jetties. In that entire length there was not a person or a house to interrupt the desolation. The top of the sand spit behind us was crowned with a jumbled mass of driftwood. Large logs, some escapees from log rafts or lumber mills, were represented; so were whole trees with stubs of branches still attached that had floated down rivers in flood and had been heaved by the largest waves above the highest tide. This bristling rampart was tangible evidence of how waves shaped beaches.

We squinted into the gusting wind to avoid the flying sand grains, and the water trickling down our cheeks tasted salty. The surf roared at us, which is to say that the wide spectrum of frequencies created by all the waves crashing, colliding, swashing, and releasing bubbles produced a high volume of white noise—a hiss of astonishing proportions broken only by the occasional crack of a single breaker slapping the water extra hard. We could feel the beach shake under our feet when a long plunging breaker collapsed.

Then we became aware of a distinctive smell. The flavor of the open sea on the Pacific Northwest is not to be confused with that of Southern California or of East Coast harbors or of tidal flats. It is a not unpleasant odor but a natural perfume, probably flavored by plankton that are churned into a froth to create an aerosol that is carried shoreward by the wind.

When the violent gusts subsided, we opened our eyes and tried to make some sense out of the mass of breaking waves churning the half-mile-wide band of water nearest the shoreline. As we huddled watching the white confusion, Isaacs decided that we inlanders were entitled to a short background lecture on what was going on.

"Most waves are caused by wind blowing across the water surface. The stronger the wind and the longer distance it blows, the bigger the waves. In the area where they are generated, waves can be large and rough, but as they move out from under the wind that created them they become long low undulations called swell.

"Swell can travel long distances with little loss of energy. Only the wave form moves forward; after it passes, the water particles end up about where they were before. Individual waves are primarily described by their period, which is the time in seconds for successive crests to pass a fixed point such as a rock.

"As swell approaches a shore, it feels bottom, and as the depth continues to shoal, the crests of the waves become more peaked. Finally, when the water depth is only about 1.2 times the wave height, the wave becomes unstable. As the water at the bottom slows, the crest rushes forward, and the wave breaks. If the top curls over an air-filled tunnel and falls into the trough ahead, the breaker is said to 'plunge'; if the top merely turns white and tumbles down the wave front, it is a 'spilling' breaker."

Our inexperienced eyes still saw only confusion, so we climbed Table Bluff to get a better perspective. Before us there was spread a grand panorama of Pacific violence, a texture of white turbulence atop the green-gray fabric of sea and sky. We could make out three broad strips of broken white water, each with green water between. Beyond, as far as the eye could see, a procession of long-crested waves advanced and furiously harled themselves onto the submerged outer bar. Then they would re-form in the deeper green water inside that bar and break a second time on the inner bar. Finally, on reaching the beach face, they would break one last time, and the last dregs of energy would be expended in a thin white swash; when this slid back down and ended in a muddy swirl called the backrush, the wave was finished.

Although John patiently explained the inner workings of the waves, we did not then appreciate such details. "A man wouldn't live long in that surf," he added. That certainly seemed to be true but who would be foolish enough to go there anyway? Who indeed?

The beach party established a wave observation post atop Table Bluff, where we could observe sea and surf conditions, time wave periods, and measure the height of breakers on the outer bar. In order to capture details of the outer breakers a half mile offshore, we used long-focal-length aerial cameras borrowed from the Navy. These were bulky and heavy, but they had excellent lenses, used negatives 8 inches square, with 200 exposures per roll. An internal clock showed in the corner of every picture, so we knew exactly when it was taken. This was very advantageous in following a sequence of events or for calculating the height of the tide. Twice a day we photographed the surf zone and recorded its vital statistics from that observation post at the edge of the bluff. Later the height of the waves could be measured on these pictures.

Sometimes, on days when the waves were large and the sky was clear, the Navy would send reconaissance planes to take a sequence of vertical photos of the surf zone. From these we could obtain water depths on and between the submerged sand bars by measuring the speed of advance of the wave front on a series of pictures taken at 3-second intervals.

Having estimated wave height and water depth from the safety of shore, we felt honor bound to check up on ourselves by making direct measurements of the profile of the sand under the surf to find out if those figures were correct. This was more than a point of pride; one of our jobs was to determine the exact configuration of the sand bars that caused the waves to break when and where they did. Did large waves in the real ocean do the same things as those in a model tank? That meant surveying the surf zone during or immediately after large waves were breaking, because when the waves changed, the shapes of the sand bars changed.

We learned a simple way for a person on the beach to measure the height of the big breakers, even though they are far offshore. Just stand at a level on the beach where your eyes are exactly aligned with both the high point of the breaker and the distant horizon; then the height of the wave is the vertical distance between your eyes and the backrush. We would often do that before risking a run through the surf, but rarely did we do it long enough to get the highest breakers, because wave heights vary so much. We found that about every 3 minutes there would be a series of three higher-than-average waves. But sometimes, because the three highest in one group were not the same height as those in the next group, we got into trouble.

The line along which we would take soundings was established by pairs of red and white banded range boards 10 feet high that were visible a mile at sea. In the system that Isaacs had developed, a dukw moved shoreward along that line while a surveyor, a thousand feet down the beach, would follow it with the telescope of a surveyor's transit. The leadsman in the dukw would heave a sounding lead out ahead and call, "Mark" into a walkie-talkie radio. The surveyor would see the splash of the lead and read the angle between it and the base line. Then, as the dukw came even with the lead and the line became vertical, the leadsman would tap the lead on the bottom, read the water level on the line, and call the depth into the radio. The recorder on shore would write the angle and depth in a notebook, and these would later be converted to profiles and contour maps.

Such surveying is easy in low waves. But in the winter breakers at Table Bluff, it was a stimulating experience, because a dukw has about 2 feet of freeboard, and the breakers were often 10 to 15 feet high, occasionally more. There was an excellent chance that the leadsman (me) would get cold seawater down the back of his neck, not to mention the possibility of drowning.

First it was necessary to get out through the lines of breakers to the relative calm beyond, where a survey run could be started. On the way out, the driver and I were forcibly reminded that the energy in a breaking wave is proportional to the square of its height. That is, a 14-foot breaker on the outer bar has four times as much energy as a 7-foot breaker on the inner bar. The driver and I often made that kind of calculation in our heads when the breakers ahead were unexpectedly large. That was because it was our considered opinion that the chances of *not* returning to the beach alive increased along with the energy of the breaker. If the dukw foundered or rolled over on the outer bar when the breakers were large, there was a very small chance that either of us could have made it ashore.

On days when the outer breakers were relatively low, which meant 10 to 12 feet, the dukw would buck its way out, usually taking two or three breaking waves head-on at each of the two bars. Since the top of a dukw cab is only 4 feet above the still-water line, driving deliberately into an onrushing wall of water two or three times higher than your head takes either nerve or poor judgment. We had a lot of both.

It was much better to take a breaker head-on than to turn at the last minute and take the chance of being hit broadside and perhaps being rolled over. As dukw and wave collided at about 20 miles per hour, there would be a resounding metallic thud. The hull would shudder, forward motion would be momentarily reversed, and the impact would fling a huge white splash in all directions. There were small holes where the canvas roof joined the corners of the windows through which water would squirt like a high pressure hose on the men inside. Green water crashing on top would pour onto the grating between the cab and the cargo compartment and the automatic bilge pumps would start throwing 3-inch jets of water into the air on both sides. It was not unusual for the canvas seams in the roof to split or for the windshield wipers and rear-view mirrors to be torn away by these impacts. The original light-steel frames that supported the canvas over the cargo compartment soon buckled under plunging breakers and were replaced with heavy pipe.

Once the dukw was outside the breaker zone and aligned with red-striped range markers, we would check to see if the radio was working and make sure the transit man was ready before heading landward. Then we would select what we hoped would be the lowest of the oncoming waves for a ride across the outer bar. Once committed, there was no possibility of turning back.

When that wave overtook us, the dukw would begin to surfboard. I would heave the sounding lead ahead and to one side into the trough; as we passed it, and the line became vertical, I could read the water depth in the trough. Then the dukw's stern would begin to rise as the crest overtook us, and the bow tilted downward. As its slope increased, our seagoing truck would begin to slide down the front of the breaker that was peaking behind us, and the driver would fight desperately with throttle and rudder to maintain its position at right angles to the wave front. While running before the wave, we were traveling at twice normal speed, often with the bow plowing the slick green water of the trough ahead, sometimes up to the windshield.

While balancing under this incipient waterfall, I would estimate the height of the wave that was about to come crashing down, add one third of that (5 feet for a 15-foot wave) to the trough depth, call the answer into the microphone, and duck. Then the reaching crest of the plunging wave would collapse on us, not quite capsizing the dukw.

Now the bow would rise, and the wave would pass underneath, lifting us on its shoulders. The dukw would become level again, its deck 8 feet or so above the surface of the trough just ahead. We found it exhilarating to ride the crest of a wave, driven forward by the pressure of the moving water against the wheels and axles that hung below. After a few seconds atop this rolling pinnacle of water, the wave would pass on, and the dukw would slide down the spent breaker's back side.

Now the idling engine would begin to roar as the driver redlined the tachometer to get clear of the bar before the next wave broke. The total time elapsed for a ride on an outer breaker was not much more than 15 seconds, but those were long seconds. By comparison the breakers on the inner bar were pretty tame.

At the end of the run as the dukw neared the beach, the wheels would touch bottom gently as the springs resumed their load. The driver would shift power from the propeller to the wheels, and with gears grinding and canvas dripping, our dukw would slosh out of the water. Once safely up on the beach the driver would threaten to quit, logically arguing that his life was worth more than a beach profile. I would agree sympathetically, check with the transit party by radio to see if they had gotten the data, and after a short breather, we'd agree that this was really fun and head back out to sea to run another line of soundings farther down the beach.

Our two principal drivers on most of the work, Rex Goodwin and Bill Lloyd, were most amiable and loyal fellows who did many things beyond what could be recompensed by a paycheck. There is no doubt that they held my life in their hands on many occasions and cussed me for risking theirs. However, since I was always in the dukw on surf runs, and each of them was only there half the time, neither could claim that his risk was as great as mine.

On days when the surf was low and the wind was down, John and I would stroll along the beach, looking at beach features, strange flotsam, and sea life. We learned that firm sand on the beach face usually indicated erosion, and soft sand beneath ones feet suggested the beach was growing. Plainly large waves cut the beach back, sometimes leaving a vertical scarp several feet high, and small waves brought new sand, but exactly where the boundary was between the two we did not then know. We noticed that each thin white swash of water onto the beach pushed a few of the larger sand grains ahead of it, and these remained to create a graceful swash mark at the highest point reached. When this uprush overruns previously dried sand, the newly wetted surface sand will trap air beneath in low bubbles or domes that are fun to step on and collapse.

On these calm days, there were only low plunging breakers on the inner bar. Lines of pelicans moving down the coast would glide parallel to a break-ing crest, gradually losing altitude until they reached the end of the collapsing curl. As it squeezed air out the end of the tube, they would be lifted by that jet and glide seaward to catch the next wave. They traveled for miles without seeming effort, using only wave power, while beyond them, mile-long files of black cormorants had to work hard, flapping their short wings, to stay air-borne. At the water's edge sandpipers, knee deep in the swash, would poke long bills through fast-flowing backrush water to snag tiny animals from the sand.

This was an entirely new and very satisfying life for a fellow who, only a couple of months before, had been working underground over 10,000 feet above sea level! In John Isaacs, I had found an entertaining teacher and a good friend, who had an instinctive insight into the ocean's workings. Months after beginning work, I first heard the word "oceanography." The explanation of what oceanographers did was vague enough and the playing field grand enough to encourage a free spirit like me to consider changing careers. The timing, it turned out, was perfect.

After surf surveying became a normal daily operation, the experience of riding the big breakers became a terrifying source of pleasure, something like mountain climbing. None of us wore life jackets, although they might have helped if our tin boat had gone down; instead we developed a fatalistic attitude.

For a few months an excitable fellow named Martin drove for the beach party, and although we sometimes had misgivings about his judgment, he was generally reliable in the surf. On one occasion when a big breaker shoved the dukw rudely to one side, buried half the windshield, and seemed about to roll it over, Martin panicked and tried to scramble out of the driver's seat, hollering something about "bailing out." With heavy-weather clothes on, in big surf a half a mile from the beach, that would have been suicide, so I, standing on the cross walk behind the cockpit, shoved him back. "Shut up and drive," I suggested, gripping his shoulder firmly and waving a fist in front of his face. That simple explanation, reinforced by a refreshing shot of cold water from the wave, was sufficient to keep him in the cockpit, and we reached shore safely. Later, another of his lapses cost us a dukw.

Our friends at the Coast Guard lighthouse and radio station sometimes watched us from the cliff, and eventually one of these fellows asked to take a ride with us through the surf. According to the account he gave the others later, he was terrified by the experience, and this enhanced our reputation as fools: "Why do you guys risk your necks every day to find out how deep the water is in a place where no one will ever take a boat?" Our explanation sounded odd, even to ourselves: We wanted to understand the physics of waves and beaches, and our only explanation was scientific curiosity. That was the first hint I had of the difficulty of explaining ocean research to outsiders. Shortly thereafter Isaacs received a formal message from the commander of the Coast Guard lifeboat station five miles away inside Humboldt Bay. "Don't expect us to come rescue you if you get in trouble." We didn't. We would have drowned long before they could have launched a boat.

Like most scientific studies, the purpose of the Waves Project was to improve man's understanding of the world around him. We were pioneers because not much was then known about the mechanism of how breaking waves move sand or how changes in wave characteristics cause beaches of various materials and orientations to build, rearrange themselves, or disappear. Hundreds of millions of dollars worth of shoreline installations depend on understanding what will happen when man interferes with waves and natural shoreline processes by building structures such as jetties and breakwaters.

THE NORTHERN BEACHES

In January, under dark skies and intermittent rain, the beach party headed north to study the coast of Oregon. It was a new world for all of us except John Isaacs, who had once been a forest ranger and commercial fisherman there. A year before he had led an expedition up this coast; this time we were resurveying those beaches to see what had changed.

Although we found Oregon beautiful in a green and rainy sort of way, with its soggy meadows and dense dripping forests, the gloomy atmosphere was a bit depressing. The small coastal towns exuded a feeling of mossy decay; those were lean times in the lumbering and fishing business.

That coast is intersected by many rivers, and until the old ferries were replaced by modern bridges in the late 1930s, no convenient coastal highway existed. The side roads that branched from that one artery were largely single-lane mud tracks with deep ruts made by logging trucks. Woe to those who encountered a loaded log truck barreling toward them while they were trying to reach a beach.

Our job was to collect data on breakers and beaches in the rough winter months and make comparative surveys that would lead to the understanding of beach processes. The Oregon coast was a good place to work, because there were so many kinds of beaches with different exposures to waves.

The beach at Table Bluff had served as my standard; indeed, it was the only one I had seen. It is long and straight with intermediate-size quartz sand, and it faces directly into the North Pacific. Now we were able to compare it with many other variations. Some of the Oregon beaches were extremely flat, others were quite steep; some were light-colored and some were dark; some were hard enough for cars to race on, others were so soft the dukws had to grind along with lowered tire pressure. In small bays between rocky headlands, low tide might expose a flat hard beach made of fine-grained light sand. Behind and above it, awash only at high tide, there would be a steep beach of dark basalt cobbles placed there by storm waves and topped with driftwood.

After months on dunes and beaches, our party became confident that a duke could go just about anywhere, but we were surprised to find hidden quicksand amid the dunes at Coos Bay. The lead dukw had easily traversed a wide flat area, and we had seen nothing unusual about the appearance of the sand. Then loud blasts from the airhorn of the following dukw attracted our attention. Looking back, we saw that its hull was flat down on the sand surface, the wheels buried completely. It had been following exactly in the tracks of the first dukw—a practice we deliberately followed to avoid such problems.

Walking back we discovered that near the tracks of the first dukw, the formerly hard sand surface now acted like a rubber membrane over a thick liquid. While our feet remained in contact with the sand, we could bounce up and down several inches, sending waves a short way across the sand surface. Under a thin cover of dry sand and organic material the sand-water mixture below was thixatropic, meaning that it was reasonably solid until vibrated by the passage of the first dukw, and it was liquefied by the vibration caused by the second one. Cussing the good fortune that brought us this unneeded piece of new scientific knowledge, we jacked, cribbed, laid board tracks, and winched from a safe distance. An hour or two after nightfall both dukws were rolling again.

A day or two later we were on the beach south of Cape Lookout where a lump of dark basalt called Haystack Rock rises about 100 feet vertically from a white sand beach. Clearly the rock is a temptation to climbers, but according to local legend, it had never been climbed. It did not look very difficult, and I considered making the attempt until Isaacs pointed out that the upper part of Haystack Rock is inhabited by thousands of nesting sea gulls—big, aggressive birds with a wingspread of 4 or 5 feet. He claimed that the last man who tried to scale the rock was attacked by gulls and driven frantic by their picking at his hands and face until he fell to his death. That story had a certain logic I did not wish to test, and so I left that bit of glory for someone else.

Then we moved on north to the entrance to Tillamook Bay, a place that illustrates man's frustrations in dealing with coastal sand movements. Many years ago this entrance, through which the bay and ocean exchange tidal water twice each day, was a collection of sand bars braided with shallow channels. There was not enough depth in any one of these for fishboats or coastal schooners to navigate except at highest tide. It was easy to see that sand being moved southward along the coast by the littoral current was responsible for clogging the channel, so the Army Corps of Engineers decided to build a rock jetty perpendicular to the shoreline along the north side of the entrance. This would dam the inflow of sand from the north and maintain a single deep channel of water with navigable depths close to the jetty.

When completed, the structure did both of those things very nicely, and for a while everyone was pleased. Unfortunately there were some other, long-term effects, and Tillamook is now recognized as a classic case of a cure that arguably made the situation worse.

As the result of building the jetty, the sandy beach north of the entrance grew ever wider while that on the south side disappeared. A summer resort, including hotel, swimming pool, and many houses were undermined and lost. Today the south shore on Kinchaloe Spit is made entirely of large basalt cobbles that form berms, bars, rip channels, and other beach features much like those seen in sand. Over the years Tillamook's jetty has been extended several times as the sand has filled its north side and then flowed around the end, causing the original problem to recur. This is an example of many similar situations around the world. In spite of years of research and experimentation with many schemes for building jetties and bypassing sand, there is no universally satisfactory solution that keeps an entrance channel open where sand is flowing along the coast.

Finally our party came within the influence of the Columbia River. This great river drains a huge basalt plateau in eastern Oregon and Washington. Before the dams were built on it at Grand Coulee and Bonneville, the Columbia in flood would release 1.25 million cubic feet of water per second into the North Pacific. This flow brought with it a lot of fine dark sand and constantly rearranged the sand already there, so the lower Columbia became well known for huge shifting bars, some named for the shipwrecks they caused. The river entrance is a place to be reckoned with. An old chart of the Columbia River entrance region shows sixty-two shipwrecks going back to the *Peacock* in 1841.

One can imagine how scary it must have been for the skipper of a square-rigged sailing ship to enter this unmarked and steadily changing maze of sand bars. It must have been equally terrifying for him to head back out to sea across the entrance bar, even when the wind was favorable. On the Columbia Bar, breakers 40 feet high have been reported in winter storms, an awesome force even for a large ship to encounter.

South of the river the beach on Clatsop Spit is a quarter of a mile wide at low tide and hard enough for ordinary autos to drive on; it is backed by dunes, ridges, and sand flats. This area has several claims to fame. First, on November 15, 1805, it was the terminus of the Lewis and Clark Expedition. Second, it is the last resting place of the *Peter Iredale,* a great iron sailing ship that had been becalmed just as it was about to enter the river in 1906 and slowly drifted on the sands. And finally, Fort Clatsop, a truly insignificant defense installation, was shelled by a Japanese submarine in 1942!

The beach at Clatsop is also a great place to dig razor clams. These critters, about 6 inches long with sharp elongated shells, live about a foot below the surface of the wetted beach. By sneaking up on a tiny hole, jamming a long narrow shovel known as a "clam gun" down alongside it, and making a fast heave, you may be lucky enough to catch one of the slippery devils. Often you are not. How these creatures can jet themselves downward through sand hard enough to drive a car on is another unsolved beach mystery.

The beach party located some of the survey stakes left from the previous year's survey near the wreck of the *Peter Iredale,* whose rusting bowsprit still stabs at the sky, and started to work. The survey of the upper beach was quickly finished, and since the surf appeared no higher than our normal maximum, we decided to make the undersea profiles at once and finish this beach so the party could push on north. Although it was late afternoon, dukw driver Bill Lloyd and I headed out through the surf. Just beyond the point of no return, it dawned on us that the outermost breakers on this very flat beach were much farther seaward than usual, and we had underestimated their height. There was nothing to do but smash our way out through the breakers, some of which must have been nearly 20 feet high, so we could turn around.

Once beyond the outer bar, Bill and I looked about to reassess our situation. We were nearly a mile at sea amid an endless procession of waves overhung by low gray clouds; it would be necessary to run a long rough gauntlet to get back to shore. Isaacs, watching us through the transit, came up on the radio, sounding a little nervous for our safety. "I can only see you when you go over the top of a big one. Be careful." It was a little late for that.

There was no wind, and a succession of textbook waves, all peaked in preparation for breaking, moved inexorably toward us. A 15-foot crest, give or take a few feet, would pass under us every 15 seconds, and as it heaved the dukw up, we could look around. Then we would slide back into a long flat valley of water that disappeared into the clouds until the next crest lifted us.

There being no reasonable alternative to a run for shore, the only question was, which wave should we bet our lives on? Every few minutes there are usually two or three waves larger than the others, and since every wave loses energy to the one behind it, our plan was to wait two more waves after the last big one in the set and then go for it. If we could time our run so that the big ones broke just ahead of us, we could ride in on the slightly smaller waves that followed. That was the theory.

The light was beginning to fade when Bill and I finally placed our bet. The wave we chose was not noticeably smaller than the others; when it reached its highest peak, just before breaking, I looked back at the great glassy underside of the curl and thought, "What the hell am I doing a mile from shore in a tin truck that's about to be crushed under an avalanche of water?" Bill's knuckles were white as he gripped the steering wheel; his eyes focused on the onrushing trough, and his lips moved silently.

We were indeed smashed harder and submerged deeper than usual by this surging mountain of cold water. Then the tin truck I had recently demeaned rose gallantly through the mass of water to its usual position, balanced atop the broken wave. When we landed, we were wet with sweat as well as seawater.

The strange thing is that although we cursed our own stupidity for having taken so great a risk for so small a reason as one more beach profile, we repeated that performance, with variations, a number of times over the next few years. Moreover, we always took soundings during those runs, reasoning that "as long as we're taking the risk we might as well do something useful."

There were times when a series of narrow escapes and drenchings made the drivers and me consider seeking some other kind of employment but we were determined not to be driven from the field of battle by danger and discomfort. On the rare calm days, the job was pure fun, and there was no reason to quit. So we stayed.

The Northwest Coast in winter can be a thoroughly miserable place to live and work. Our baggage, motel rooms, and nearly everything we encountered were constantly damp. Although we wore long johns and well-insulated heavy-weather clothes, these invariably got wet, and it was rare to feel warm and dry at the same time. We sympathized with the local lumbermen and fishermen who had to live with constant sog.

Years later John Isaacs was driving across California's Mojave Desert in the summer. He stopped to refuel at a gas pump in front of a lonely shack set about by scraggly mesquite bushes, all parched by the 110-degree heat. The operator of this oasis was plainly happy with his lot, and John couldn't resist asking why.

"Well," says the desert rat, "I worked twenty years as a gill-netter on the Columbia River, and I'm just starting to get dried out." Everyone in our group would have understood.

In subsequent years the dukws returned many times to the Oregon coast to repeat beach surveys, install wave recorders, and study coastal changes. Mostly this work was fun, but there were exceptions.

The coast highway winds along the side of Neahkahnie Mountain, a rugged headland about 30 miles south of the Columbia River. From this vantage point, several hundred feet above the sea, there is a spectacular view of the coast to the south, including the surf on Manzanita Beach. Since our job was to record the appearance of breaker zones under various wave conditions, the party stopped each time it passed to photograph the wide surf zone below. I would be driving a university car that contained the cameras, trailed by the two dukws driven by Rex and Martin. When I stopped, the dukws would park and wait patiently until I had the photos.

On one occasion the lead dukw parked on the road shoulder just behind the car, and the second dukw stopped immediately behind it. Below the shoulder the road fill slanted steeply for a few feet into a gently sloping meadow, 100 yards wide, knee-deep with ferns and flowers. This pastoral beauty ended abruptly against the sea air; below was a vertical cliff 200 feet high.

After getting the pictures, I returned to the car, put away the aerial camera, and waved the party onward. Rex signaled OK, and his dukw began to move; so did the second one. A moment later we realized it was driverless, for as it moved, it began to turn to the right, instinctively heading for the sea. It rolled down the embankment, gathering speed, and raced, up to its hull in flowers, across the meadow. By the time it reached the top of the cliff, it was doing at least 40 miles an hour, enough to launch it well out into space in a spectacular final leap.

At that moment driver Martin came dashing up gasping, "Did you see it?" I had, and I was mad. Nearly every day we had risked losing that machine in the surf and were prepared for that possibility, but losing it on (or off) a mountain would be hard to explain. Martin claimed that he hadn't had time to adjust the hand brake so he had parked it against the other dukw and gone off to pick flowers or something.

We followed the tracks down the meadow to as close to the edge of the cliff as we dared. In the midst of the boiling white water below, the only identifiable object was a black and yellow marker buoy that had been aboard the dukw. There was no reasonable way to get to the foot of that cliff for a closer look, so after five minutes of fuming and speculation, we tramped back up the road, glad that our valuables were carried aboard the other dukw.

From the next town I called Berkeley to tell Isaacs the story, When I told him we'd lost a dukw, John quickly said, "Was anybody hurt?" I said, "No," and continued with the story. At its end John said, "Are you sure you didn't hurt him?"

A familiar line in John Masefield's "Sea Fever" about "the flung spray and the blown spume" has a special meaning on the coast of Oregon.

On Tillamook Rock, several miles at sea off the southern end of Clatsop Spit, stands a lighthouse that was built long ago to guide mariners to the Columbia entrance and keep them off the dangerous coast. The base of the lighthouse is 90 feet above the sea, and the light itself is 50 feet higher. Spray goes routinely over the light, and during heavy winter swell, people often cluster on the nearest headland to watch the show. When a great wave smashes into the base of the cliff, the impact is like an explosion; the island shakes, and great sheets of white water go high in the air. This spray is not made up merely of water droplets; it carries with it rocks snatched by the waves from the cliff below. After the glass of the lantern had been broken several times, the Lighthouse Service installed steel gratings 135 feet above the sea to protect it from the "flung spray." Oregon spray is a thing to be reckoned with.

Spume is sea foam, caused by minute organisms in the water; in the churning surf these can act like a detergent to create suds. Early one morning while driving the coast highway behind Boiler Bay, we encountered a surrealist scene. Where the road circled inland around the bay, perhaps 50 yards from the beach, the forest floor was covered with soaplike bubbles of spume 2 to 5 feet deep. Except for the roadway itself, where preceding cars had cleared it away, greenish white suds extended from the beach into the woods beyond the road. We could not resist wading beachward through this diaphanous blanket, groping with our feet amid suds over the knees not knowing where the humps and hollows were, leaving behind a trenchlike trail. Eventually our shins identified the driftwood hidden near the water's edge, and we had a close-up view of the churning bay, with wispy masses of spume being torn from the wave tops by a brisk breeze. We took several photos of this curious billowy topography while discussing what a great soap commercial it would make if Venus were shown rising through this vast mass of suds holding a box of Duz.

Later we learned that a truck driver had discovered this odd phenomenon at 3 A.M. when, driving at 40 miles per hour he suddenly found himself amid spume up to the windshield. No doubt his story was heard with some suspicion at the next truck stop.

Surveying beaches and measuring waves was a daytime occupation. We got used to the virtually constant rain, the drenchings in cold seawater, the surplus army C-rations for lunch while huddled around a driftwood fire, and the miserable meals and motels that were available in the remote towns. Those went with the job and the territory. But in the evenings we noticed the absence of female companionship.

There were few opportunities to meet young unattached ladies in the small towns along the Oregon coast. But at the mouth of the Columbia River sat the 10,000-person metropolis of Astoria. It had been set up by John Jacob

Astor in 1811 as a fur trading station and continued to survive on shipping, fishing, and lumbering. It was not exactly a thriving cultural center, but it did have a movie theater, three Finnish steam baths, a dozen bars and pinochle parlors, six pool halls, three restaurants where the food was passable, and several "houses of ill repute" (whose reputations were reasonably good). Everything in Astoria was about as up-to-date as Kansas City in the musical *Oklahoma!* So we beach surveyors washed off most of the salt and sand, put on our best clothes, and went out for a big night on the town.

The social status of one place towered above all the other attractions. Amato's Supper Club had an orchestra, a dance floor, and waiters who wore black ties. That was class. It was expensive, but it was the best; drinks were 50 cents and dinner was $2. For one night we would not reckon the cost.

It chanced that there was a table full of young ladies who were also out on the town, courtesy of a cosmetic company to whom I shall be eternally grateful. Among them a slender blonde, Rhoda Nergaard, caught my attention.

We danced the evening away, she daintily skipping on the air above my clumsy shuffle, as we talked about past, present, and future. Her parents had come from Norway as young people, bought a small farm, and started raising chickens. When Rhoda batted her big blue eyes, life on a chicken farm along the Columbia with pigs and horses for pets and a two-room schoolhouse, sounded romantic. She was bright, bouncy, and beautiful. That should have been enough but her farm-style cooking set the hook. She has been my constant companion ever since.

In a few days the beach party moved north into Washington, crossing the shifting sand shoals of the wide Columbia on a ferry whose ornery captain carried a rifle on the bridge, where he traded shots with enraged fishermen, similarly armed, after he ran their nets down.

Facilities in the tiny coastal towns were excessively modest. Motels often consisted of a short row of tiny dripping cabins with sagging beds, soggy mattresses, and cracked mirrors; often the paint was peeling on the sign that said "modern", which in the local usage meant electricity and sometimes indoor toilets. It did *not* mean heat (you built a wood fire in a small cooking stove for that) or blankets or towels. It was possible to borrow blankets from the manager if you left something valuable for security, but it was futile to ask about fancy accessories like sheets or pillowcases or soap.

The cafés matched the motels. On arriving in a new town, we would check out the eating places by peering through steamed windows, observing the curl of the chicken-fried steak and smelling the grease. Then we would sigh and say, "Well, it's only twenty miles to the next town." John concluded that the fine food mentioned on the signs meant that it was ground into tiny particles. The higher class joints advertised "Dancing on Saturday nights," and, for amplification, "No calks [spiked logging boots] allowed" and "Tables for Ladies." We wondered what kind of ladies would dance with men wearing calks if they had been allowed.

Finally our party reached the northernmost beach we would survey, under the lee of Point Grenville. The sand was flat and hard; puddles of rainwater reflected the tall firs and large boulders jutted through the sand. A thick mist hung close about us as though a rain cloud had been snagged and held by the cluster of huge stumps that ranged above the high tide mark. The dismalness and quiet of this surreal fogbound scene fitted somewhere between an abandoned cathedral and a scene from a Dracula movie. As we began to lay out survey stations, a black bear huffed down from a nearby fir and padded off into the gloomy forest muttering something about outsiders invading his beach.

The Washington coast is Indian country, a land of dense rain forests, deep glassy lakes, and brimming streams. It is a dark green world, pressed by a leaden sky, but endowed with timber, game, and fish. Because the deep dripping brush of this northern jungle covers much of the ground, local people travel by dugout canoe when they can. It is a bit of a shock to see a red object flash through the dense forest and realize that it was the headband of an Indian in an outboard-driven dugout, following an almost-hidden stream. Even with very small motors, these long slender craft move at astonishingly high speeds that make one appreciate native boats.

From Cape Elizabeth north cliffs rise vertically from the sea; dark low rocks poke up through boiling water; offshore there is desolate Destruction Island. These are features that make a modern seaman wary, and they must have seemed menacing to early explorers in square-rigged ships. Those who did make it ashore found no hospitality; for centuries the Indians fought off English, Spanish, and Russians. Japanese who reached this coast after drifting across the Pacific in dismasted junks were enslaved and sometimes sacrificed.

But by 1946 the Indians had become more hospitable. When, late one moonless night, the dirt logging road led us to Neah Bay on Cape Flattery, the northwestern-most town in the continental United States, we found space in a hotel that offered twenty small cells and two bathrooms. Hundreds of pinpoints of light spread out before our windows, but we were too tired to investigate what appeared to be a large city and flopped into bed. Next morning, looking out the same windows on a deserted bay ringed by a single line of houses, we realized the "city" had been fishing boats, and this was salmon season; all had sailed before dawn, leaving us the only inhabitants of the hotel and the only outsiders in the village. For breakfast our Indian hosts offered fresh salmon splayed on slender wooden spikes over an open fire, a meal I would like to relive, the best possible reminder of their ancient civilization.

Although we did not recognize it at the time, the waves and beaches of the deserted North Pacific coast were one of the frontiers of oceanography. Along its rough, dark shore we learned much about the ways of the sea in its wintry mood that would be useful later on.

THE SCHOOL OF HARD ROCKS

As we traveled the coast, driving through the rain, walking the beaches, and living in remote motels, John and I talked about our lives and adventures before the Waves Project. Although our backgrounds were different, many of our feelings and hopes were similar. We were still unaware of our steady drift toward lifetime careers in oceanography, even though we were literally immersed in the subject; both of us had vague expectations of going on to something else. By nature we were interested in a great many subjects, and over the next few years became generalists in ocean matters. A generalist is one who has enough background information in several sciences to bring the findings of other fields of endeavor to bear on the subject at hand. I learned from Isaacs that a few months or years of concentrated study in another field in one's spare time was enough to learn to talk intelligently about it and make use of its concepts and findings. Why not try to learn something useful about all ocean-related subjects? One benefit would be a better perspective on the world ocean.

I had grown up poor in Bronxville, an affluent suburb of New York City, but was much more influenced by the lives and adventures of great explorers than by those of the wealthy stockbrokers who lived nearby. As a small child, when King Tut's tomb was discovered in 1922, my earliest ambition was to dig for ancient treasures in the desert. But a few years later, after Charles Lindbergh photographed jungle-covered Mayan temples from the air, I considered a career as a jungle explorer. But archaeology had to compete with the adventures that were implied by the marvelous animal exhibits with romantic backdrops that were on display in the American Museum of Natural History. Perhaps a naturalist-geographer.

The museum was also home base for famous adventurers, who assembled at the Explorers Club just around the corner, and membership in that club became a boyhood dream. Exploring and adventuring seemed the absolute apex of the good life, and as a start in that direction, I became an enthusiastic boy scout. The climax of that career came when Robert Lord Baden-Powell, founder of the Boy Scouts, happened to be in Bronxville and personally pinned on my Eagle Scout badge.

I was a bright kid with little interest in school, preferring to play hooky and lead a Tom Sawyerlike life in the marshes and forests along the Hudson River. Luckily for me the New York school system required only that students pass a state Regents examination to move up or graduate, so a month of study at the end of each school year was sufficient for me to get by, and I graduated at age sixteen. My father had left, and my most remarkable mother supported brother Bob and me by running a small private school called The Children's Workshop and writing a newspaper column.

During the great depression, jobs of any kind were hard to find, but I managed to get a stifling position as a gas station attendant for the respectable sum of $16 a week. This was hardly adventure, but when the tunnel came to town, my life changed.

The Delaware Aqueduct, intended to bring water to New York City, would be the longest tunnel in the world. It would run 85 miles under the Hudson River and several reservoirs; its finished diameter at the New York end would be 32 feet. The tunnel would be "driven" in each direction from the bottoms of twenty-three shafts about 3 miles apart, whose depth ranged from 200 to 800 feet. The important part for me was that Shaft 23 was in Bronxville. As soon as the first wooden construction shacks were set up at the shaft site, I was around asking Big Bill Merrill of the Dravo Corporation for a job. "What the hell do you know about sinking a shaft?" he asked gruffly.

"Nothing," I answered. "I want to find out."

He grunted from the bottom of a huge belly, tilted his felt hat back, wiped his forehead, and looked into the distance. Finally he said, "OK, you can be an assistant timekeeper."

In the early stages when the shaft was no more than 40 feet deep, Big Bill would stand at the rim and look down into the rocky pit some 25 feet in diameter, where six or eight jackhammers were making a terrible racket. If he wanted something, he would give a bellow, and in spite of the deafening roar of the drills, every man would look up. I became Bill's general flunky for the three shafts he bossed, but after a few months topside, I wanted to be involved in underground operations and wear a glamorous aluminum safety helmet. Part of my interest was in higher pay, but some of it stemmed from an idealistic, romantic view of participating in building a better America. Bill Merrill scoffed at such damn foolishness but agreed to let me start as a mucker just to see how long I could stand it.

A shaft mucker shovels the broken rock he's standing on into a steel bucket about 4 feet across and 3 feet deep. There is a steady drip of water down his neck from springs above, the loud whine from an air-driven pump, and an empty bucket weighing half a ton hanging above his head. Digging broken rock is numbing work but, except for a 15-minute lunch break, the mostly black crew and I kept at it for eight hours a day, looking up only when the buckets were changed. The job became an endurance contest as well as a miserable way to make a living. After a couple of weeks, it was a relief to be called out of the hole in midshift and fired; my mother had phoned to remind Bill Merrill I was below the legal age.

Bill used to impart chips of wisdom to me, and one of these was, "Never hire a man who rolls his own cigarettes or wears his boots outside his pants." Foolishly I asked, "Why?"

His answer could have come from an astonished John Wayne: "Why, if he's not stoppin' work to take a stone out of his boots, he'll be rollin' a cigarette."

A few miles north another shaft that had just "bottomed out and turned tunnel" gave me a job as a nipper. For the princely sum of a dollar an hour, a nipper moved sharpened drill steel into the "heading" (the rock face being drilled) and took the dull steel away. These drill rods weighed about 8 pounds a foot and ranged from 2-foot-long "starters" to 10-foot lengths used for deep holes. Carrying a few of those at a time over a rough rock floor was another muscle builder, but it gave me a chance to watch the miners work.

Miners are the underground elite, who run big horizontal compressed-air drills called liners. These heavy machines were supported by a steel frame mounted on wheels called a jumbo, which moved in and out of the heading on railroad tracks. In the big tunnel there were usually eight drills and eight miners on the jumbo, each with a helper called a chuck tender, all under the supervision of a foreman who also did the blasting. Each miner was responsible for drilling eight or ten holes about 10 feet deep in the rock face. Those in the center of the jumbo drilled vertical lines of "cut" holes in at a 60-degree angle in such a way as to outline a vertical wedge of rock in the center of the heading. These were backed up with other lines of holes a foot or two away at increasingly steeper angles. The other drills made a line of holes around the circular perimeter that diverged outward slightly, plus numerous holes perpendicular to the rock face. This was all done in an atmosphere that was noisy, wet, grimy with rock dust, and slimy with drill oil.

Each miner would jam dynamite primed with delay blasting caps into his completed holes and pound it hard with a long wooden pole until it squished out tight against the sides of the hole. Then the blaster would connect the pairs of wires issuing from each hole to central firing wires. The jumbo would be moved back, the men would take shelter from flying rocks a few hundred feet back behind the muck cars, and the blaster would send an electrical current through the firing wire. Instantly the cut holes would explode, flinging a deep wedge of rock out of the center; then with successive one-second delays, the lines of reliever holes would fire, breaking rock into the opening left by the center wedge. Finally, the outermost holes along the sides would break in, the "lifters" would raise the bottom, and the "back" holes would bring down the uppermost rock. If all went well, there was a clean round hole full of broken rock waiting to be moved by a mechanical mucker. If it didn't, the blaster was fired. Since each of the three shifts started with a muck pile and ended with a blast, the tunnel advanced about 24 feet a day.

The effect of the shock waves from this series of shots on the atmosphere in the tunnel was fascinating. The air was always cool and moist and, after the drills had been running, enough mist hung in the tunnel so that one could see only a few hundred feet down the tunnel. But the successive compression and rarefaction waves from each of the six or eight delay shots would instantly create a dense fog in the tunnel that would last a second, followed by super clear visibility for another second that would allow one to see for a quarter of a mile. My interest in the laws of physics that governed this odd behavior eventually led me to school again.

In the meantime my ambition was to become a driller or miner and make the cosmic sum of $10 a day, which was enough to buy room and board for a week in those days. Once a committee of New York taxpayers, outraged by the city's expenditure on such high wages, made an inspection visit. After what they considered to be a narrow escape from being run over by a train of muck cars, a scary trip down in the "cage," and a few minutes' exposure to the noise and wet, they scurried off to report that the city was getting good value for its high wages.

To become a miner it was necessary to serve an apprenticeship as a chuck tender on one of the big drills. My opportunity came when a chuck tender on the other shift was squashed by a rock fall. The next day I took his place, and the man who replaced me as nipper on the previous shift was killed by another rock fall. This gave my fellow workers and my mother the notion that I was being saved by some higher power for a fancier death.

Because of the impossibly high levels of noise, all communication in the heading was by hand signals. No one had explained these to me before the drills roared into action, so on my first shift I stood like a dope while the frustrated driller held up one finger and shook his fist at me. I thought, "What's this guy mad about?" But in fact he was sending a signal meaning, "tighten the number one bolt." After a couple of shifts, I became useful, and in a week we had a winning team.

When my proficiency at running a drill was at least average, I quit and went to another shaft to rustle a job, knowing that miners were hard to find, and there was a big turnover. Lugging a gunnysack containing boots, wet clothes full of rock dust, and a beat-up hard-boiled hat, I lurked in the change room. When the shift boss, a deeply lined old *Polak* named Yakko, came by, I swaggered over and and slurred out of the side of my mouth, "How ya fixed for miners, boss?" There were no other rustlers around that night, so he gave me, a fuzzy-faced kid, a baleful stare and said, "I'm one man short. I give you a try."

Running a big drill in a tunnel seemed to be an adventurous thing to do. The combined smell of fresh broken rock, dynamite fumes, wet pine lagging, and machine oil mist gives a hard-rock tunnel a characteristic scent that old miners claimed "gets in your blood." After the first week went well, I was given the job of drilling the more difficult cut holes, and from then on in the daily race between the eight drillers to finish our holes, I came off the jumbo first about 90 percent of the time. Yakko sat on a muck pile nearby, coughing and wheezing with a bad case of miner's consumption from the rock dust in his lungs. To save him some effort I took over many of his blaster's duties and, in a couple of months, became the de facto shift boss.

At age twenty-two I awoke to the importance of a formal education. Some of the years that should have been spent studying had been squandered on manual labor, but they were not entirely wasted; at least they had converted a useless teenager into a man who liked hard work and could get along on rough construction jobs. The time spent in the big tunnels was not without cost; the intense sound of rock drills running in a confined space destroyed some of my ability to hear high-frequency sounds.

In 1938, full of unwarranted optimism that the mining industry would be eager to have such a talented prospect as myself, I headed west to attend the Colorado School of Mines. The school had a great reputation, based on the many hard-rock mining engineers it had sent to mines around the world. Once there, I soon found out that the strength of a Mines education was in its sensible concentration on basic engineering, not on mining methods. After nearly four years of alternating between work in the tunnels of the East Coast and the classrooms of Colorado, I was on the verge of graduating when an altercation with the president over my maverick attitude toward the school caused him to expel me a few months before graduation. However my grades were good, and forty years later the school presented me with its Distinguished Achievement Award.

At the time I did not care very much. The United States had just entered World War II, and I had an extracurricular job searching for tungsten, a scarce metal needed for war production, in behalf of Hassie Hunt, son of the wealthy H. L. Hunt of Dallas, Texas. Our most promising prospect was the Yukon Tunnel, deep in the San Juan Mountains near Silverton, Colorado. The dump outside the tunnel had a good showing of huebnerite, a reddish tungsten mineral not previously mined there, and it was a reasonable place to start. The tunnel ran over 2,000 feet back into the mountain, and somewhere near its deepest penetration the previous miners had encountered the vein containing the tungsten. Unfortunately, the oxygen level in the tunnel air was so low we could not get in to examine whatever was there.

With a small crew of men who were unfit for military service or much else, it took a couple of months to raise the roof of the old blower house that had been caved in by snow, get the blowers running again, and tape over hundreds of holes in the old air ducts to reach the vein. After the tunnel had aired out for a few days, and we had started sampling, the state mine inspector arrived to check on safety conditions. He and I walked in together, and by the time we got to the working area, he was winded. "Let's take five while I have a smoke," he puffed, and struck a match. It glowed and went out. So did the next three.

"Cheap matches," I suggested helpfully.

He was still attempting to light his pipe when the correct explanation dawned, and his energy returned. "There's not enough oxygen for a match to burn. Let's get the hell out of here." By the time we were outside, the inspector's breathing was good enough for him to order work stopped until we installed a better air supply. Instead, because the tungsten showing was not very promising, we abandoned the property.

Unlike most of Colorado's old mining towns, which were made of shacks, hastily stuck to the steep sides of ravines amid a boom, Silverton had well-built houses along wide streets neatly laid out on the flat bottom of an ancient lake bed. The town is at an elevation of 9,300 feet and usually snow-topped mountains rise to nearly 14,000 feet around it; the Animas River flows across the flat on one side of town to plunge into a narrow canyon it shares with a narrow-gauge railroad. The town was pretty much isolated from the outside world, especially in winter when snowslides (avalanches) blocked the narrow dirt road through the mountains for a week at a time.

Silverton became my base for exploring the rugged surrounding mountains, as I searched for mineral deposits that might have new economic potential because of advances in metallurgy or wartime demand. Treeline in those parts is just under 12,000 feet; above that there are steep naked rocks and broad alpine meadows that serve as summer grazing ground for great herds of sheep. There were hundreds of old mines and prospect holes, half a dozen ghost towns, and virtually no people. Except in the worst of winter, I traveled the desolate high country on snowshoes when necessary, accompanied by Muffin, my collie-shepherd dog. Snowslides were a constant danger, and on several occasions we narrowly missed being buried or swept off a mountain. It was also dangerous going into old mines, whose loose rock entrances were dubiously supported by old, rotten timbers that might collapse if disturbed. In that event we might have become permanently entombed; no one would have known where to look for us.

Resampling old mines, especially alone in the mountains, could be dangerous, and there were a lot of close calls such as one at the long-deserted Caledonian Mine. It had two tunnels into the mountain at different levels, drifts along the vein, and far underground, a 200-foot vertical shaft in the vein between the two drifts. The shaft contained a wooden ladder that hadn't been used for over twenty years and a rusty pipe for compressed air, both held in place by transverse timbers. With the intention of sampling the mineralized vein between levels, I started down the ladder from the upper level. At first the timbers seemed solid, but about 30 feet down, some subliminal indication of trouble caused me to turn my back on the steeply sloping ladder. At that moment it tore loose from its supporting timber; I grabbed the air pipe and hung on for dear life while the ladder crashed downward through the blackness toward the level below, carrying rotted timbers with it. My position was roughly equivalent to that of a blind man clinging to the downspout at the sixteenth floor of a twenty-story building. Somehow I shinnied back up the pipe to Muffin who was waiting in the dark drift to slap his big wet tongue over my face a few times as I emerged from the black hole.

The following winter I took a job dewatering a long-deserted mine about 11,000 feet up on a mountain behind Alma, Colorado, an old mining camp between Fairplay and Breckenridge. The shaft was about 150 feet deep and filled with water; the owner wanted to take samples in the tunnel at the bottom, where there were alleged to be high values in lead and silver. A high-voltage power line near the mine could be tapped for power, and three of us spent a hard day's work skidding a 200-pound electric pump up the trail to the mine. My two helpers stuck around a couple of days until the pump was in the shaft, running; then they decided the location was too remote and quit. So, with only Muffin for company, I stayed on the mountain for several weeks, sleeping in a cabin next to the shaft, eating the traditional "bacon and beans" of the Forty Niners, and roaming the beautiful deserted mountains in the daytime.

One day it started to snow heavily, and by nightfall the wind became violent. About 2 A.M. I awakened and, above the sounds of the storm, noticed a change in the sound of the pump. Slipping into heavy socks and rubber boots, and putting a heavy jacket over my long johns, I went out to investigate. The pump motor now only gave a hum, meaning that one of the three electrical leads to it was open. After winching the pump a few feet higher above the water in the shaft, I started to check out the circuits.

There were big knife-switches on the high-voltage pole where the mine power took off from the main line, and it was apparent that one of them had been blown open by the wind. The power pole was a hundred yards up the steep hillside behind the cabin, and with only a small flashlight, I slogged up to it through snow over my knees. The light beam cut through the swirl of white, and I could barely make out three switches about 12 feet above the ground; one of them had fallen open. By scuffing under the snow, I located the long pole used to throw the switches and, with the flashlight held tight against it so that the light shone on the hook, with its tip I managed to snag the ring at the end of the foot-long switch and push it closed again. The pump immediately kicked in and resumed its task.

On climbing back into my crude miner's bunk, I asked myself why it was necessary to joust with a 17,000-volt power line at night during a blizzard. Was my life equal in value to the pump or the mine? Of course not; that was not the point. A man who chooses to live along a frontier must respond when his courage is tested. Although no one else will ever know how brave or foolish his actions, a man is accountable to himself.

Between short engineering jobs, I worked as a "ten-day tramp." That was a standard phrase used to describe the restless hard-rock miners who knew how to do the dozens of complex operations connected with getting ore out of the ground. Because there was a chronic shortage of experienced men, there were always jobs open to these tramp miners, who were constantly on the move from one mine to another, rarely working for more than a couple of paydays at any one job. It was a way to learn as much as possible about how mining was done.

So I roamed from mine to mine throughout the Rockies, learning about shrinkage stopes, cut-and-fill operations, and block caving. Each of these not only required special skills but were often dangerous and miserable. Driven by the reckless enthusiasm of youth I had no fear and was willing to try anything.

During that period I was temporarily blinded by cement dust from a gunnite operation in a burning tunnel where the temperature was 120 degrees F, carried out for dead from the Big Thompson tunnel, 4 miles underground, rode aerial tramways for miles in subzero weather while wet from mine water, became involved in a union war underground, where some had sworn to "get" me, and literally scared a rider to death when an ore truck I was driving ran away on a long steep mountain grade (he was stiff and blue by the time I got it under control). Gradually the idea penetrated my mind that there must be a better way to make a living, in a warmer climate. So eventually I pushed on to California.

In all the conversations I had with Isaacs on a great many subjects, I don't recall we ever talked about making money or getting rich or anything connected with business. Those matters were of no interest to us. Instead we spoke about science, the endless frontier, of exploring the earth, and of living interesting lives filled with new ideas about the world around us. My ambition, in addition to knowing something about each of many subjects, was to make some important invention or discovery that would result in a major advance in science or engineering. I wanted to go to remote places and do seemingly impossible things. The fun for me would be to do things *first*. Somewhere, somehow, someday I would.

II

Coastal Technology

TIDAL WAVE

Early in March 1946, while the beach party was passing through Astoria, Oregon, a cryptic message arrived from the university: "Get your vaccinations updated; you will soon be going out into the Pacific to make some special measurements of waves." John and I got the shots and then moped around the motel with sore arms and upset stomachs trying to figure out why. The answer showed up in a local newspaper: "President Truman announced today that the United States will test nuclear explosions on some old warships at a remote Pacific atoll." A few days later the paper printed a sketch of Bikini lagoon, and I recognized that its shape was the same as the white plaster oval in the old swimming pool at Berkeley. The steel plate those fellows had been dropping was intended to simulate a nuclear explosion and the waves it would create. Maybe they weren't so crazy—unless you count being in the wave business as crazy.

Then it was decided that Isaacs would go to Bikini to measure the explosion waves, and I would take over as leader of the beach party. He left at once, and on April 1, 1946, the day I returned to Berkeley with the beach party, the first question we got was, "Did you see the big wave?" It was not an April Fool's joke. A tsunami, popularly called a tidal wave, originating in the Aleutians had struck at various places around the Pacific on the first day in four months we hadn't been on the beach. Not being there to greet it was either good or rotten luck, depending on exactly where you were when it arrived and how much you enjoy such things.

30

Isaacs had been en route to Bikini when the A-bomb tests were delayed, and he had been held up in Hawaii. So he was assigned to check out the effects of the tidal wave in the islands; my job was to study the effects of the wave on the California coast, accompanied by a representative of the Corps of Engineers.

For several days the two of us drove along the coast road making notes and taking photos of damaged buildings, stranded boats, and measuring high-water marks indicated by driftwood. Mainly we interviewed people about what they had seen. The first thing those on the beach reported was the arrival of the trough of the wave, looking very much like an extreme low tide. People on the beach had followed the receding water seaward to pick up sea life and inspect a bottom never before exposed. Then, after a minute with the water at its lowest ebb, the wave crest arrived. It rushed shoreward as a breakerlike wall of water, chasing people from the muddy bottom and carrying anything that would float to well above the highest high tide. At Half Moon Bay, fishing boats anchored offshore were deposited on the crest of a coastal road 15 feet above sea level.

After sorting through our notes and collecting the accounts of distant observers, we put together the following picture: Sometime during the night an underwater landslide in a deep undersea trench south of the Aleutian Islands had started a train of long-period waves that radiated outward in all directions. In the deep Pacific these waves moved at speeds of over 400 miles an hour, but their height was low, perhaps only a foot. As they moved onto the continental shelf, they were refracted (bent) to conform to the shape of the bottom. At the same time, the shoaling water caused the wave height to increase greatly. Thus the largest waves struck headlands pointed toward the Aleutians. Depending on the nearness to the epicenter, the underwater configuration offshore, and the orientation of the coast, the tidal wave had a large breaking front or a gentle rise and fall in the sea surface. Often tidal waves act something like a rapid, much exaggerated tidal change in which 12 hours has been compressed into a few minutes.

The largest waves that caused the most dramatic damage were those that encountered land nearest the wave's origin. At Scotch Cap, Alaska, an island that marks one side of Unimak Pass, the Coast Guard had a lighthouse and radio station in a two-story reinforced concrete structure whose main floor was 52 feet above high water; nearby, 103 feet above high water on the rim of a mesa, a radio mast stood on a solid concrete base.

When this station did not come up on the radio next morning, an aircraft was sent to investigate. It found only ragged traces of the lighthouse foundation and a lump of concrete where the radio mast had been. Besides the damage to structures, five men were missing, and vehicles atop the mesa had been moved a considerable distance by an overtopping rush of water. It was concluded that at least one wave more than 100 feet high had arrived about 2:40 A.M. and swept everything away.

Much more has been learned about tsunamis since then, but our group made some contributions to what was known at the time. We pointed out that the first evidence of such a wave is a drop in sea level as water is sucked back from the shore to form the oncoming wave crest. Second, we discovered that, except for headlands pointing into the wave, embayments facing exactly opposite to the wave direction were likely to be most affected. Half Moon Bay and San Simeon Bay, both of which face south, lost boats and buildings. Years later we learned that Taiohae village in the Marquesas Islands at the head of a deep, narrow bay facing south was damaged even though it was 4,000 miles from the epicenter.

In the Hawaiian Islands points on the north shore facing into the wave and bays on the south coast facing away from the wave were hardest hit. The captain of a ship anchored offshore in Hilo Bay was astonished to see the city destroyed and a pier wrecked by waves that passed unnoticed under his ship. Dozens of small wooden houses on the two streets nearest the water's edge washed inland to become a line of driftwood at the high water mark.

Isaacs and some associates compiled a report on the effects of that wave on the Hawaiian Islands, and John, as usual, came up with one of the best stories. He reported that at the Navy's Kaneohe Bay base on the east side of Oahu, a road ran out on a point of land to some storage bunkers that had been mounded over with dirt. Early on the morning of April 1, an enlisted man on guard duty was driving innocently along this road when he observed the sea level inexplicably rising, about to cover the road. Quite sensibly he headed for the highest ground and drove the jeep as far up the side of a bunker as it would go. Then he scrambled to the summit and watched with amazement as the sea rose around his little island. Presently thick white clouds enveloped him and reduced visibility to a foot or two. It seems that he had taken refuge on a bunker containing a stockpile of smoke-screen chemicals that were activated by contact with water. For hours after the tsunami had subsided, this poor guy sat on his hilltop in dense white smoke, thinking he was still surrounded by water and afraid to move.

Largely as a result of the damage and confusion generated by that wave, a Seismic Sea Wave Warning System was set up in the Pacific. It was quite clear that a somewhat larger tsunami could do a lot of damage to the growing coastal population.

OPERATION CROSSROADS

Neither the original test of an atomic weapon in the desert at Almagordo nor the subsequent destruction of two Japanese cities produced much information about the effects of such weapons on military targets at sea. Questions remained, such as, What would an atomic (nuclear) explosion equivalent to 20,000 tons of TNT do to large naval vessels? How much would this new kind of weapon influence ship design and the spacing of ships in fleets or convoys? What would the effect be of a nuclear burst on airplanes or tanks at various distances?

A large supply of used equipment left over from World War II was available for experimental purposes, including at least three nuclear bombs and capital ships of the German, Japanese, and United States navies. Therefore Joint Task Force One was created to "carry out the atomic bombing of a target array of naval ships." Operation Crossroads was to take place in the lagoon at Bikini atoll, Marshall Islands, 4,150 miles southwest of San Francisco. The tests would be commanded by Vice Admiral W. H. P. Blandy and include more than 200 ships, 150 aircraft, and 42,000 men as well as a thousand civilian observers, reporters, and representatives from many countries.

The effects of exploding nuclear weapons at two quite different heights were to be measured. The first, called Able, was to be an air burst detonated some 1,500 feet above the center of the test fleet. The second, called Baker, was to be an underwater burst suspended at middepth in the 180-foot-deep lagoon. The positions of the seventy-odd battleships, cruisers, submarines, landing ships, and the aircraft carrier *Saratoga* were carefully planned in advance so that the maximum information could be obtained.

John Isaacs was concerned only with measuring the waves produced by the explosions. He followed the plan we had discussed months before and set up long-focal-length aerial cameras atop two camera towers on Bikini island. The towers were 2,000 feet apart, and the cameras would take simultaneous photos every 3 seconds (the time required for the camera to advance the film) starting half a minute before detonation. Long rolls of film in these automatic cameras permitted many large photos (9- × 18-inch negatives) to be made of each shot. When the camera was aimed to include the entire fleet, which was 3 or 4 miles away, the origin and progress of the waves to the beach on Bikini island could be followed. As it turned out, the presence of Isaacs and the existence of this set of wave photos were an important factor in the success of the tests. This had to do with the physics of explosions, basic principles of seamanship, and bombing accuracy.

Explosive effects radiate outward from the point of detonation in an expanding sphere. As the distance increases, the heat, nuclear radiation, and pressure effects of the blast decrease rapidly as the energy spreads evenly over that huge surface. A small increase in distance could mean a large decrease in blast pressure, for example. Therefore, in order to be able to relate the amount of each kind of energy received by each part of a ship to the damage it sustains, it is necessary to know the distance from the center of the explosion quite closely.

Ships at anchor do not remain in one place; they constantly swing about their anchors in what is called a watch circle. A ship with a 5-to-1 scope (five times as much anchor chain as the depth of the water) in Bikini lagoon could conceivably swing a thousand feet in any direction. So the test ships, most of which were anchored weeks before the first shot, constantly changed position as the wind shifted. In order to locate ships at the moment of burst the official plan had been to take lines of aerial photos of the test fleet the day before each shot and assemble these into a mosiac. But when this was attempted, it was found that the ships moved so much from one photo run to the next, only a few minutes later, that no satisfactory mosaic could be made. The uncertainty in the position of each ship at the time of the explosions could be great enough to make some measurements useless.

Third, in spite of the Air Force's standard claim that it could reliably "drop a bomb in a pickle barrel," on Able shot the carefully selected team of bombardiers missed the orange-painted battleship *Nevada* by over 1,500 feet.

The result of all the above was that, after the shots, no one knew exactly where the ships had been relative to surface zero (the point on the water directly beneath or above the bursts). When that became apparent, Isaacs proposed measuring the position of the bow and stern of each of the test ships on his pairs of large photos and, by triangulation, computing their precise positions. This offer was immediately accepted, and for months after John returned to Berkeley he and several helpers measured photos by securing them to a steel platform above which a traveling low-power microscope moved along graduated steel tracks. Its position could be read in tenths of millimeters. When a photo was fixed to the platform in a standard way, the position of the bow or stern of each ship could be measured precisely. Then, by means of simple trigonometry, ship positions could be determined within a few feet. The result was that thousands of weapons effects measurements that would otherwise have been nearly useless now become valuable.

The waves created by the Baker shot made a couple of breakers on the beach at Bikini island about 8 feet high. This was a relatively minor finding, but it reminded everyone how puny the bomb energy was by comparison with a Pacific storm that would send several thousand breakers higher than that to the northwest beaches. The photos made for wave measurements of the mile-wide column of water rising into a mushroom cloud and surrounded by a 600 foot high "base surge" that towered above the test fleet became the most reproduced pictures of the test series.

CANNERY ROW

One of the people I interviewed who observed the tidal wave was Dr. Rolf Bolin, a bouncy, bespectacled professor of marine biology at Hopkins Marine Station at Monterey, California. When the wave arrived, he had been netting squid from a small boat just inside the Monterey breakwater. Bolin noticed that unusual currents moved his skiff around, but he observed no change in sea level, and it did not occur to him that this was a major ocean event. Therefore he was astonished to learn later that at the same moment in Pacific Grove, only 1 mile west but in a cove facing south, the water had risen 16 feet to wet the dangling hand of a man napping on a concrete bench.

Bolin and I quickly became good friends, and when Dean O'Brien sent the beach party to the Monterey area, he arranged for us to park the dukws on the grounds of the Marine Station. As a return courtesy, we would take his friends and students on biology collecting trips. On these occasions we would strip the canvas covers from the cargo compartment, load up with biologists, and roll down Hopkins' tiny beach into the calm waters of Monterey Bay for a day of dredging, trawling, and fishing. In the days before SCUBA made it convenient for marine biologists to dive, nearly every haul brought up sea life that few had seen before.

One of the regulars on these trips was a marine biologist named Ed Ricketts who had been made famous by John Steinbeck as the Doc of *Cannery Row* and who was well known to biologists for his book, with Jack Calvin, *Between Pacific Tides.* Ed was a kindly, easygoing fellow who cared little about money but a great deal about the finer things in life, including women, beer, Gregorian chants, and sea animals. His base, the Pacific Biological Laboratory, was a tiny, two-story, unpainted clapboard building squeezed between two of the thirty-five canneries that were then operating. Ed lived on the second floor and had his laboratory at ground level; out behind it on pilings, exposed to the sky above and the bay below, were concrete tanks that held the specimens of marine life he sold for instructional purposes.

One reason the beach party had been sent to the Monterey region was because Dean O'Brien thought we should work the northern beaches in the winter and the southern ones in the summer, an idea that some might regard as perverse. Another was that he had spent his summers on the beach at Carmel for several years and had made informal surveys of the coming and going of sand there. Because Carmel Beach is a closed system (that is, the sand there is prevented from leaving or being replenished from outside sources by rocky headlands at its extremities) it was a good place to study how sand is shifted from onshore to offshore, or laterally along the beach in response to changing wave action.

The beach party laid out a grid of survey lines along the curving shoreline of Carmel Beach and began making profiles that recorded the growth of the summer berm and the shift of the widest beach from north to south. In time these surveys expanded to include six beaches around Monterey Bay and including the steep, coarse-grained beaches at the Soldier's Club at Fort Ord, the mouth of the Carmel River, and the rocky north beach at Point Sur.

Summertime surveys made it possible for us to map the weekly growth of the berm and measure the quantity of sand brought ashore by the lower waves. From May until the first great storm of October, the berms on some of the beaches widened over 100 feet. Then the attack of higher, steeper waves of winter began moving the sand offshore again onto underwater bars. We were compiling basic numbers, and our reports on the details of how the changes took place and what caused them added much to the general knowledge of beaches.

On the first attempt to survey in the surf at Point Sur, the dukw had barely left the beach face when Rex Goodwin, who was driving, saw the glistening tip of a black rock in the white water directly ahead. When he throttled back and turned left to miss the rock, the dukw swung parallel to the onrushing waves, and the next breaker carried it back onto the beach face. As the dukw's wheels touched bottom its sidewise momentum caused it to tip over and land with its flat left side against the sand and its six wheels pointing out to sea. I was tossed from my customary position on the grating behind the driver onto the beach, but Rex was still in the driver's seat lying on his side, under the canvas cover. I scrambled to my feet, slapped the windshield to get his attention, and shouted, "Stay there, the next wave will roll you back." If the next wave had been much larger than average it could have rolled the dukw on top of him, but he stayed, and a rather ordinary wave set the dukw back on its wheels. The engine was still running, and Rex, a little shocked but wearing a triumphant grin, drove it up the beach. That was our last attempt to obtain beach profiles at that location.

Cannery Row and the fishing fleet that supported it were vital forces in Monterey life. After a few weeks in the area and charmed by the local atmosphere, Rhoda and I bought a tiny, run-down house on the rim of the hill just three short blocks above Ed Rickett's laboratory for $4,950. The down payment on that huge sum required some skillful financial maneuvers, which involved selling my car, borrowing on my university expense account, and selling Isaacs' outboard motor while he was still in Bikini. Despite these sacrifices the view of Monterey Bay beyond the line of steaming canneries made it worthwhile.

To borrow a line or two from Steinbeck, "When the purse seiners have made a catch the cannery whistles scream. People all over town scramble into their clothes and come running. It is a stink . . . a dream." We agreed. The outrageous stench of cooking sardines penetrated our sinuses, insulted our nasal passages, and overloaded our taste buds. It was palpable and could not be ignored. But we, like the other denizens of the area, would stand on our front porch, breathe the dense air deeply, smile, and speak of the "million-dollar smell." It was the smell of money, jobs, prosperity. We loved it because it was the essence of life in Monterey, and we were part of it. The people who went running to work ran past our house. Had we guessed that Cannery Row would be idle and its indelicious smell gone forever within only two years, we might have shown even more appreciation. As it was, I was glad to spend most of the day on a distant beach while Rhoda worked in a Pacific Grove drugstore. If an official of today's EPA were to sample that atmosphere, he would doubtless create a tizzy of regulations, but in those days anyone who objected would have been doused with sardine oil and run out of town on a cannery cart.

Now our work took on a new dimension; we would try to measure and record the shapes and periods of waves. In the surf at Table Bluff, the beach party had installed a sensor on the bottom beneath the surf that would detect the changing water pressure as waves passed above. It was mounted on a tripod from which a light neoprene cable led ashore to a recorder. For a few hours this instrument had recorded waves, until the surf, enraged by this man-made intruder, threw the tripod back on the beach, its cable wound around it. Considering the difficulties of that location and our innocence in ocean matters, that was a relative success.

Now, in the calm of Monterey Bay not far off Cannery Row, we decided to try again. Roscoe Hughes and Frank Snodgrass brought an improved wave recorder system from Berkeley; we lowered it from a dukw in what seemed to be a suitable location and ran a slender cable ashore. The recorder's pen immediately traced low, graceful curves on the chart, representing waves, and we began celebrating. Before the celebration was over, the darn thing quit recording; on retrieving the sensor, we found that even in that calm location the cable had parted, perhaps after being snagged by a fishing boat. There was more to wave measuring than we had thought.

Not long after our little house on the crest of the hill became livable, John Isaacs returned from the Crossroads tests, and when he visited us in Monterey, we had a party for a dozen friends including Ed Ricketts. There was no furniture, so everyone sat cross-legged on a rug before a fireplace that blazed with driftwood, telling tall tales and drinking. Well into the evening, Ed Ricketts heaved himself uncertainly into a nearly vertical position and said, "I'm running an important experiment, and I must make my midnight titration." As he lurched in the direction of the front door, Isaacs and I, who were somewhat steadier, volunteered to go along and help.

Cannery Row was just down the hill, but we banded together for stability and navigational purposes. Eventually we reached Ed's laboratory, and with the aid of a fortuitous street light, he selected the right key. At first the lock would not hold still enough for it to be inserted, but we discovered that if two of us steadied the building and the other two steadied the key, the door could be unlatched. Once inside, after intensive groping for a switch, a burst of light revealed a table covered with laboratory glassware including neatly arranged beakers, reagent bottles, and a slender glass titrating column. It was evident that one clumsy move would break a lot of glass and destroy the experiment, whatever it was; anyone with good judgment would have concluded that we'd all better get out before that happened. But Ed staggered forward and solidly planted his feet before the delicate array of glass. At that moment he became cold sober. He selected the proper beaker, read the meniscus on the column, stirred with one hand while cracking the glass valve with the other one, and saw the liquid turn color. Then he wrote the numbers in his book and fell back in our arms, exhausted by the concentration. Together we staggered back up the hill to proceed with the party.

Except for Ed Ricketts, the Chinese groceryman, and John Steinbeck himself, the rest of the cast of Cannery Row were gone before we arrived. But somehow those marvelous characters still made their presence felt.

For a few months we had to leave Monterey, and just before we departed, Ed Ricketts gave a party to welcome Steinbeck back from Mexico. His parlor was barely adequate for the dozen friends who sat cross-legged on the floor drinking beer and eating steamed mussels that had been liberated from nearby pilings earlier in the afternoon. The guests were all regulars, summoned by rumor and perhaps some subliminal hint that a turning point in our lives was approaching. As the evening advanced, the group's enthusiasm began to subside into mellowness; the music and the random chatter subsided. It was time for John to tell his tale.

Outwardly Steinbeck was not a handsome man, but he had a rugged, kindly quality about him that seemed inherently truthful. His special talent was for penetrating insights into the workings of human nature. Somehow he could see right through whatever screens a man's soul and reduce what he saw to the written word. This night, dressed in careless clothes and with a stubby beard, he sat on a day bed beneath a bloated caricature of himself that he and Doc loved, but I hated. In a slightly hoarse voice he told of life in the small Mexican village where his book *The Pearl* had been made into a movie. Steinbeck was a very sympathetic observer of the quiet people of Mexico, and his oral version of their lives allowed him to add emphasis with changes in voice level, pauses, gestures, and facial expressions. The audience ringed before his dais was so intent on every word that the night vanished.

So, within the next few months, did the life of the Cannery Row we knew. The sardines did not come back that year; Ed Ricketts' car was struck by a train only a block away from the lab, and he died of the injuries; we did not see John Steinbeck again for nearly fifteen years.

Our daughter Anitra was born in Oakland, California, a week before we all moved to Carmel. There, with money from my mother's estate and help from my brother Bob, Rhoda and I built a new house in the form of an artist's studio around a monumental fireplace in Del Monte Forest, part of the exclusive seventeen-mile drive.

MEASURING WAVES

Early in 1947, no suitable instruments existed for continuously recording the periods and heights of waves approaching a coast, so the Waves Project decided to work on that problem intensively. It would be necessary to hold a sensor that converted pressure changes into electrical impulses firmly on the bottom under rough water and bring the electrical signals to a recording device on shore. Electronics was in its unreliable infancy, and many seagoing engineers, including me, preferred rugged intruments that had "less than one vacuum tube."

The design group in Berkeley under Dick Folsom had been steadily improving sensors, circuits, housings, and underwater seals. But they knew little about the practical problems of maintaining cables across wave-impacted shores or installing instruments at sea. It was evident that the slender, lightweight cables we had previously tried would not survive in ocean waves, but the project could not afford the cost of proper cables. Then by chance, on a visit to Mare Island Navy Yard, I spotted dozens of large spools of heavily armored submarine cable in a surplus yard. This cable had 10 conductors, a heavy neoprene sheath, jute wrapping, and 20 strands of number-9 wire around the outside for armor. It was practically perfect for what we wanted, and it was about to be incinerated to salvage the copper.

The Navy chief in charge said, "Ya want it? Where?"

The next week Navy trucks delivered twenty spools to the university. The cable had the main characteristics we needed; it was rugged, heavy, and had lots of conductors. The first use of it would be at Heceta Head, a picturesque rocky headland that jutted boldly into the Pacific from the Oregon coast. Strange as it seems, we decided to test a new instrument, cable, and installation method all at once in very rough waters hundreds of miles from Berkeley. But the Oregon coast was where the big waves lived that we wanted to measure, so the dukws and a couple of miles of heavy cable headed north, prepared to do battle with the Pacific's winter waves. Heceta Head was already under assault by a northwest storm.

In order to keep occupied until the storm passed, I decided to measure long-period waves at Depoe Bay where an open outer bay and a small inner harbor are connected by a crooked rocky channel. This channel, only 25 feet wide and 150 feet long, is spanned by a bridge on the coast highway. When its fishing boats return amid big swell, the highway bridge is lined with spectators who like to watch the death-defying run through the narrow opening below.

Each boat maneuvers just off the entrance to the channel, where reflected waves from the rocky cliff add to incoming waves. The skipper watches over his shoulder and picks his moment; then he hits the throttle and aims for the slot, knowing that if his hull touches on either side it will be holed, and he is likely to lose both boat and livelihood. Exceptionally lucky boats, not quite aligned, have been lifted over the rock on the north side by a large wave and dropped gently in the channel. But the record of safe passages is very good, and nearly always a cheer goes up from the watchers on the bridge above.

Those very characteristics—large waves, narrow opening, and quiet water inside—made Depoe Bay an ideal place to measure long-period waves. The narrow channel that prevented a rapid flow of water between the inner and outer bays acted like a filter. We reasoned that, if we measured and recorded the water level in the small inner bay, from which the 15-second swell is excluded, we should be able to see the small rise and fall caused by longer period waves. That proved to be correct, and we were able to record groups of waves with periods of about 3 minutes and a height of a few inches. This was much longer than the natural period of sloshing of either bay.

In two days the storm subsided, and we returned to Heceta Head. The cable spool, weighing about a ton, was mounted on an axle in the cargo compartment of one dukw, and the heavy steel tripod holding the wave sensor dangled from an A frame over the stern. The cable was connected to the instrument, and the armor firmly secured to the tripod. The plan was simple, and our luck was good. The dukw ran off the beach and out to the site selected, about half a mile off the headland. There we lowered the tripod to the bottom in 40 feet of water, buoyed it, and started shoreward, unreeling cable as we went. Once safely ashore, we unspooled the rest of the cable and dragged it up the cliff to a small building near the lighthouse and connected the recorder.

Large winter storms often follow each other in quick succession, and we had been lucky to get two days of calm between storms to make the installation. Barely had the recorder begun operating when another violent storm at sea tested this first serious wave meter with a long sequence of high waves.

The new system worked beautifully, and we felt we had solved the wave recorder problem. In accordance with the O'Brien dictum, if it worked here, it would work anywhere. We were euphoric as the recorder pen swept back and forth almost the width of the chart paper recording trains of steep 17-foot-high waves. But any imagined triumph over the ocean is probably premature.

At that time, not long after World War II, Japanese spherical moored mines, which were supposed to sink any ship that touched them, continued to break loose from their anchors and drift across the Pacific. Some had been sighted off the Oregon coast, and this, understandably, made local mariners nervous. The buoy we used to mark the tripod looked a little like a mine case, and although the words "University of California Wave Meter" and our telephone number were painted on it in yellow, few boatmen get close enough to mines to read any fine print that might be on them. Someone reported our buoy to the authorities, and a few days later the Coast Guard proudly reported sinking an enemy mine off Heceta Head with gunfire. The buoy was not an essential part of our instrument, but it indicated to an observer which wave was being recorded. Without it the tripod would be hard to recover.

The big storms moved several feet of sand from the area of bottom on which our tripod rested. When smaller waves followed, they returned the borrowed sand and buried our now unmarked tripod. The rubber bellows that had expanded and contracted in accordance with pressure changes was now under sand and could no longer expand. As a result, the recorder pen barely moved when a large wave passed instead of making a great sweeping arc. Nor could we locate the buried tripod or grapple for the cable that was under a couple of feet of sand. After a month or two we abandoned that site.

A second spool of cable had been shipped to the north coast as a spare, and it was decided to install it at La Push, Washington, a tiny Indian fishing village in the rain forest on an otherwise deserted stretch of coast.

The pressure sensor was to be planted just seaward of a small steep island crowned with firs, behind which the Quillayute River flowed into the sea. There was a beach to its south from which the dukws could operate, and along its upper margin, well above ordinary high tide, there was the usual line of large white driftwood logs that seemed to have been there for years.

We set up the spool of cable on the dunes and, with storm flags flying, dragged the cable directly out to sea and lowered the tripod. The shoreward end of the cable was led under the driftwood and into the Coast Guard Station. As before, the instrument recorded waves from the first moment, and we went to our bunks that night hoping for a set of storm waves so there would be some special records to take back to Berkeley.

We were not disappointed; the next morning opened with strong wind gusts, driving rain, and breakers on the beach. Inside the snug station we sipped coffee with several Coast Guardsmen and watched the pen sweep back and forth across the strip chart in accordance with the waves a half mile offshore. Suddenly Rex Goodwin, who was peering out the rain-splattered window through the murk hollered, "Look! Look at the cable!"

We sprang to the windows to see and were equally astonished. Our 1.25-inch armored cable that weighed a couple of pounds a foot was stretched through the air above the breakers. A huge driftwood log being rolled about on the beach by waves had snagged it and was pulling hard enough so that several hundred feet of cable was airborne, vibrating like a fiddle string and radiating a halo of spray. For a few more seconds the recorder pen continued to trace the passing waves. Then the cable, with a breaking strength of around ten tons, went slack, and the pen centered itself on the chart. We groaned and cussed the orneriness of nature.

We persisted. The next wave recorder was installed off Point Sur, 30 miles south of Monterey, California, on a rocky bottom. This time we mounted the cable spool on a surplus steel life raft and towed it out to sea with a dukw, unlaying as it went. This meter lasted for many months, long enough to make us hope that our luck had turned and that from here on all would be well.

Point Sur was also used by the Navy, which had several cables leading to offshore hydrophones. These were intended to listen for SOFAR bombs from downed aircraft and, incidentally, to the sound of passing ships. But the most interesting sounds they received were made by small sea animals that lived near the hydrophones and were identified whimsically as the whistler, the carpenter (sound of sawing), the snapper, etc. One day when I stopped by the receiving stations to listen to some of these fascinating sounds, the sailor on watch said casually, "A Russian submarine just went by."

"How do you know?"

"Well," he said, "I can see the horizon, and there are no ships out there. None of our subs are operating in this area. Besides, ours usually travel on the surface, and the sound signature I got isn't like any of our boats."

That sounded reasonable, so I asked, "Did you report it?"

"Why should I? I don't report any other ships."

That's the way it was in 1948.

About that time the Navy's SOFAR group ran a submarine cable from far out in Monterey Bay almost to the beach, before it occurred to them they had no way to get the end ashore through the surf. Their cable ship, the *Glassford,* was hove to just beyond the breakers, rolling heavily, waiting for somebody to do something. The Navy commander in charge of the project found us on another beach and asked if our dukws would run a line from their ship to the beach so they could haul the end of their cable ashore. In an hour we were on the beach at the mouth of the Salinas River.

We drove out through breakers that were about 10 feet high on the outer bar, but when we arrived alongside the *Glassford,* no one aboard heard our shouts or noticed we were there. Dukw and ship heaved and pounded against each other in the passing waves; impatiently, I grabbed a dangling line to pull myself up to the rail. Halfway up the side, that line gave way, dropping me into the churning water between dukw and ship. Somehow I dived clear and came up spluttering to one side, having avoided getting smashed between the two. These antics finally attracted the attention of those on deck. It was easy enough to run their towline ashore, but honor required that my banged shin bone and dunking be repaid by an evening for Rhoda and me at the Del Monte Officers' Club, courtesy of the commander. He was glad to oblige.

The most elaborate of the wave meters was placed off Point Arguello in 1949. This headland, at the southern end of the westward-facing coast, joins with Point Conception in forming a great double cape that marks the beginning of the Southern California bight. A lighthouse and foghorn at the tip of one of Arguello's long rocky fingers warns mariners coming down the coast to stay well offshore when they turn eastward into the Santa Barbara channel. It was a fine location for measuring waves if we could figure out where and how to install a wave meter.

A mile of heavy cable, mounted on an axle and skids, had been delivered to a railroad siding nearby. From there a dukw dragged it down a long dirt road to the top of a rocky slot in the south side of the point, 40 feet above a tiny cobble beach.

On that part of the coast the cliff is almost unbroken, and there was no good place nearby for a dukw to enter the water. However, at the Coast Guard boat station about a mile away, there was a small protected beach where the cliff was only 12 feet high. On learning that the Coast Guard owned a small bulldozer, I slightly exaggerated my experience in running such machinery and got the loan of it. When the cut was half made, I inadvertently drove the 'dozer off the almost vertical edge, and it came within a whisker of rolling over on me. There was a bit of a problem getting it back up, but by the end of the day we had a short, steep dukw road to the beach.

Once in the water, the two dukws ran close under the slot in the cliff and anchored. There we shackled them together side-by-side, heaved a line from the cliff, and pulled the end of the cable aboard. The conductors were then spliced to the sensors, and the tedious business of transferring most of the cable from the spool on the cliff to the dukws began. We hauled it aboard and "figure-eighted" the cable with alternate loops in each cargo compartment so it would pay out without fouling.

Once the cable was aboard, the dukws headed due south from the point, dragging another thousand feet of cable off the spool to get it clear of the rocky zone around the point. Then they turned westward, out to sea, paying out the cable that was carried aboard. Each heavy loop chonked and clumped as two of us flipped it into pipe guides and braked it with muscle power. At its seaward extremity we lowered the tripod and got a fix on its position. That installation was decidedly a success and would have been more so if I had not cut halfway through my thumb with a hacksaw when the job was almost finished. But, when everything works, one can ignore such minor setbacks.

Everything, at Point Arguello, meant two kinds of pressure transducers (the new type was a thermopile inside a rubber bellows that measured adiabatic heating caused by pressure changes) and a wave direction indicator invented by John Isaacs. The latter consisted of a Rayleigh disk—a flat sheet of stainless steel about a foot in diameter supported on a rotatable vertical rod connected to a selsyn motor. This disc acted like a crosswise weather vane. That is, instead of pointing into a wind or current, It took a position perpendicular to the direction of flow. For orbital currents that reverse direction with every wave, the disk was ideal. Signals from the selsyn master under the sea were repeated by a slave selsyn motor ashore that drove a recorder pen. With these instruments every wave's height, period, and direction were measured for some months.

Three distinguished visitors came to look at the installation: John Isaacs, who had moved from Berkeley to the Scripps Institution of Oceanography in

La Jolla, California, Jim Snodgrass also of Scripps, and Allyn Vine of the Woods Hole Oceanographic Institution.

After examining the wave recorder, which was performing beautifully, we all spent several hours collecting excellent fossils from the bedded limestones along the coast. These rock slabs went into the trunk of the car we used to drive Al Vine to the airport at Santa Maria. Those were the days of DC-3s and barracklike terminals where one clerk comprised the entire staff of a plane. All of us looked salty and scruffy after a day climbing about on the cliffs, and Al was not dressed as a dignified air traveler. The clerk eyed us warily as he weighed Vine's suitcase and announced, "Thirty-two pounds".

"How much weight am I allowed?"

"Forty-four pounds."

"Just a minute," says Al, "I'll go get some more rocks," and he started out the door.

Shadows of doubt and dismay crossed the clerk's face. He called wistfully at Vine's back, "It's not necessary to have the full weight, sir."

By this time we were sure that every possible wave recorder problem had been met and solved, and the installation would be permanent. Wrong again! A few months after we left, the lighthouse crew decided to use the rocky slot where our cable came up from the cobble beach for a trash dump. Eventually the trash caught on fire and all the insulation and soft parts of our cable where it hung down the cliff were consumed. We scratched Arguello.

The Berkeley instrument design group then sensibly decided that the next wave meter should be installed at a place that was much easier to get to, such as the tip of the Monterey peninsula. This required the cable to be laid out through a small bay near Asilomar. It was a simple job; soon we had a dukw loaded with cable spool, the splice to the instruments made, and the shore recorders in place. We waited and waited and waited, but the waves were too high to install the tripod.

Frank Snodgrass came down from Berkeley to see what the holdup was on a day that the drivers and helpers were away. When we reached the site, the sea looked flat. In a moment of inspiration I said, "Let's you and I run the cable. Right now."

He looked dismayed. "Just the two of us?"

"Sure, it'll be easy. Then the job's finished."

With some trepidation he agreed. We climbed aboard the dukw, and I drove it out through low breakers. We lowered the tripod, slowly laid the first few hundred feet of cable, and tested the primitive wooden brake we used to keep tension on the cable as it was laid. All seemed well when we started toward shore through breakers only 4 or 5 feet high. Then a wave notably larger than its fellows lifted the dukw, and it started to surfboard. As the dukw's speed tripled, the cable spool began to spin. We were still OK; the cable was going in fine until the wave passed us; then the dukw slowed markedly, but momentum kept the cable spool turning. It was evident that the rotational inertia of the mass of cable was beyond the capacity of our simple brake and Frank's muscle; the spinning spool started to throw off large loops of cable. After the first of these heavy lassos encircled the driver's compartment Frank and I scrambled out on the forward deck where we could fend off the coils of this mechanical python that could crush us if we got caught in a loop. With the next wave the dukw moved ahead again, and for a short while it looked as though all would end well. Then a loop hung up on something, bringing the dukw to a sudden stop. No longer able to rise with the wave crests and surfboard ahead, the dukw took a breaker over the stern, filled with water, and sank on the spot. Frank and I swam ashore, concluding that maybe somebody up there didn't want waves recorded.

SHIPS AGROUND

One foggy morning in July 1949 the beach party arrived at the Point Arguello to find the lighthouse crew there in a state of excitement. "There's a ship on the beach just north of Honda Point." That point was already famous for shipwrecks because a flotilla of Navy destroyers had run on the rocks there in 1923, and some of their rusty bones could still be seen. We piled in the dukws and dashed off to see the latest victim.

From the road, the hazy outline of a freighter showed through the fog. It was hard aground on the outer bar with breakers smashing into its exposed flank and white water going higher than the stack. Coast Guard crews were huddled around fires on the beach unable to get out to the ship; their surf boat waited offshore, unwilling to risk running the outer breakers, which were about 12 feet high. This was a job made for dukws, so we volunteered to take Coast Guard officers out to find out what ship it was and what had gone wrong.

In the hours since the ship first grounded, it had been pushed around by the waves until it lay almost parallel to the outer bar, broadside to the breakers. As each wave slammed into its 422-foot-long side, a force of hundreds of tons drove the ship a little farther landward and threw a great sheet of water into the air that landed on the upper decks and drained off in a series of waterfalls. We circled the ship once to look it over and take photos; plainly its situation was desperate. Then we brought the dukw into its quiet lee and climbed a rope ladder to the deck.

We were aboard the *Ioannis Kulukundis* of Syra, Greece, a Victory class ship loaded with 9,000 tons of Canadian wheat bound for South Africa. Smoke still came from the stack, but the main engines were down because the cooling water intake was filled with sand. Discipline had vanished. The ship was dying, and the crew had already moved their sea bags full of personal possessions to the deck. Captain Mastrominas sat disconsolately on a coil of line with his head in his hands. He told us that either the mate had not laid the course properly, or the helmsman had not paid proper attention. His voice almost broke when he told us that if the ship had steered a course one degree further to the west it would easily have cleared Point Arguello and the sands that held it.

The Greek crew hung over the rail consuming the ship's remaining liquor and begging the dukw drivers to take them and their belongings ashore. As a friendly gesture they tossed down half a bottle of brandy, an act that started a rumor on the beach that we were helping smuggle the ship's cargo of whisky ashore. Later the captain asked us to ferry the crew to the beach, and the Coast Guard agreed, since there was no future for them on the ship. This rescue mission started another rumor that we were aiding illegal immigration.

When the underwriters and salvage experts arrived at the site, we ferried them out in return for the right to listen to their comments. "What happened?" they asked the captain.

As soon as the ship had struck, he awoke and ordered, "Full astern." Instead of backing the ship off, the foreward wash of the propeller moved sand toward the grounded waist of the ship that settled around the hull and gripped it even more firmly. Now the bow hung over the sandy trough inside the bar, and the stern was afloat outside the bar, the ship flexed with every wave; it was only a question of time before its back would break.

We followed the experts as they groped through the hull. It was dark, of course, without generators, and the meat in the refrigerators had already begun to rot. The passageways had the aroma of long dead Limburger cheese, and each impact of a wave set the hull vibrating in unusual modes. We all descended cold greasy ladders and poked around the dripping engine room by flashlight. It was eerie to watch the ship's steel skin wrinkle and move in and out like an accordion as the hull flexed with each passing wave. But all five cargo holds were still dry, and we sat on our haunches ankle deep in the great piles of wheat while bright lanterns threw deep shadows across the creaking hulk. In this eerie conference room, the salvage experts conferred and gave their opinion that the ship itself was not worth the cost of salvage because there were so many similar ones on the market, cheap. The wheat, however, was valuable if it could be removed before it got wet, expanded, and burst the compartments that held it.

On this sad note we returned the visitors to shore and left for Berkeley. A month later we learned that continued pounding by the waves had broken the hull in two at the engine room and driven both the bow and stern sections, with their holds still dry, closer to the beach, allowing much of the wheat to be salvaged by aerial tramways.

A year later a news flash about a Japanese ship that had run aground near Fort Bragg, California, drew me to see how it would be handled. After a two-hour dash from Berkeley, I found myself on a rocky cliff with some of the same salvage men who had been at Point Arguello, peering out through the fog at another beached freighter. The ship was about 600 feet off the cliff, and we were standing about level with its bridge. The *Kenkoku Maru* had ridden up on some rocks at an angle of 45 degrees with the coast, and its painted water line was far above the level of the sea. Its captain was rumored to be considering hari-kari (ritual suicide). Some of the professional salvors took a short look, decided the ship would be there forever, and left. One did not.

The Smith-Rice Company knew what to do and had taken a no-cure-no-pay contract to get it off the rocks and into a San Francisco dry dock. Mr. Rice had taken personal charge, because he was using his own money. If he did not get the ship off (no cure), he would not be reimbursed for the cost of the attempt (no pay). I carried a movie camera and stuck close to him to see what he would do, making myself useful by calculating tides and carrying messages. The *Kenkoku Maru* carried no cargo, and it had been making 10 knots when it ran up at high tide. This was the worst combination of conditions. There was nothing to remove to lighten the ship, no higher water to float it, and the friction of a hard grounding to overcome.

As soon as a ship runs aground, it blocks off wave action in its lee. This zone of quieter water allows the suspended sand churned up by the waves on either side to settle in a sand bar called a tombolo on its landward side that will eventually connect the ship to the beach. As this pile of sand grows and the sand under the ends of the ship is eroded, the chance of the ship breaking its back increases, especially if the hull is allowed to move.

There were also rocks of unknown size and shape directly under the hull that might project up through the ribs and make it virtually impossible to drag the ship off. One thing was in the salvor's favor: The *Kenkoku Maru* was a small ship (about 6,000 tons) with a relatively rigid hull and a stout double-bottom; although the ship's outer skin had been ripped open by rocks, the upper side of its double bottom was still watertight.

First Mr. Rice ordered water pumped into the cargo compartments to hold the ship hard against the bottom so waves could not move it or flex it while he got ready for the big pull. Next his men ran stout steel cables from the bow and stern to "dead men" (buried logs) on the cliff. These would help hold it steady while an aerial tramway was rigged to move men and equipment aboard. He and I were the first to ride over the surf from cliff to ship in an open metal basket; after us came eight of the largest self-powered air compressors the tram wire could support. The plan was to float the ship on a bubble of air and pull it off using "ground tackle" during the highest tide of the month. Rice was edgy; if he couldn't get the ship to dry dock, his company would be out the cost of this effort, about $200,000, a sum which was roughly equivalent in 1988 buying power to $2 million. It was a big gamble with the outcome much in doubt.

The highest tide of the month was a week away at three in the morning, and every action focused on that moment. By that time two large salvage tugs offshore would have the ground tackle in place. A tug pulling only with its propeller does not have much capability for putting tension on a line to a fixed object; a 2,000-HP tug might pull as much as 40,000 pounds, or about the same as a 200-HP tractor on land. But if it puts out an array of heavy anchors (ground tackle) to hold itself firmly and then pulls with a deck winch and multiple pulleys, it can pull several hundred thousand pounds. Rice would need every bit of that.

Before the sun set on the appointed night, the tugs were ready, and the lines to the ship were rigged. Aboard the *Kenkoku Maru* the pumps began unloading the water they had previously pumped into the hull. Hoses from the air compressors had been fitted to the tops of the double-bottom tanks so that they could be kept full of air even though it steadily bubbled out through the rock-punched holes in the outer hull. Finally, the cables to shore were released, except for one on the bow that was attached to a bulldozer on the cliff. It would pull to one side and help swing the hull perpendicular to the shore.

The timing of each step was carefully planned so that the maximum pull and the most buoyancy would come at the highest tide. At midnight the tugs started taking a strain and just before the highest tide the *Kenkoku Maru* came free, but not without a lot of grinding and screeching as the steel slid across the rocks. Bubbles belched from the torn bottom as it was towed triumphantly to a San Francisco drydock. Mr. Rice cured, got paid, and threw a helluva party.

AERIAL PHOTOGRAPHY

As the Waves Project became increasingly involved in the scientific study of coastal processes and shoreline changes, it found that good aerial photos were very helpful. We especially wanted portraits of tidal entrances and headlands in various wave conditions. Since it was hopeless to try to describe exactly where pictures were needed and what they were supposed to show, it was decided that I would fly with the Navy and take the pictures myself. Our preference was for oblique photos, looking down on the shoreline at an angle from an altitude of 5,000 feet.

The Navy was very cooperative and made Catalina PBY flying boats, Beechcraft, and blimps available as well as cameras, film, and photo labs, all of which were surplus. The Catalinas were especially useful because they were reliable and slow; and the large Plexiglas "blisters" on the cabin aft of the wing could be opened to permit a clear, wide view of the coast. In fact, they were so large that the photographer using them was attached to the plane by a light cable to make sure he was not bounced out when the air got rough. On the other types of aircraft, it was necessary to remove a door to get oblique shots, and it was not unusual to sit for hours on a makeshift seat just inside the door opening holding a heavy aerial camera, ready to fire at suitable subjects. It was cold, uncomfortable, and hard on the arm muscles because these large cameras weighed 20 to 40 pounds in still air. When turbulent air bounced the plane, the effective weight of the camera increased considerably, and there were times when these vertical battering rams almost got heaved overboard to keep them from smashing my knees. These cameras all had pairs of rugged handles, each capable of lifting at least a ton, so the wry instruction issued to Navy photographers was "Don't worry about the camera. Just bring back the handles."

For a period of three years in the free time between other jobs mentioned, I flew in dozens of Navy aircraft from several squadrons, eventually photographing much of the Pacific coast several times from various altitudes and angles. Because the Project was interested in storm waves and their effects, it was not unusual to fly in bad weather.

Sometimes we pushed our luck a little too far. On one occasion a Catalina flying boat assigned to the project took off from Astoria Naval Air Station into a near gale, after the pilot had been assured by the Navy aerologist that "the weather topside is a lot better." The big plane careened down the runway like a drunken goose, hopping from one set of wheels to the other. A lucky bounce got us over the tree tops at the end, and with the pilot fighting the manual controls all the way, we climbed to about 1,000 feet.

My seat was a wooden platform under the port blister, to which I was secured by a wire tether. Even on a quiet day, this was not a particularly comfortable place, but on this flight I was tossed about like a cat in a tumble dryer. It was necessary to jam the heavy camera in a corner and brace both feet against it to hold it there. When turbulence dropped the plane, I would float for a moment in midair and then be jerked back by the wire into a crash landing on the hard seat, all the while fighting to keep the camera from coming adrift and smashing itself or some vital part of the aircraft. My navy friends were green with air sickness as the violently bouncing plane banked steeply over the Columbia, where we saw a Coast Guard PBY like ours land, disabled, in the river beneath. That was enough.

The pilot sensibly completed his turn and landed again on the strip from which we had just taken off for what may have been the shortest flight ever made from Astoria airport. Once back on solid ground, he regained his proper color and went to discuss the weather topside with the aerologist once again.

I was intrigued by other kinds of waves that could not be seen or measured in the usual manner. One way of studying long-period waves in harbors is to take time-lapse pictures of groups of anchored or loosely moored boats with a 16-mm motion picture camera. Pictures taken at the rate of one frame a second and viewed at twenty-four frames a second speed up the action and convert the seemingly random motions of the boats into very distinctive patterns. Usually all the boats dance in one direction; then they all reverse and dance the other way with a period of about three minutes. I had often taken time-lapse pictures of the boats in Monterey harbor from a low hill, but when such pictures are taken at a low angle, the relative amount of boat motion is hard to measure. It seemed possible that an aerial camera mounted on an aircraft hovering directly above the harbor could take a series of pictures that could be converted to a time-lapse movie.

For this the Navy made a blimp available. Blimps are streamlined balloons with a gondola hanging below driven by a couple of propellers. The pilots sit in a Plexiglas bubble at the bow, and behind them there is a cabin lined with windows containing a few seats and a small galley. We had used blimps before to take oblique pictures of the shoreline—a fairly easy task requiring one simply to open the aftermost window in the cabin and point the camera while the blimp flies slowly along the coast. But in order to get a sequence of vertical photos of the fishing boats in Monterey harbor, the blimp had to hold still directly above the center of the harbor, and the camera had to be held outside it with the viewfinder cross hairs trained on a specific point below. A ten-minute sequence (200 exposures at three-second intervals) was needed at an altitude of about 5,000 feet.

The plan was for the blimp's pilot to maintain the airship's position above the harbor as well as possible. I would straddle the window sill with one leg and most of my body outside; Don Mc-Adam, our phlegmatic field party engineer, would sit on the cabin deck with a firm grip around my other leg. This arrangement felt reasonably secure, so the blimp eased into position above the center of the harbor.

I aimed the camera and started the intervalometer that automatically fired it, concentrating so hard on keeping the cross hairs on the target that I became oblivious to all else for about five minutes. It was plain that the blimp was losing altitude because the boats in the viewfinder were steadily growing larger. Then I became vaguely aware of a series of clanks and clumps in the cabin and noticed that my body had twisted considerably to keep the camera vertical and aimed properly.

Finally the pilot's voice came over the intercom, and it had a frantic quality: "Mr. Bascom, I can't hold us anymore." At that I looked about for the first time and saw that the blimp was in an almost vertical position, tail down, at an altitude of only 1,500 feet and settling fast. Don McAdam, still clutching my leg, was flat on his back against the after wall of the gondola and piled around him were all the loose articles that had rained down as he clung to my leg, including utensils from the galley, books, jackets, etc. As the pilot gunned the engines, and the airship began to level off, I thought how lucky it was for both of us that nothing heavy had fallen on his head and knocked him out.

Temporarily we became notorious as the guys who stood a blimp on end over Monterey Harbor. A little too late we recognized the huge difference between flying very slowly and stopping completely. At rest, without the dynamic stability that keeps blimps on an even keel, a blimp is delicately balanced; the weight of two of us far aft of the center of buoyancy had upset it. But the pictures we got were unique.

It was a lot of fun flying with the Navy as a civilian, free to come and go. They generously allowed us to use the photo lab equipment to develop Waves Project films and cooperated in many other ways. One result of this joint effort over several years is a large collection of aerial photos of the Pacific coast in many moods at the Water Resources Center Archives at Berkeley.

AMPHIBIOUS OPERATIONS

The work on beaches and in the surf led to an invitation from the Navy to observe and comment on practice amphibious operations. The first was at Oceanside, California; early on a fall morning in 1947 Isaacs and I set up an observation post on the bluff above the beach, where waves of landing craft were scheduled to discharge men and machinery. The weather was beautiful, and the largest breakers were only about 6 feet high, so we assumed the amphibious forces would breeze through this routine exercise. Our plan was to photograph whatever happened with long-focal-length aerial cameras that showed the time of each picture.

First it was necessary for us to be able to identify the various kinds of landing craft, ships, and vehicles that came and went on the beach below us. Mainly these were small, wooden LCVPs (Landing Craft Vehicles and Personnel), 65-foot steel LCMs (Landing Craft Machinery), and 115 foot LCTs (Landing Craft Tanks). The landing ships were LSMs (Landing Ship Machinery) and LSTs (Landing Ship Tanks), the distinction between the two classes being that craft were small enough to be transported on a ship. A couple of miles offshore, larger amphibious cargo and personnel ships moved slowly, waiting to offload into subsequent waves of craft.

We had hoped to assemble some statistics on the effectiveness of each kind of craft, including how they responded to the breakers and how long it took to discharge machinery and retract. But we were surprised by the complexity of even a small peacetime operation; the landings that started in an orderly fashion soon became too chaotic to record. A lot of things could go wrong even with low waves on a fairly hard sand beach.

The small craft were easily turned sideways on the beach face by low waves and currents; when they tried to retract, the wash of a boat's propeller would move sand under the hull, sticking it harder. Loose lines inevitably wound up in the screw; ramps would become jammed down, or up; and so on.

Many of the larger landing ships grounded on the offshore bar, dropped their ramps, and discharged their trucks into the deep trough inside the bar. These would immediately disappear from sight and be marked only by drivers swimming above the cab or standing on top of their sunken cargo. Engines were supposed to have air intake and exhaust pipes that extended above the cab so the truck could run under water but these often leaked, causing the engine to stall. Of course, any vehicle that stopped blocked all those behind it on the ship.

When the tide went out, the boats that had reached the beach face could not recross the bar, which was now too shallow, so they milled around in the pond between the bar and the beach face. Ships aground on the bar were stuck until the tide came in again, often with their seawater intake ports out of water so there was no water for cooling engines or fighting fires. By low tide the beach had become one big junkyard. These problems were not unexpected by

Navy officers who had been through such landings before, but I think we may have been the first to provide reports illustrated with timed photos showing what went wrong and when.

We asked ourselves, if amphibious forces have this much trouble landing on an easy beach in good weather, what would they do on a steep coarse-grained beach, or on a very flat beach with a wide surf zone if there were big breakers? It was not long before we learned some of the answers.

After watching such practice operations, we began to develop some insights. In 1949 Operation Miki was planned as a two-stage amphibious operation with practice landings first at Oceanside followed by "real" ones in Hawaii. The Wave Project's ground observations would be made by Professors Joe Johnson and Bob Wiegel; I would take pictures from the air, flying back and forth above the landing zone to get oblique photos of the beach operations.

This time there were more ships and men involved, but the results at Oceanside were about the same as before. There was to be a review of that operation by the senior officers on the following morning at the Amphibious Base in Coronado, so I spent most of the intervening night in the photo lab developing and printing a roll of large aerial photos. These pictures with clocks in the corners captured the sequence of events, leaving no question about what happened or at what time.

Admiral Doyle, then Commander of Amphibious Forces, Pacific, presided. He would listen to a commander give his version of how well his group had performed, and then he would select a picture that showed an array of boats stranded crosswise or otherwise in trouble and say, "Captain, are these your boats still on the beach at ten o'clock?"

Captain, looking horrified: "Oh, no, sir." Then, after inspecting the photo more closely, he would hesitantly recant, "Well, yes, sir, those *do* look like our boats."

At the end they all agreed that the operation in Hawaii would go much better and that much was to be learned from aerial photos with clocks in them. "Too damn much," one bruised ego suggested out of the side of his mouth.

In the Hawaiian phase of Operation Miki, the Blue Team would make the landing. That was us, the invading good guys, who were establishing a beachhead on some friendly country seized by bad guys known as the Red Team. In the landing, which would be made somewhere on the island of Oahu, the Red team would be composed of United States forces stationed there, also largely Navy. The attack was to be made on unspecified beaches within an announced five-day period, but the time and location were kept secret by those making the landings in the hope of surprising the defenders. My job was to take aerial photos of the actual attack from a Catalina seaplane based at Barbers Point Naval Air Station, west of Honolulu, deep in Red territory.

Because I knew the times and places of the landing, but these were not known to our "enemies" at Barbers Point with whom I ate and bunked, it was necessary to schedule the plane every day, as though that was D-Day, in order to avoid giving away the actual date. Early each morning I would load the cameras into the plane and take off; once in the air with cameras loaded there was no reason not to use up the film, so I set about photographing the most interesting coastal scenes of Hawaii, Molokai, and Maui from the air. Finally, after three days of this deluxe sightseeing, it really was D-Day.

The operation began as usual with an Underwater Demolition Team going in just at dawn to check out several beaches on the west coast of Oahu near Waianae for wave conditions and obstacles to landing. The swimmers saw only the beginnings of a beautiful day with low swell offshore and elegant beaches of white sand that would some day attract tourists. There were no breakers except those that broke directly on the steep beach face, and the swimmers had fun riding the swash of their uprush 50 feet or so up the coarse sand, where they would dig in and hold on for a moment until the water drained down through the sand leaving them high and dry.

On returning to the ships, they reported that the beach was free of obstacles, and the first wave of landing craft started shoreward. But beaches that are fine for swimming can be disastrous for boats and vehicles with a job to do. What the UDTs did *not* report was that the beach was made of coarse round grains of coral sand that were lubricated by the swash of every wave. They should have foreseen that it would be quite impossible to drive jeeps and trucks even the few feet required to get them off the beach and onto the hard ground behind it.

The first line of boats reached the beach, expecting to touch at the water's edge, drop their ramps, discharge marines, and back away. Instead, they were swept up the beach face by the thick uprush, turned sidewise, and left high and dry on the soft sand at the crest of the berm. The marines, glad to be ashore under any circumstances, left the instant the ramps dropped and slogged their way to the road through ankle-deep sand in which their vehicles immediately struck. It was almost impossible to beach a boat properly, discharge cargo, and retract again. Although a few of the larger waves would break directly on the beach face, and lose their forward momentum, more often the waves simply collapsed into a thick swash of water that would carry a boat upward.

The second and third lines of boats had similar experiences, except that some of these craft rammed the ones already there; soon the situation became hopeless. The surf that had looked so low and innocent to the swimmers smashed some of the wooden LCVPs to kindling in an hour. Other craft intended to be used as tugs would move in close to throw a towline to those already stuck and would themselves be picked up by a wave and tossed up on the sand. Atop the berm there was a line of stranded boats and stuck trucks surrounded by sweating marines.

It was obvious the Blues had picked the wrong beaches, and the Navy's high opinion of its competence in amphibious warfare derived from its World War II successes suffered considerably. We suspected that landing on these beaches was not much worse than on the coarse lava-sand beaches of Iwo Jima, but there the risks from enemy fire had been so great that the problems caused by sand and waves seemed relatively insignificant.

A couple of days later several hundred of the officers responsible for Operation Miki met in an open-air theater at Pearl Harbor to critique it. A tough commander ran the show. Precisely at 2 P.M. he convened the meeting as follows: "Gentlemen, we have two hours' time and seventy-two speakers. That allows one minute and forty seconds each. When the bell rings, your time is up. The first speaker is Admiral So and So," and he cocked a photo timer.

The admiral was halfway through his introduction when the buzzer sounded and the commander cut in, "Thank you, sir. The next speaker is _____," and so on. After the first few talks self-destructed, those assembled heard crisp, direct comments that rarely took more than 90 seconds. The critique was over promptly at 4 P.M.

About a year later, two Navy officers showed up unexpectedly at the door of my house in Del Monte Forest. The commander began, "I have brought the LSD *Gunston Hall* to Monterey with an amphibious unit aboard." The lieutenant added, "I have an Underwater Demolition Team for beach reconnaissance." In unison they said, "Admiral Doyle wants you to test us."

It is unusual for the Navy to send a ship to visit a civilian even with some sort of advance warning, but evidently a message from the admiral had gone astray. In any case the immediate question was how to test a collection of boats and a UDT team for the next few days. We went down to Monterey harbor and rode the captain's gig out to a large gray ship that dominated that corner of the bay.

An LSD, for Landing Ship Dock, is over 500 feet long and displaces 6,800 tons when its well is dry. The forward part of the ship is mostly machine shops and living quarters. The aftermost half of the ship is a docking well that can be "flooded down" to take on or discharge the amphibious craft it carries for long voyages. When the stern ramp is closed and the water is pumped out, the craft are in dry dock, easy to work on. LSDs are valuable work ships, as necessary to the amphibious fleet as a carrier is to the fighting fleet.

The Underwater Demolition Team had about thirty men aboard, bursting to show off their special skills. I asked the lieutenant what they would do if an amphibious landing were to be made on the nearby beach at Fort Ord. "Well," he said, "we would swim in at dawn and make sure there were no beach obstacles, estimate how well wheeled vehicles could move on the beach, and inspect the shoreward exits from it. Then we would swim back out through the surf to be picked up by our boats so we could report what we had learned."

"OK," I replied. "Do that tomorrow morning."

When they arrived at 6 A.M., I was waiting on the beach with a movie camera running. There was a gray mist on the sea when their LCPs with rubber rafts tied close alongside came zooming along just outside the breakers. The swimmers were macho types, some with wartime experience, but all had become acclimated to the warm waters around their base in Coronado. Half of them had on rubber suits; the rest wore only swim trunks. They entered the water while the boat was underway at 10 knots by bouncing one at a time onto the rubber raft whence they rolled into the water. Although the breakers were no more than 8 feet high and the distance to shore only about 600 feet, they had a hard time of it. Blowing hard and shivering, they hauled out on the sand like aging sea lions, their enthusiasm of yesterday gone. They were in no condition to measure anything or even to swim back out to their boat. There was no need to explain to that group that better protective suits and additional training in cold surf were needed if they were to go up against a real enemy.

The LCMs launched from the *Gunston Hall* did not do much better; they broached on the steep beach, missed the rip channels, and grounded on the bar. All of us learned. We recognized that there must be much better ways of landing men and machinery where no ports exist. Years later, on senior advisory committees of the Navy and Marine Corps, I would work on those problems.

Five years of research under Dean M. P. O'Brien was about to come to a close. The Waves Project had made considerable progress in understanding the interactions of waves and beaches, in developing wave-measuring devices, and in describing some aspects of shorelines development. Unfortunately project members did not promptly write up the results in formal scientific journals, and some of our findings were not made available for several years. Perhaps the best known account of our work is in my book *Waves and Beaches,* which became a widely used beginning textbook.

Although Mike O'Brien did not get to share directly in the fun and adventures of the beach party, he was certainly one of the leaders of oceanography in the study of the effects of the waves on shorelines and man's coastal structures.

He was a civil engineer who focused on solving specific problems and he liked answers that could be reduced to orderly equations; his questions were pragmatic: What wave conditions cause sand to move along the near-shore bottom? How can better wave forecasts be made? How does one calculate the forces exerted by waves on structures?

I much admired Dean O'Brien's style, especially his ability to juggle a lot of projects at the same time and to cut through complications with fundamental principles. Once we asked his opinion of some data that showed that the water in a line of holes across a beach stood at irregular heights. He answered in a grand professorial manner, "Water always flows downhill," and turned away, leaving us to figure out the details.

III

Military Matters

MINE WARFARE

During the summer of 1950, the Undersea Warfare Committee of the National Academy of Sciences conducted a summer study at the Navy Electronics Laboratory in San Diego. Under Dr. Gaylord Harnwell, president of the University of Pennsylvania, a group of sixty physicists and engineers drawn from university and Navy laboratories around the country assembled in some barracklike wooden buildings on the waterfront to see what could be done to counter a new kind of antiship mine that could make harbors unusable and many waterways impassable. I was a member of the group, on leave from Berkeley.

Mine warfare as waged in World War II derived from World War I experiences in the shallow water around Europe, which in turn came from those of the Russo-Japanese War of 1904 where "fields" of moored mines were first used. These mines consisted of a spherical float about three feet in diameter containing high explosive and bristling with "horns" about 8 inches long. Anchored to a weight, they floated upward as far as a wire-rope tether would allow. When these mines were laid, a clever mechanism adjusted the wire length so that the explosive sphere rose to a few feet below the low-tide water level. If a passing ship's hull chanced to brush against a horn and break it off, the explosive would ignite, and the shock wave would either sink the ship by caving in its bottom or disable it by breaking pipes and machinery. Sailors standing on deck, well above the explosion, would sometimes have their ankles broken by the sharp jolt. The existence of a few lines of such mines across waters used by ships, or just the possibility that a line of mines might be there, was a substantial deterrent to marine traffic. Often the first indication that an area had been mined was a sinking ship.

The countermeasure for such mines was to "sweep" a channel through the suspected area by towing a wire that was held out to one side of the sweeping ship by a paravane, something like a guided torpedo. This wire would cut the mine's mooring cable and when it bobbed to the surface it could be exploded by gunfire from a safe distance. Those were the easy ones.

In World War II the "influence" mine had been invented. These mines were cylinders about 6 feet long and 2 feet in diameter filled with high explosives, and dropped like bombs from aircraft or launched from ordinary-looking cargo ships with hidden underwater openings. They would lie quietly on the bottom until triggered by the magnetic field of a ship, or by the pressure signal generated by ship-made waves, or by both signals simultaneously. Some of the mines also counted ships and exploded when, say, the seventh ship activated the firing mechanism. Or they could be triggered randomly by a firing device known as a "roulette" wheel, or by a clock, or by combinations of the above. These devices made the psychological threat much greater. Who knew if there were mines on the bottom and when they might be armed and ready to explode?

It is relatively easy to sweep purely magnetic mines by dragging a magnetic-field generator through the suspect waters, but very difficult to generate the pressure pattern of a large ship's bow wave without using a large ship. So, the problem our group worked on was this: What is the best way to counter the threat of these bottom mines that could deny American ships the right of passage in many straits and harbors?

Against sophisticated triggering mechanisms, which seemed impossible to sweep, one strategy is to find and attack each mine directly. But how does one find mines in a huge harbor littered with metallic junk that has collected over many decades—especially when the junk has magnetic and sound-reflecting properties not unlike those of the mines? Moreover, many harbors have bottoms made of deep soft mud in which these mines bury themselves but still retain their effectiveness against ships.

My job was to learn about mines on and in the bottom, so I was given several dummy mines and a group of Navy divers to experiment with. The study needed answers to a number of questions: How far would an air-dropped mine sink into the mud? On a hard bottom what velocity of current would move a mine or bury it? What were all those other objects on the bottom that confused the search? And so on.

At first we sent down Navy divers dressed in "Jack Brown" shallow-water diving suits, to examine various parts of the bottom of San Diego Bay; when they surfaced, I would interview them in detail about what they had observed. The fellows found it hard to describe what they saw when the visibility at the bottom was near zero, or to translate groping in the murk into a useful description. So, after a few minutes of rather casual instruction from Lieutenant Buster Tribble, I put on his suit and went down to feel for myself. The Jack Browns were safe and simple. Air was pumped down from the surface into a large face plate from which the excess constantly bubbled away. The suit kept the diver reasonably dry; it protected him from cold, mud, and sharp objects; a lifeline could drag him back to the surface if necessary, and a tender constantly watched for signs of trouble. Diving was a straightforward way to learn about bottom conditions down to 60 feet or so if the currents were moderate. Sometimes the diver could see a couple of feet.

In order to find out what happened to mines in different kinds of bottoms, we dropped test mines at various places in San Diego Bay and marked each with a small buoy so it could be easily located. In the central bay the layer of soft mud was only a foot thick, and the upper half of the mine remained "proud of the bottom," but the mine farthest up the bay, some six miles from the entrance, sank several feet below the surface in soft mud. Once a week I would inspect each mine by feel, the visibility being zero, diving down into the mud to determine if its position had changed. Such diving is not for the fastidious.

As for checking the mine at the entrance to the bay where tidal currents can be quite strong, the Navy divers recommended against the light diving dress. "When we dive there on the degaussing range, we use hard-hats," they warned, "or we get blown away by the tidal currents. Ya better call Mr. Sylvester." So I did, and Mr. Sylvester, a long-time chief master diver, agreed to help. The next day he showed up at the Navy Electronics Lab's pier on a tuglike diving boat, ready to go. Sylvester was a big, easygoing fellow who had been running diving jobs for over twenty years; like most other divers who have survived that long, he was meticulous about safety matters.

We anchored up near the south side of the channel where the water depth was about 40 feet. As a boy I had read Commander Edward Ellsberg's books on submarine salvage and pored over photos of diving operations, but I had never actually seen a deep-sea diving dress before. Now a regular Navy diver dressed in one while I watched, and Chief Sylvester explained its complexities.

The suit, made of rubber-covered twill, is entered through the top. The only other openings are for the hands which are sealed by long rubber wristlets that are so tight a lot of slippery soap and effort are required to get the hands through. Then a bronze corselet ring goes over the head, sealing the upper edge of the suit. Lead-soled shoes weighing 20 pounds apiece are lashed to the ankles, and a weight belt with another 40 pounds of lead is suspended between crotch and shoulders by heavy leather straps. Finally the dresser brings up a spherical helmet made of tinned copper with three round glass viewing ports, each protected by tiny bars. He sets the helmet in place on the corselet and, with a rotary quarter twist, locks it securely in place. Already attached to the helmet are the airhose and telephone line; now the lifeline is secured to the main straps, and the diver climbs down the ladder until he's underwater and weightless again. Then his tender lowers him to the bottom on his lifeline.

Fascinated, I watched the operation, questioned the diver below by telephone, and learned the signals to the tender on the lifeline. It was a darned short course in hard-hat diving, but in an hour Mr. Sylvester had been persuaded to let me try it out. I wanted to get the feel of being in a deep-sea suit on the bottom, although in these very easy circumstances, this dive was far from being adventurous. A diving dress is a substantial barrier against the outside world; when the dresser closes the face plate and cuts off direct conversation with the others on deck, the isolation is instantly palpable. With the force of gravity on the diver's feet nearly twice as great as usual, it is a considerable effort to slog a few feet across the deck to the ladder.

Mr. Sylvester picked up the microphone, and his casual voice came over the speaker inside my helmet. "Y'all ready? Just climb down."

As I descended the ladder, the buoyancy of the suit gradually reduced the weight on my feet, and by the time my helmet went under, I was just barely afloat. "Dump a little air—not too much" was the next instruction. Inside the helmet there is an air-dump valve actuated by a touch of the right side of the chin. As the air escapes and the suit volume becomes smaller, the diver sinks. Slowly I settled to the bottom, where it was possible to stand solidly on two feet and look about in reasonable visibility.

The weighted suit that had been a terrible burden on deck was a pleasure on the bottom. It was warm and dry; there was an endless supply of air and a telephone. The only problem was that the constant hiss of air entering the helmet made it impossible for me to understand words coming through the telephone speaker. In order to hear what those on deck had to say, it was necessary to reach up to a valve on the outside of the helmet and shut off the incoming air. There was plenty of air in the suit for a few minutes of breathing but, wary of blacking out, I would soon open the air valve again.

The point of examining ground mines in various kinds of diving dress was that few if any of the others involved in the mine study had dived under these circumstances. Later I was able to make some sensible comments about proposals for using divers to find, indentify, and salvage or detonate mines. In the final report I wrote some sarcastic words, approximately like the following, about using divers to find and identify mines: The diver is lowered into cold inky blackness until he sinks over his knees into soft mud riddled with fragments of junk. He cannot see his hand against his face plate, or a compass, or a detection instrument. After about a minute a voice from above comes over the telephone: "Don't just stand there. Go find a mine." There had to be a better way.

Several British mine experts were members of the group; their harbors would be obvious targets for a mine blockade in some future military confrontation. One of those officers sidled up to me one day, made sure no one was watching, and sneaked a photo out of his wallet. "This is top secret," he confided. It was of a television screen, a rare piece of hardware in 1950, and it showed the single word, *Affray,* in letters of steel.

A few months before a British submarine, returning on the surface from a peacetime training mission, had been accidentally rammed by a freighter and sunk in an area where there were dozens of other wrecks. The sonars then in use could locate wrecks in shallow water but not identify them, so the Brits had built the first underwater television system. A hard-hat diver equipped with a huge primitive camera had been sent down with orders to photograph some object that would positively confirm or deny that one hulk was the submarine they were looking for. With marvelous luck, considering the circumstances, he got the clinching evidence, the name of the sub, *Affray.* It was the birth of a technique that is still developing.

The visitors also brought some World War II mine-hunting reports from a Royal Navy group named ORGASM. This whimsical acronym attracted more than average attention to the Operations Research Group for Anti-Submarine Measures; they were equally known for the last page of their reports. It was a carefully posed picture of the group's commanding officer trussed to a chair behind his desk while a Hitlerlike interrogator is about to inject him with a horse-sized hypodermic needle. The caption was, "This is a SECRET report. Don't tell anyone." It was very effective; everybody smiled, and everybody remembered.

The British Commonwealth also contributed a marvelous little man in his seventies with dancing blue eyes and a crown of white hair who effervesced enthusiasm. Sir Charles Wright had started his career with Shackleton in the

Antarctic looking for the lost Scott polar expedition. After that he had become a telegraph engineer in the frontline trenches of World War I, a navy problem solver, and head of England's National Physical Laboratory. After work he and I would revive ourselves in his rooms with whiskey and sea stories. The problem our study group had with Sir Charles was that, no matter what ideas we thought of, he had already tried them, and they didn't work. He could remember in great detail how the similar earlier experiments had been done and why the outcome was unsatisfactory. No one doubted that he was right, but the others needed to try their own ideas and make their own mistakes. Sir Charles, always cheerful, understood and often contributed suggestions that helped reduce our chances of failure. Best of all, he never said, "I told you so" when our results turned out to be as poor as he had foretold.

An efficient search method requires a high "signal-to-noise ratio." That means that the response (or signal) from the object being sought should be large compared with the response from its surroundings (the noise). For example, it is difficult to search for mines with a magnetic detector if there are a great many other magnet-attracting objects in the region. Thus, it would be easier to hunt mines anywhere except in a harbor that is heavily littered with pieces of iron junk.

There was speculation on how much such junk there was on the floor of San Diego Bay and how to measure it. Finally some bright fellow suggested we secure a hydrophone to a short piece of pipe, tow it along the bottom, and listen to the sound it made. This we did and were rewarded with a steady clankety-clank as the pipe was dragged up and down the main channel. The bay sounded like paradise for a scrap dealer but an obstacle course for a mine-hunting physicist. In addition to interfering with magnetic instruments, the junk also reflected sound, depending on its size, shape, and orientation. Briefly the group considered whether one helpful countermeasure would be to rake or dredge the bottoms of certain bays to clean up the junk.

Some good ideas came out of that summer's work over 35 years ago but mines of the type described are difficult to counter and can be a serious threat. In the Persian Gulf even the pre-World War I type mines became a problem because this unglamorous part of the Navy's operations had been neglected.

THE SCRIPPS INSTITUTION OF OCEANOGRAPHY

There were many overlapping interests between the Navy Electronic Laboratory's mine hunters and the nearby Scripps Institution of Oceanography, so I asked a few questions about the origin of the latter. In 1903 Dr. William

Ritter, head of the University of California's Department of Zoology, arrived at San Diego Bay with a small group of students to study the marine biology of the region. On the sand near the sumptuous Hotel del Coronado, they moved into "furnished tents" that could be rented for 50 cents a day and started collecting plankton. The largest contribution to that first summer's work was a gift of $500 from wealthy newspaper publisher E. W. Scripps. He subsequently contributed more money and his yacht, and, best of all, suggested that Ritter contact his sister, Ellen Scripps, a wise and wealthy woman, for support.

The following year the group moved north about ten miles to a small wooden laboratory building in a public park at La Jolla, built with funds subscribed by some of the local citizens. Professor Alexander Agassiz of Harvard, a world-famous naturalist, was brought in to comment on the new lab and the possibilities it opened. He was delighted by what he saw and not only gave a supporting lecture but personally contributed books and scientific equipment. This enthusiasm helped encourage Ellen Scripps to donate $50,-000 to support the work for three years. The next year the lab purchased 170 acres on the beach north of town for $1,000, and Ellen contributed another $1,000 for a road to reach this remote site, a mile away.

By 1912 this independent laboratory had achieved enough stature to become The Scripps Institution for Biological Research of the University of California, and in 1925 the present more general name was adopted. After World War II, influenced by Directors Harold Sverdrup and Carl Eckart, whose interests were primarily in physics and mathematics, the shift from biology to physical oceanography and geophysics continued. By 1950 the Institution consisted mainly of a pier, the sailing ship *E. W. Scripps,* three small laboratory buildings plus some tiny wooden houses for the faculty, and a staff of fifty persons.

During that summer of 1950, a party was given by Dr. Russell Raitt, a seismologist and one of the leading faculty members, and his wife Helen. The idea was to bring together some of the visiting physicists from the San Diego mine-countermeasures study with the Scripps staff. So, on a warm August evening, the two groups congregated to swap sea stories and eat a large pot of bouillabaisse made of sea animals gathered earlier that day from the nearby reefs and beaches. We were informally scattered around Helen's patio, eating a delicious concoction of mussels, clams, and fish, when some sharp-eyed physicist exclaimed, "These clams seem to have letters written on them." The rest of us inspected our dinner more carefully, and sure enough, nearly all the small white clams had strange glyphs on their shells.

A choking sound attracted attention to the purpling face of a Scripps biologist. His fingers dived into the bowl on his lap, and he held up an example, wailing, "These are *my* clams, and that's *my* code. They live in the beach by Scripps pier. Every week I dig them up and weigh them to see how fast they're growing. You're eating my experiment!"

There was a moment of silence as we mourned the dear departed. Then we reconsidered and resumed eating; after all, the clams were tasty, and we were in our proper place at the top of the food web. However, as fellow scientists with our own share of experimental problems, we restrained our sympathetic laughter when he moaned, "This will set me back a year!" and tried to comfort the fellow by arguing that his measurements were of little value anyway. He refused our consolation and we were forced to conclude that some biologists have no sense of humor.

At the end of the summer Dr. Roger Revelle, the acting director, invited me to join the staff at Scripps, intimating that many interesting projects were just ahead. The Waves Project was tapering to a close, John Isaacs had been at Scripps for two years, and it was time for me to try something new. The shallow water coastal work had been fun, but I still had not really seen the open ocean or glamorous distant shores. After five years my education in oceanography had scarcely begun. So I accepted the offer.

Not long before moving to La Jolla from the pine forest at Monterey, a couple of friends from the Navy's SOFAR Project stopped by, inviting me to take a relaxing overnight cruise on their cable-laying ship—the same one for which the dukw had run a line through the surf. Guilelessly I accepted, and a couple of days later was sunning myself aboard the *Glassford* off Point Sur, watching with indifference while the crew grappled for a submarine cable thick as a man's wrist, brought it aboard, and opened it up. I was casually leaning on the rail when the cable officer approached and said, "We're ready for you to make the splice now."

For *me* to make their splice? Until that moment no one had mentioned that I was supposed to work for my holiday at sea. When I looked at the maze of conductors and strength members to be dealt with, it was clear that the complex splice required was beyond my competence.

There was no point in giving away that secret so I replied coolly, "If I weren't here, what would you do?" The officer indicated that a chief petty officer aboard could probably do the job. With just the proper touch of loftiness and condescension I suggested, "This is a great chance for him to get the experience." From then on, my contribution to the splice consisted of making appreciative remarks about the chief's work.

While we waited, the *Glassford* rolled easily in light swell on a glassy sea surface. There was not a breath of wind in the early afternoon, and the air temperature went from warm to uncomfortable. Scanning the offshore horizon for a weather front or a breeze, we observed a dark line, low against the water a few miles away. We tried to guess what sort of object would have that appearance and hazarded the opinion that it was a string of barges moving northward, somehow without a visible tug. But as we watched, the line expanded vertically, and soon it looked more like a box; we changed our guess to a new kind of aircraft carrier without a superstructure. Both the shape and the change were puzzling, so I trained a movie camera on whatever it was and filmed away. As we continued to watch, the "carrier" shrank back into a thin line of "barges" again and then finally, as it continued to move, resolved itself into an ordinary cargo ship. We had been watching a mirage, but the explanations we discussed at the time were rather poor guesses; later I read up on the subject.

Those on the deck of the cable layer had seen a rare "mirror mirage" caused by a density inversion in the air not far above the ocean surface. The low dark line was the part of the ship's hull we could see directly beneath the inversion layer; the upper dark line (the deck of our "carrier") that made the image look rectangular was an upside-down mirror image of the low line reflected by the underside of the atmospheric layer. The decks and superstructure of the mirrored ship somehow merged to fill in the space between the two lines.

The chief's splice was a success, but there was one annoying result of that trip; while I was at sea, a lymph node in the valley between my right leg and abdomen became greatly enlarged. A few days later doctors at Carmel Hospital removed the node but neglected to mention the pathologist's report. I remained blissfully ignorant of things to come, and we moved to La Jolla.

In order to obtain an office at Scripps it was necessary to apply to Dr. Martin Johnson, a senior professor who was chairman of the space committee. My informant smiled cryptically, "You'll find him on the second floor of the old building at the end of a passage behind the men's room."

Following the instructions, I groped my way down a narrow hallway toward a crack of light at the end, and knocked. There was a muffled, "Come in," and I opened the door to a revelation. Martin might well have grinned and said, "Gotcha!" The domain of the chairman of the space committee was perhaps 8 feet square, less the piles of books and papers that reached almost to the ceiling on all sides except the window and the door. His desk and chair barely fitted in the remaining space.

"What can I do for you?" he said as I stood at rigid attention for fear of touching one of the stacks of papers that hemmed me in and triggering an avalanche. I was awed by the genius of the system. Who would have the guts to ask this guy for space—or complain about whatever he gave them? I threw myself on his mercy.

Dr. Roger Revelle, the acting director, was 6 foot 4, handsome and personable, a veteran of cruises on the old *E. W. Scripps* to study carbonate deposits on the ocean floor. Not long out of a Navy uniform, Roger "knew the ropes" in Washington, which meant he was familiar with the most likely sources of ocean research funds. We had often talked before when we met informally in the garden behind the old Cosmos Club on Lafayette Square across from the White House. There, on sweltering summer evenings in the nonair-conditioned wartime city, oceanographers would gather to hold the sticky air at bay with the club's famous mint juleps.

When I presented myself before his desk at Scripps to find out what he wanted me work on, my eyes fixed at once on an orange-red poster behind his head: A radical slogan, "Fan the flames of controversy" floated above the fire of revolution. Until then I had supposed that a director wanted as little controversy as possible. This was a new attitude. Scripps, I thought, will be a good place to work; here one can reason aloud from the evidence and express opinions that may upset the establishment.

For a starter Roger gave me a letter to answer. It was from somebody who wanted to bring water to Southern California by building a thousand-mile-long undersea pipeline to transport the outflow of the Columbia River directly to the Los Angeles region. This was not the engineering problem it first seemed but one that required an understanding of the fundamentals of water economics. Obviously the cost of such a project and the technical problems of building and maintaining such a line would be tremendous, but the scheme was not a good idea for a different kind of reason.

The principal demand for water is not for drinking water or household use but for irrigation. That means a *lot* of water is needed at a *low* price and a *high* altitude. Water is a very inexpensive commodity; the principal cost is in moving it about, especially pumping it uphill. If the ocean were freshwater, few California irrigators could afford to lift it to the elevations where it is needed for agriculture. Most farmers can only afford to open a gate in a canal and let the water run downhill onto their crops. By the time I sent off an answer to the letter, I had developed an interest in California's water problems and had learned that water is priced according to its use. At that time, household water was selling for about $160 an acre foot, industrial water for $80, water for irrigating oranges cost $30, and farm crop water sold for $6 to $10. This astonishing spread in price was for the same water from the same source. Desalinizing seawater was so expensive as to be an unacceptable solution except perhaps for emergency household use in a wealthy city at sea level.

Others on the Scripps staff also had some interest in the matter and would occasionally discuss what oceanographers might do to help with the water problem. One long evening when John Isaacs, Walter Munk, and I sat cross-legged on the floor of Walter's old cabin on the lower campus with a jug of wine between us, John brought forth his great iceberg plan. He suggested moving an Antarctic iceberg to California.

Each year huge chunks of frozen fresh water break free from the Ross ice shelf; often these are five to ten miles long with flat tops a hundred feet above the sea and a bottom perhaps 600 feet below it. Some single bergs contain more water than all Southern California uses in a year. If one could tow such a berg with the help of known current systems, perhaps water shortages could be avoided. The Peru or Humboldt Current would carry it north along the coast of South America to the west-flowing Equatorial current and winds that would move it beyond Hawaii. Then the tugs would nudge it into the North Pacific gyre, which would bring it back across the North Pacific to the Vancouver region and then into the south-flowing California current. When the berg reached Southern California, it would be dragged into the lee of one of the channel islands, perhaps San Clemente, where it would be parked. Once there, a floating dam around the ice would collect the melted fresh water (which would float on the surface above the denser salt water) and allow it to be pumped to the mainland. John calculated that there would be more fresh water condensed by the cooling of air passing above the berg and falling as rain than would melt from it. In a manner of speaking, we changed a jug of wine into a lot of water in one evening.

It was a cockamamie scheme that we later learned had been tried in the mid-1800s when a sailing ship towed a small berg to Callao, Peru. But it is the kind of idea that scientists like to play with, because it stimulates thinking on the subjects of heat flow, ocean currents, and the unusual problems of towing very large objects. The whole thing should have died there, but it received some publicity and, with variations, was studied and promoted by others as a practical solution for northern hemisphere water problems, which it certainly is not.

Next Director Revelle asked me to meet with some representatives of the Hanford, Washington, nuclear energy group, who wanted to talk about disposing of nuclear waste in the ocean. From the figures they had, it seemed reasonable to consider dropping lumps of concrete made partly of high-level radioactive materials into a seven-mile-deep Pacific trench. At that time, none of us knew much about either deep trenches or radioactivity, but as it turned out ocean problems were not particularly relevant.

Assuming that a method could be developed for making concrete lumps from radioactive waste, the big problem was how to move them from Hanford down the Columbia River to a place where they could be loaded on ships. That would entail considerable risk because an accident could seriously contaminate the Columbia River, bring transportation on that route to a stop, and generally disrupt life in the region. We closed the conversation by encouraging the Hanford people to come back as soon as they had solved their onshore transportation problem; they never returned.

At cocktail time, which inevitably comes when the sun drops below the yardarm, visitors to Scripps were usually made aware of a sunset phenomenon known as the green flash. "Have you seen it?" the faithful would ask, and if the initiate hesitated for an instant, he would be assured that there *really is* a green flash, and if he were to watch the sunset over the ocean every night, he would eventually see it. Other persons overhearing this exchange would flick their eyes heavenward as they would if a snipe hunt had been proposed.

The story is this: At the moment the upper rim of the sun sinks below a clear, distant horizon, a green flare or flash appears above the horizon for a fraction of a second. Anyone seeing this flash has no doubt about the evidence of his eyes, but many people have looked dozens of times and not seen it; understandably, they remain suspicious. The reason the phenomenon is rarely seen from a densely inhabited coast is that the tiniest amount of haze or cloud will obscure the flash of green. However, it is often seen from ships well out at sea where the horizon is clear.

Dr. Carl Eckart, once director of Scripps, often explained the physics of the green flash to visitors. They would stand on the cliff looking out to sea, and Carl was used to saying, "Well, I guess tonight wasn't the right night. Just keep looking." Then one night Carl saw it himself. "My God!" he said. "There *really is* a green flash."

By day I used some of my time to design instruments, work with graduate students, and think about mine countermeasures. One of the problems of influence mines is that they can be set off by ship-made waves, and no one knew how to build a ship to sweep them that would not make such waves. Hydrofoil and hovercraft schemes were considered and rejected for various reasons. But in 1953 John Isaacs thought of a new hull shape that looked as if it might do the trick. He reasoned that since the main waves a ship makes come from the bow pushing water aside, why not turn the hull inside out and capture these waves inside the ship?

He proposed splitting an ordinary-shaped hull down the middle, transposing the two sides, and leaving some space between them. A flat bottom, open at bow and stern, would be built between the keels of the half-hulls with the propeller above it at the midpoint. With this arrangement, all outside surfaces would be straight and could not deflect water or make waves. The flaring bows would push water toward the center of the ship, but, because there was a bottom beneath connecting the two keels, its effect would not be transmitted downward, and a mine below would not detect a change in pressure. As might be expected, Isaacs had a hard time either explaining the concept or selling his proposed hull arrangement to doubters in the Bureau of Ships so he built a model to experiment with.

La Push was a plywood box 4 feet square and 8 feet long, open at the ends. Inside the box on each side was half-a-hull 1 foot wide and 4 feet high that curved into the outer walls at each end. An outboard motor hung through a hole in the center of the upper surface with its propeller close to the bottom in the 2-foot-wide channel between the hulls. We moved this strange craft up to a small reservoir behind San Diego and began to explore its possibilities. I was the pilot; John kept notes and watched the instruments that recorded pressure on shore.

It was easily predictable that the water entering the 4-foot wide bow opening would have to move twice as fast when it went through the 2-foot wide channel between the hulls, but two other aspects of *La Push*'s performance surprised us. As soon as the propeller began to turn, the craft not only surged ahead but rose vertically in the water; although the deck remained horizontal, 1 foot of freeboard at rest became nearly 3 feet when the craft was underway. This was because a region of high velocity flow is also one of low pressure; the racing water in the channel had the effect of increasing the upward force on the bottom, and the boat lifted accordingly. Second, we now saw that the water sucked in at the bow by the propeller and drawn through the central channel at high speed, to be expelled like a jet below the surface of the surrounding lake, would necessarily be overrun from behind by water flowing in from the lake surface. The result was a constantly breaking wave inside the boat, just behind the propeller, equivalent to a "standing wave" over a rock in a fast-flowing stream.

After getting to know the craft's idiosyncrasies, we set about making definitive measurements to see if *La Push* really did capture its own pressure signal. To find out I would drive it between lines of slender bamboo poles jammed into the lake's muddy bottom. This kept the craft a standard distance from the pressure sensors distributed on the bottom. At the model equivalent of a full-scale speed of 10 knots, there was no appreciable pressure signal beneath or to the side of *La Push,* although it still pushed a low (Bernoulli) wave ahead. Although these experiments appeared to confirm John's concept, no larger scale version was ever built, possibly because no one in the Navy wanted to command a giant box.

John and I still spent a good deal of time together off the job, and we had many private jokes. Usually these were based on a common knowledge of Mark Twain stories or Kipling poetry or some subject we had often discussed. With that background, a single key word or gesture between us would transmit a whole concept. For example, John had read somewhere that the Prussian general staff classified officers in four categories that were a combination of smart, stupid, lazy, and energetic. Category 1 was smart and energetic, and Category 2 was smart and lazy; those were preferred for officers. It was also easy to deal with Category 4, stupid and lazy; the worst problem was how to handle Category 3, stupid and energetic.

On various occasions when we were acting as advisers to Navy projects, John and I found ourselves among senior officers sitting around large conference tables listening to enthusiastic briefings on hopeless schemes. John would surreptitiously hold up three fingers, so only I could see and wait for a barely perceptible confirming nod.

FROGMEN ON THE ATTACK

Late in the summer of 1950 some rugged Navy demolition divers took me along for a careless afternoon on a rubber boat in the ocean off Coronado. They had brought along a new device for "free diving," meaning one could dive free of any lines to the surface. It consisted of a bank of three air bottles, each only about 3 inches in diameter and 18 inches long, surmounted by a "demand" air regulator called the Gagnan-Cousteau device, from which a tube led to each side of the mouth. Internal and external pressures were balanced so that the diver could suck air in through one tube and expel it through the other.

"Nothing to it," they said, strapping the bottles on my back and shoving me overside. "Just breathe through your mouth." That was all the instruction they gave me. While sinking, I jammed the rubber flange against the outside of my lips, bit into the rubber grips tightly, and tried to inhale normally. Every breath of air was accompanied by a dose of saltwater, and after a few minutes I surfaced, coughing and spluttering.

My Navy instructors guffawed. "You're supposed to put the rubber flange *inside* your lips, dummy. No wonder you breathed salt water."

Encouraged by this expression of sympathy, I went down again, and this time breathed well enough to enjoy the freedom thoroughly. For the first time in a horizontal position, instead of standing in a diving suit, I looked closely at the bottom and its small animals. From then on I was an ardent enthusiast for free diving.

This self-contained equipment that used ordinary air was a wonderful invention; with it a man could breathe normally while swimming about underwater. But for most professional divers it was a new concept; they were wary of it because they were accustomed to working in miserable conditions, cold, dirty harbor water with near-zero visibility. For them diving had mostly been dangerous, uncomfortable work; it meant manual labor such as assembling pipe lines, building bridge abutments, or salvaging ships, where the guy on the bottom needed all the help from the surface he could get. Few of the pros imagined that anyone would dive for recreation. They were not about to trade safe, well-tested diving dress for a SCUBA device with a limited air supply, no telephone, and no safety line to a tender just to get greater freedom of movement.

During World War II the U.S. Navy Underwater Demolition Teams rarely used breathing devices; they held their breath while swimming down to inspect tank traps or attach explosives to obstacles. Other navies, especially the British and Italian, used "frogmen" equipped with some form of underwater oxygen supply to sneak into enemy harbors at night and attack ships. Some of these extraordinarily brave fellows would straddle self-propelled torpedolike packages of explosives and, with only their heads above water, steer these toward a target before bailing out at the last moment. Alternately, they would ride small submersibles to the target, dismounting to place magnetic limpet-mines against the hulls of anchored enemy ships. These mines had tiny propellers that would be turned by the forward movement of the ship. After the ship had gone ten miles or more, these would automatically fire the explosives. This made it hard to tell if the ship had been torpedoed and resulted in its sinking in deep water where little help would be available.

The breathing devices these underwater warriors wore used oxygen instead of air and recirculated the exhaled gas through CO_2-absorbing lime. The advantage of oxygen-breathing equipment was that no bubbles of exhaled air came to the surface to give away the presence of a swimmer beneath. The disadvantages were that divers could get oxygen poisoning at depths below 40 feet and CO_2 poisoning if the chemicals failed to absorb carbon dioxide. The U.S. Navy had several versions of these pure oxygen devices, but they were considered to be too dangerous for routine operations and were rarely used; two divers practicing with them died in the swimming pool at the

Amphibious Base. Usually the problem was that the CO_2-absorbent chemicals had been exposed to the air and used up their capacity before the dive started.

About that time Roger Revelle became chairman of the Panel on Underwater Swimmers, a subdivision of the NAS Committee on Amphibious Warfare of which Dean O'Brien was chairman and I was executive secretary. The panel members were mainly scientists, and few knew much about how swimmers had been used in military operations so I dug into World War II history and found illustrative examples in several books on the subject, especially *The Midget Raiders* and *Sea Devils*.

At Algeciras, Spain, on the Strait of Gibraltar, Italian frogmen used an ordinary-looking anchored merchant ship, ostensibly prevented from leaving by an engine breakdown, as a base for their operations. They cut a hole in its hull below the waterline large enough for men and craft to come and go unseen, and from this clandestine metal cave launched "human torpedos" to attack ships in the British base at the Rock of Gibraltar, only a few miles away across the bay.

A British underwater unit using tiny submarines called X-boats penetrated nets and other defenses in a Norwegian fjord and attacked the well-defended German battleship *Tirpitz* anchored there. This daring night operation in near-freezing water managed to place eight tons of explosives on the bottom under the *Tirpitz* keel. When it exploded, that ship was immobilized for the duration of the war. British swimmers riding torpedolike "chariots" sank a large dry dock, mined the Kiel and Corinth Canals, cut submarine telephone cables, and sank a cruiser in Singapore as well as nearly 100,000 tons of merchant shipping.

The Italians under Prince Valerio Borghese returned these favors with a flourish. Their submarine *Scire* arrived off Alexandria, Egypt, at night and launched six men astride three manned torpedos to attack the British battleships *Valiant* and *Queen Elizabeth,* as well as a tanker that lay at anchor in the harbor. Only the heads of the attackers projected above the calm water as these odd vehicles approached the harbor's protective antitorpedo nets. The British were wary that some kind of an attack would be attempted; their searchlights scanned back and forth across the dark waters, and boats cruised about dropping explosive charges to prevent swimmer operations. As the Italians watched, the nets across the entrance opened to let some British destroyers enter; when they did, the torpedo riders followed the ships in and the nets closed behind them. Once inside each torpedo headed for its assigned target.

While passing over the top of a second line of protective nets that had been installed close around the battleship *Valiant,* one of the torpedos picked up a strand of wire that fouled and stopped its propeller. At the same time it encountered fresh water flowing from the nearby mouth of the Nile and, being ballasted for seawater, promptly sank. One of the riders separated from it, but the other, a very courageous and determined fellow named De la Penne, was able to drag the disabled torpedo across the bottom through absolute blackness, guided only by the sound of a pump, until he bumped his head on the *Valiant*'s hull. By that time he had spent 40 minutes on the bottom and was not only exhausted but suffering from oxygen poisoning. Even so, he was able to set the firing clock before he surfaced.

De la Penne was immediately picked up and taken before the *Valiant*'s commanding officer, who asked exactly where the explosive charge was located and when it would explode. When the Italian refused to answer, he was placed in a lower compartment of the ship where he would surely be among the first to die if there was an underwater explosion.

Ten minutes before six, the time set for the torpedo to explode, De la Penne asked to see the captain again and was brought on deck. Once there he explained that it was much too late to protect the ship but ten minutes remained to get its crew to safety on deck. When the explosion came, it heaved the ship violently upward; then the lights went out, and the hull filled with smoke as it settled the few feet to the bottom. While the *Valiant*'s officers clustered on the afterdeck trying to decide what to do next, the *Queen Elizabeth,* 500 meters astern, also blew up and settled to the shallow bottom. Both ships remained on an even keel and looked about the same as before, so the British decided not to admit that anything had happened; they continued to stand watches, raise the flags, and carry on as usual. As far as the outside world knew, the British fleet was ready to sail, and its sea threat was maintained for several months.

The Swimmer Panel was impressed by those accounts, and decided there was no point in the United States starting from scratch to learn a game if there were experts available to teach it. So it was arranged for the Navy to invite representatives of the countries that had learned the hard way during World War II to come to the Amphibious Base at Coronado, bringing their ideas and equipment. The Navy would fly them, as well as its own officers from various commands, to the meeting. Everyone would hear directly from the men whose exploits we had read about in a grand underwater swimmer symposium. A *Swimposium!* Walter Hahn of the NAS staff in Washington would make the arrangements at that end; I would handle them on the West Coast.

Preparing for the Swimposium required increasing contact with the Underwater Demolition Teams stationed at the Amphibious Base. The boss of the teams was Commander Douglas Fane, a stocky, soft-spoken, freckle-faced redhead with a reputation for daring wartime exploits. When I first walked into his office and admitted to working for the University of California, he almost dismissed me as "one of those longhairs." Then he reconsidered my hair, which was shorter than his own, and laughed at his accidental joke; but from then on I was known as "the longhair."

Fane was just about to screen some underwater movies he'd shot in the Virgin Islands. "C'mon, I gotta show a picture I took of a fish." He had spent years underwater in the tropics and must have seen many thousands of fish and I wondered what it would take to impress him. The scene opened with the camera in the black insides of a sunken ship, looking out through an open door and porthole at clear light green water beyond. Presently the porthole was eclipsed, and a shadow fell across the door opening. Then the head of a huge grouper appeared; it paused for a moment to look at the photographer and consider whether it could fit through the doorway. Deciding that it could, with its main engine turning over slowly and the fins moving ever so slightly, the great fish came on. Like the *Queen Mary* edging into a tight berth, it eased through the doorway with just enough clearance for a crack of light on each side. There was a gleam in the creature's eye and a crooked smile on its huge mouth as it joined the photographer inside the ship. "I sure hoped it wouldn't inhale," Doug Fane commented. "If it had, the camera and half of me would have been sucked in. That guy must have weighed over 300 pounds. You don't see 'em like that very often." Probably few such fish still exist; they were such easy targets for spear fishermen intent on setting records.

Commander Fane was enthusiastic about the proposed Swimposium and about the chance of getting some technical help for his swimmers, who would need better equipment to carry out some of their changing missions.

I tried to get him to be more specific. "What can the panel work on that would be most helpful?"

He didn't hesitate. "That's easy. We're going to the Arctic in July to swim among the ice floes, and we don't have a decent suit. See if you can get us something that will keep the men warm and dry." I said I'd try.

After we talked about swimming and diving matters for a while, he mentioned that some members of the team went out to the Coronados Islands almost every weekend for fun. Did I want to go along? That was too good an offer to refuse; the following weekend Doug and I with our wives and kids (he had four, all redheaded), and some of the team's top swimmers were aboard a couple of landing craft, headed for a diver's holiday.

This was the first of several trips we made to these Mexican-owned islands where the deserted kelp beds and beaches were a natural playground for lovers of the sea. We would anchor the boats in a protected cove with the ramps down so it was easy to get in and out of the water; then we would spend the day poking and probing in the fascinating world of undersea plants and animals. The kids would paddle about in face masks, peering down through the clear water at the kelp and fish and rocks below. Rhoda and Anitra especially loved squeezing babies out of viviparous perch and watching these swim for the first time. No diving gear was used on these expeditions; the Underwater Demolition Team divers could hold their breath long enough to get several abalones per dive and catch lobsters with their hands that were too big to fit in any ordinary pot.

Doug Fane brought along his friend Roberto Frassetto of the Italian nobility, a member of the Italian wartime swimmer organization. Roberto was a handsome, cultured man, knowledgeable in several arts and sciences, who had not quite recovered from wartime wounds but was full of ideas and energy.

During World War II the Italian Navy decided to attack British warships that were at anchor inside the Grand Harbor at Valletta, Malta, an island just south of Sicily in the center of the Mediterranean. The whole island, and especially this harbor, was heavily fortified, and the main ship channel to the inner harbor was guarded by nets, guns, searchlights, and patrol boats. There was no chance of forcing an entry there, but there was a small secondary entrance spanned by a bridge from which torpedo nets had been hung as a screen against intruders.

The Italian plan was to send demolition swimmers in on a moonless night to destroy the bridge and its nets with explosives. With the nets thus cleared away, small high-speed boats loaded with TNT fused to detonate on contact would race through the opening into the harbor where British warships were anchored. The drivers of these surface craft rode at the extreme rear; they were instructed to aim the boat at its target, flip over backward, and swim for their lives.

This risky operation did not go as intended. Although the demolition charges brought down most of the bridge and its nets, they also alerted the British defense artillery, which reacted immediately. Led by Frassetto the Italian boats raced in single file toward a small gap between a supporting pillar and the wreckage of the downed bridge. Frassetto, in the first boat, decided to use his boat to widen the gap for those behind; he aimed at it and flipped out. The explosion cleared the way, but the boats following him were wrecked or blown up by the defender's guns and never made it through the enlarged opening.

Frassetto was one of the few survivors of that attack. Seriously wounded by the numerous explosions, he was picked out of the water by a British patrol boat and spent a couple of years recovering in British and Italian hospitals. After the war he became a special consultant to the Office of Naval Research and was brought to this country as an expert adviser on Mediterranean matters. (He now lives in a palace on the Grand Canal in Venice where he is director of the project to prevent storm flooding.)

On one of the Sunday sorties to the Coronados Islands, I took along a new underwater movie camera, a relatively rare piece of equipment at that time, in search of subjects. Doug Fane and I were exploring the rocky beach below a cliff when we came on a cove where a herd of baby sea lions had hauled out. Dozens of shiny black pups not much over a foot long, probably waiting for their mothers to come ashore with more milk, scampered ahead of us over the slippery rocks. I made an end run around them, trying to get in position to photograph the oncoming pups and, while concentrating on the babies, almost ran into a huge bull elephant seal. Annoyed by the disturbance and embarrassed to be caught snoozing, he suddenly reared up, bending his neck backward to heave his head as high as possible, all the while snorting and blubbering through his prominent proboscis. This fellow must have weighed a thousand pounds, which was indifferently packed into a thick loose skin. He was uncertain and indignant, and he made several threatening lunges to show his unhappiness with the intruders on his beach.

Then Doug Fane had one of his great ideas. "If you want a good shot of this guy underway, get down at the water's edge, and I'll herd him toward you. When he sees his bluff won't scare me away, he'll run for the water."

It was worth a try. Kneeling on the beach below the two contestants with the camera rolling, I hollered, "Start herding" and concentrated on the scene in the viewfinder. At first the bulky critter resisted the suggestion that he move; then, as Doug had forecast, he turned abruptly and started for the water on a dead run. He had a fascinating gait that made him look like a smaller animal running inside a skin about six sizes too large, with the space in between filled with Jell-O. With every step of his front flippers, waves ran fore and aft in the blubbery material that supported his hide. His trunklike snout bobbed up and down as he gallumphed toward me, growing ever larger until he filled the viewfinder. Then this mass of blubber rolled over me, pressing my body hard against the cobbles.

It was an act without malice; I was foolish enough to be athwart his shortest route to the water, expecting him to pass on either side; he thought

I'd have sense enough to get out of the way. Being run over by a sea elephant is about like having some large, loosely packed bags of wet sand roll over you. There's a lot of weight but not enough in any one spot to do much damage.

Gleefully Doug hollered, "I'll bet you got some great pictures; now get some underwater. Just follow him out."

Again I obliged. Sea elephants may be embarrassed by their clumsiness and vulnerability on land, but in the water where they can move rapidly and gracefully, they are not to be bullied. This one didn't really want any trouble, and as I swam toward him, he graciously retreated a little way. Then, pushed a little too far, he would raise his head out of the water, glower, harumph, and lunge in my direction. Out of respect for his feelings, I retreated an equal distance. After several such advances and retreats, I concluded he did not wish to have his picture taken, and our underwater pavane ended.

A few weeks later Dr. Gifford Ewing, a physical oceanographer at Scripps, who was an expert on internal waves in the ocean and who had whimsically proposed the plankton-powered torpedo (if a giant tuna can run on plankton, the Navy should be able to build a torpedo that would do the same) invited me to fly with him in his new airplane. He brought along a wartime friend, Cresson Kearny, a retired Army major. After flying around over the La Jolla region for an hour, we landed, and Giff got out his logbook. "It's important for pilots keep track of our time in the air." That figured; the major and I looked over his shoulder to see how many hours he had. Except for one previous entry of an hour, the book was blank; we had flown half of Giff's flying time.

Cresson Kearny was a most unusual chap. He had served with the OSS in China, where he had picked up some rare virus infection that sometimes gave him goose bumps when it was hot and made him sweat when it was cold. Because of his experience in jungles before the war, the Army assigned him to Panama, where he met Giff Ewing who was then captain of a destroyer escort based there. On taking over that command Giff, a sailboat sailor from Yale, claims that he called in his executive officer, a Naval Adademy graduate, and said, "A terrible mistake has been made; I don't know much about ships. So you run this ship as you see fit while I study farming. Just keep me informed." Apparently this arrangement worked.

Kearny's job in Panama was to adapt the best civilian jungle exploration equipment for infantry use, especially to help soldiers avoid drowning in tropical rivers and swamps. One of his ingenious devices was mud shoes, something like small snowshoes, for crossing soft mud flats. Flat pads would stick to the mud, making it hard to lift one's foot; mud flowing over the edges of the pad would accumulate on top. So Kearny's version hinged the pads, making it easy to break the suction and helping the mud on top to fall off. He was just the right guy to think up a better kind of protective suit for swimmers in the Arctic.

So, with a small budget from the Amphibious Warfare Committee and using a plywood table that took up half of my cramped office under the stairs of the old library, he went to work to develop the warm, dry suit that Commander Fane had requested. All suits were intended to be dry suits in those days, and they all used woolen long-john underwear for insulation. As long as the underwear stayed dry, the diver kept warm, but eventually the suits leaked, the underwear got wet, and the insulating value was lost. So Kearny insulated some long johns by cementing quarter-inch-thick pieces of closed-cell neoprene rubber to them that would not lose its insulating value when it got wet; he also made some improvements in the shape and closure of the outer gum-rubber suit to keep the swimmer as dry as possible.

In about a month, he had a suit ready, and we took it to the swimming pool at the Amphibious Base for the prospective users to test. Warm Southern California is not an ideal place to test swimwear made for use among ice floes, so we should not have been too surprised by the result. The testers, including Walter Hahn and me, would don the suit and jump in the pool but after a few minutes we would be uncomfortably hot, literally cooking in our own sweat. Whoever tried it would soon shed the rubber outer suit and swim in the neoprene-coated underwear. Gradually the obvious dawned on us: Swimmers don't mind being wet as long as they're warm (but not too warm). The outer, waterproof suit was not needed. Thus Kearny's insulated underwear became the first "wet suit." It was as much discovered as invented, and after it was shown at the Swimposium, the possibilities in neoprene wet suits were widely exploited.

Once the technical program for the Swimposium was set and the invited speakers had accepted, our Navy counterparts at the Amphibious Base began to organize a reception for the attendees. From this point on, the tail wagged the dog. Matters relating to the underwater swimmer program were submerged under a flood of decisions to be made about the reception: who was to be invited; what they should wear, drink, and eat; what music would be played; how parking would be handled; the order of the reception line; the size and lettering on the badges; and other such vitally important matters. The Navy's capacity for organizing and presenting a cocktail party was impressive, and after having observed the meticulous detail with which this reception was carried out, we felt much more confident of its ability to plan an amphibious operation. It was a glorious reception, second only to the

Coronado Arts Ball, with a profusion of uniforms, medals, and social gal-
lantry that set the tone of international camaraderie needed for the technical
sessions to come.

The next day our experts assembled to talk about the triumphs and
disasters of World War II and to speculate about what could be done with
underwater men if there was ever another such war. Erect British colonels
with red decorations and bristling gray mustaches described bold schemes for
air-dropping swimmer-raiders and their boats behind enemy lines. Italians
with dark glasses on their foreheads, jackets draped around their shoulders,
and eloquent gestures suggested elegant but improbable attacks on anchored
fleets of warships.

Nearly everyone got the impression they would like to fight World War
III with equipment they had not finished perfecting during World War II.
French, Germans, Canadians, and Americans also contributed ideas, but
somehow they were not quite as dramatic as the authentic war heroes. Afternoon sessions were held around the admiral's swimming pool, where such
things as underwater helicopters and pedal-powered swimmer-delivery vehicles, as well as dozens of kinds of suits and breathing devices, were demonstrated.

One of the high points of the pool demonstrations was that by Cresson
Kearny of the 6-pound, 12-bladder, breath-inflated nylon boat he had devised.
These one-man craft had been used by General Joseph W. (Vinegar Joe)
Stilwell's armies for surprise mass crossings of the rivers of Burma during
World War II and by Attila the Hun over a thousand years before.

While demonstrating combat equipment in Washington during World
War II, Kearny proposed that every man on certain kinds of warships should
be furnished with such a boat so he could save himself if the ship were sunk.
When a French admiral ridiculed this as an impractical idea, partly on the
grounds that no man could be expected to take a long jump into cold water
and then keep himself afloat while inflating the raft with lung power, Kearny
insisted on demonstrating. In the midst of a Washington winter, he practically
dragged the objectors to the 14th Street bridge over the Potomac. Then, in
full uniform, he jumped into the river among chunks of floating ice, blew up
the twelve separate bladders, and paddled ashore. In so doing he turned blue
and gasped for breath, causing the admiral to remark, "Well, you did it. But
I believe French sailors would prefer to drown."

Now, for the benefit of a hundred expert observers who ringed the deep
end of the pool, he gave a similar demonstration. This time the water was
warm, and his clothes were more suitable, but he provided an additional

obstacle that partly made up for the missing chunks of ice; he gave a simultaneous lecture about the advantages of his nylon boats. The crowd was not quite prepared for his presentation.

Kearny, standing near the edge of the pool with a small package attached to his belt, began with a loud, "Abandon ship!" and jumped into the pool. When he surfaced holding an amorphous object, he shouted, "This is the way a man would inflate this boat at sea," and expended his remaining breath through a tube into one of the bladders, after which he promptly sank. In a few moments he had fought his way back to the surface, taken a deep breath, shouted another sentence, blown into the tube, and sunk again. In about twenty sentences, the raft was sufficiently inflated to hold Kearny, who then climbed aboard. The assembled experts cheered and applauded this oratorical tour de force; however it is doubtful that any of the watchers was converted to the general use of such rafts.

THE RUMBLE OF A DISTANT DRUM

Behind the scenes at the University of California there were rumors of work on man-made thermonuclear reactions. In 1951 some physicists believed that it was possible to duplicate the process that produced the energy of the sun, at least for a microsecond; others were dubious. This reduced to the question of whether light elements, specifically hydrogen two (deuterium) and lithium six, could be raised to a high enough temperature by a fission reaction to start a fusion reaction. Preliminary tests at Enewetak, code-named Greenhouse, had confirmed Dr. Edward Teller's "new concept," and indicated it was possible. Now the push was on for a full-scale test of a thermonuclear explosion. It would be named Mike, and it would be shot at Enewetak as part of a test series to be called Ivy. The Scripps Institution of Oceanography was invited to measure the waves it generated.

We began by estimating the effects of the proposed explosion, extrapolating from the Crossroads tests six years earlier. There was a big difference; this one could be several hundred times larger, if it exploded at all. One number that was bandied about with some confidence was the probable energy release: ten to the twenty-fourth power (10^{24}) ergs. Ergs are small units of energy; each increase in the power is ten times as great as the next lower number. According to a scale John Isaacs assembled many years later, it takes 10^5 ergs to strike a typewriter key, the average daily diet for humans is 10^{14}, a tornado is 10^{22}, and the sun's energy for a year is 10^{41} ergs. Thus the energy in the proposed explosion would be about equal to a hundred tornadoes.

Somehow the explosion's energy would be partitioned between heat, radioactivity, and blast effects, the last being divided between the amounts of motion transmitted to air, ground, and water. If we were to measure the ocean waves produced, it would be necessary to make an advance estimate of the amount of energy going into waves and to guess at the number of waves, their period, and the height of those waves. In addition to the possibility that the estimate of explosive energy could be wrong by a factor of ten, we had to forecast the mechanism that would form the waves.

Estimating waves caused by an explosion at the surface in deep water would have been relatively easy. The explosive pressure would push the water downward and outward, sending out a small wave and creating a hemispherical crater extending hundreds of feet below the general sea level. When that pressure passed its peak, the water in the bottom of the crater would be heaved upward rapidly by hydrostatic pressure from beneath, while the watery rim of the crater would rush down and inward. This would cause a peak of water, probably over a thousand feet high, to be thrown up in the center.

The local reduction in sea level caused by the inrush of water would be transmitted outward as a wave trough. This would be followed immediately by the collapse of the great center peak that would send the largest wave crest outward. Then, while the water level at the center of this erstwhile crater rose and fell several times as it came to equilibrium, a series of lesser waves would be generated. The extreme wave heights close to the crater would decrease rapidly as the wave moved outward and its energy was distributed along an ever-expanding perimeter.

However, the shot code-named Mike, whose waves we had to forecast, would not be in deep water but in the middle of a flat coral reef over a mile wide. Trying to estimate how a nonhomogenous coral reef full of internal water-filled spaces would be effected by a thermonuclear explosion was another matter. No one doubted that a large crater would be formed, but there was no really sound basis for estimating its size. There would be no rebounding water beneath to send out a really large wave, but now we faced what might be an even more critical question: Could a crater be created that would be large enough to breach the outer edge of the reef?

If that were to happen, a large wave might be directed northward into the open Pacific where it could conceivably cause damage to distant coastal dwellers. Even more worrisome was the possibility that a large chunk of the outer edge of the atoll would be broken off and slide into the depths, causing a major tsunami, as the landslide in the Aleutians had done only five years before. If that happened, much additional energy would be added to the explosion-caused wave. This caused considerable nervousness among Task Force leaders until it was learned that the outer slope of most Pacific atolls, including Enewetak, is only 22 degrees, and coral rubble will stand on a slope about 10 degrees steeper. They were much relieved to learn a submarine landslide would not add to the problem.

Without much to go on, we turned for guidance on explosive effects to a situation that seemed to be roughly comparable—the eruption of the volcano Krakatoa. The energy expended in its main explosion had been estimated at 10^{25} ergs, only about ten times more than Mike was expected to produce. I spent New Years Eve at the beginning of 1952 abstracting of the Report of the Royal Society of London on that event; some excerpts from that report hint at the measurement problems that faced us.

The volcano Krakatoa in the Dutch East Indies erupted with the most violent explosion of recorded history. First the throat of the volcano ruptured, allowing cold sea water to freeze a crust on the rising molten magma. Then, as a boiler with the safety valve tied down, the pressure began to build up. On the morning of August 27, 1883, this crust let go. The entire northern portion of the island disappeared and in place of ten square miles with an average elevation of 700 feet, there was a depression whose bottom was more than 900 feet below sea level.

Over four cubic miles of rock were blown away and two new islands rose in the Sunda Strait. A hundred miles from the center of the explosion the sky was so dark that lamps had to be used at midday. Dust and ashes rose 17 miles; the finer particles were carried around the earth by stratospheric winds and caused red sunsets in London. The sounds of the explosion were heard 3000 miles away. But the most damaging effect of the eruption was the initiation of seismic sea waves averaging 50 feet high that inundated the whole of the foreshores of Java and Sumatra bordering the Strait. The villages of Merak and Telok Betong, at heights above sea level of 115 and 72 feet respectively, were carried away by a great wave. More than 36,000 people were drowned and many vessels were washed ashore, including the government steamer *Berouw* which was carried 1.8 miles inland and left 30 feet above sea level. Moving west these waves reached South Africa, Cape Horn and Panama.

That hinted at the possibilities, and although Mike would surely not be as big a blast as Krakatoa, it could possibly generate some fairly large waves. As project engineer on the wave measurements, I felt a little like the mine-hunting diver standing in the cold blackness, over his knees in mud, who got the order, "Don't just stand there. Do something."

It was evident that different kinds of instruments would be needed at different distances from the explosion and different measuring locations. Those in close to the explosion could be expected to record a few large waves in a short time period; those at a distance would be exposed to low waves over a longer time. At all locations the explosion waves had to be distinguished from normal waves and tidal changes; considering that Enewetak is in the trade-wind belt, where 12-foot-high wind waves are normal, that might not be so easy. Moreover, the depth of water at each instrument and the general submarine topography would influence the way wave heights and periods were measured. The principles that the instruments would utilize were clear enough, but the question remained: Exactly what kind of wave meters should be built for the very different kinds of locations where they would be installed? While I was considering the matter, another problem arose.

HOW BLUE THE SKY

While walking home from Scripps after work one day, I suddenly felt a pain in my pelvic region so intense that I fell to the sidewalk and writhed about for a minute or two before continuing. It was caused by a hard knot of tissue half as big as my fist on the upper surface of my right pelvic bone. It had been much smaller when first noticed several months before, but a doctor employed by the university's health plan insisted he could feel nothing and suggested that a couple of aspirins would take care of any discomfort.

Now I tried a doctor in La Jolla, who felt the lump and took some X-rays showing my urethra making a wide detour around it. "We'd better have a look," he said, and a few days later I was rolled into a Scripps Hospital operating room for a biopsy.

When it was over, I was moved back to my room and, still in a stupor from the anesthetic, began reciting Kipling's "Rhyme of the Three Sealers." Rhoda, anxiously waiting to see me after the surgery, chanced to enter the room just as I reached the line "So go in grace with Him to face, and an ill-spent life behind. . . ." Not unreasonably, she thought that I had been told I would soon die.

Then the doctor arrived to give his opinion: "Better notify any close relatives. There is an inoperable tumor, a lymphosarcoma, on his pelvic bone, stopping the circulation to his right leg. He may have six months, at best a year to live." Rhoda recoiled in shock, and the Isaacs family graciously took her and Anitra in for a couple of days until I was released from the hospital, not much damaged by the operation.

At that time the chances of recovering from that particular form of cancer were not very good, perhaps 10 percent. The only known hope was radiation treatments, so I signed up for a series of these at Scripps Hospital and returned to work. After a couple of weeks of watching and listening, I became sufficiently suspicious of the radiologist's competence and attitude to ask some pointed questions. The answers revealed how little he knew about cancer and radiation. The perfunctory radiation treatments were given without hope of saving my life; his modest objective was to delay my incipient death for a few more months. But I had no intention of dying. Until this time my feelings had been mainly those of annoyance at the inconvenience of losing a few hours from work during the treatments; now alarm bells sounded. I immediately quit that hospital and set about looking for somebody who at least seemed to know something and might take a serious interest in my problem.

My next appearance was center stage in a medical theater in Pasadena. I was flat on my back on a gurney, naked under a white sheet, medical history inscribed on a clipboard at my feet. A masked man wheeled me into a circular arena, where a few stark vertical figures in white waited with dour expressions. Around and above, circles of solemn doctors in three-piece suits leaned forward to pronounce sentence on the prostrate figures brought before them.

I pictured myself a poor pilgrim dragged before a Druid council, a white-robed Christian pushed out onto the colosseum floor as lion fodder, a foolish scientist brought before the Inquisition for having claimed the earth moves about the sun. My life hung in the balance, awaiting the verdict of this mass of medical brain power. The chief inquisitor intoned my life history and pathology; then an assistant pulled down the sheet, prodded my belly and described the lump he felt. My Druid captors somberly wagged their heads, muttered incantations, shuffled their sandals, and dissolved into Roman spectators blind for blood. When the Emperor called, "How say you?" a roar went up around the colosseum. "Death!" they shouted, and turned their thumbs down. The breath of the lions was hot against my throat.

"You may have another six months to live," said the inquisitor. "Pax vobiscum." His bony fingers made the sign of the caduseus in the air.

In this mad dream I leaped naked to the floor, shook my fist, and shouted defiantly at the judges, "What's the matter with you guys? All I've got is one lousy lump. Don't you have any smart suggestions?"

They did not, but my negative attitude was duly recorded, and they made it quite clear that I was expected to be quiet and die with dignity by the date they had set. If they had sent a report, it would have said, " . . . and don't come back again if you're going to behave like that."

I scrambled into my clothes and got the hell out of there. Apparently this cancer thing was going to be more of a nuisance than I'd expected, and there was a lot to do to get ready for Mike, now only eight months away. My best bet seemed to be the Los Angeles Tumor Clinic, which specialized in high voltage X-ray treatment.

In 1952 high voltage meant 250,000 volts. By today's standards that is very low voltage, the main distinction being that low-voltage rays are diffuse and hard to focus so there is a lot of side damage to noncancerous tissue, but then it was the best available. In order to be near the clinic, I moved into a cheap hotel nearby and began six weeks of treatment with large doses of radiation penetrating my pelvic region, all aimed at the lump inside. When I insisted, the technician shielded my testes with a lead pad, considering that to be a strange whim that was without merit.

Spare time in this period was spent in studying what was then known about nuclear weapons effects, especially radiation. The material then in print was cautious, vaguely warning that all radiation is dangerous to cells but not distinguishing between its varieties or describing any practicable actions to take in various circumstances. It was of little value either to a patient or a nuclear test participant.

One of the leading radiologists at the LA Tumor Clinic was Dr. Jim Nolan; we soon found a common interest. In 1945 he had been an M.D. in the Army, stationed at Los Alamos. His primary jobs were to deliver babies and deal with any radiation effects arising from the early experiments with radioactive materials. Fate gave him a more important part in the beginnings of the nuclear age when he was selected to accompany the first A-bomb, a uranium device, from Los Alamos to the frontline air base on the island of Tinian.

This device, also called Little Boy, was transported from the mountains of New Mexico to Albuquerque in an ordinary van and thence by air and truck to Hunter's Point Naval Station in San Francisco. There, a mysterious box was loaded on the cruiser *Indianapolis,* accompanied by a tight-lipped "Field Artillery" officer, Captain James Nolan, who wore his crossed cannons upside down and refused to carry a .45 revolver on the grounds that it was "a dangerous instrument." The ship left Hunter's Point almost exactly at the same moment that the first plutonium weapon (Trinity) was exploded in New Mexico, making the bomb he was accompanying obsolescent before it left the United States.

During the voyage to Tinian, this "artilleryman's" knowledge of the Army's techniques for firing large guns was of great interest to the Navy gunnery officers, but Dr. Nolan had never seen a big gun fired, so he spent much of the voyage in his quarters feigning sea sickness to avoid questions. The delivery safely completed, Nolan remained at the front and was one of the first radiologists to visit the bombed Japanese cities a few weeks later. The *Indianapolis,* having done its duty, steamed off to be torpedoed and sunk with great loss of life by a Japanese submarine that had not learned the war was over.

It's surprising how blue the sky can seem and how colorful flowers can be when one is living under a death sentence, even though that outcome is only half-believed. Somehow one's senses are sharpened, and there is an extra awareness of the surrounding world not felt by persons who expect to see many more years. I understood and appreciated that intenseness, and yet I was not greatly bothered by the hovering forecast of doom. After six weeks of radiation treatment, the tumor had melted down, and the pain had ceased; radiation sickness had been a minor annoyance. With the cancer business out of the way, I could turn my attention to more interesting problems.

IV

From Cancer to Capricorn

PREPARING FOR IVY

Many questions needed to be answered before final plans could be made for measuring the waves that would be created by Mike, the first thermonuclear explosion, during the test series code-named Ivy: What instruments should we build? Where should they be installed? How much logistics support would each of the various locations require? Because the kind of instruments depended largely on where they would be used and because each location could be vastly different from the others, specific information was needed about reefs and channels and wave action on mid-Pacific islands that was not readily available in the United States. For example, although the official charts of Enewetak showed wide coral reefs rimming the atoll's lagoon, there was no hint as to whether a bulldozer could operate on those reefs or a swimmer enter the ocean from their outer edge. It was also necessary to keep up-to-date with the constantly changing plans of Joint Task Force 132, so I began commuting between Scripps and Los Alamos, New Mexico.

Los Alamos, located in a beautiful secluded forest on the rim of the Rio Grande gorge, operated as a special campus of the University of California; this remote area in another state had been selected during World War II to minimize contacts with the outside world. Before it became a weapons laboratory, the buildings had been a boy's school, and the old main building was still used as a dining hall, but most of the new buildings were permanent temporaries. This unlikely boom town had originally been set up to accommo-

89

date sixty people; then it grew to 600 and to 6,000, and was still growing when I last saw it.

My principal contact there was a Navy captain named Duncan Curry, a very insightful and cooperative fellow. After I described all the things that needed to be done, he suggested two things: "First, you should personally visit every place you may want to put a wave station and see for yourself what it's like. Second, I'll see if I can arrange to get you an executive officer who knows the Navy ropes and can serve as a liaison with the Task Force. You're going to need a real gung-ho helper." He was certainly right about that. A few weeks later a set of orders arrived for me to fly around the Pacific islands on available Navy planes, touching down on Hawaii, Johnson, Kwajalein, Enewetak, Bikini, Guam, Midway, and Japan.

The planes on those runs were DC-6s, slow and noisy by today's standards but with long sea legs. The military version was windowless and equipped with bucket seats. Today a bucket seat means a body-fitting, soft leather seat for a sports car, but in 1951 bucket seats on airplanes were parallel pipe bars about a foot apart with canvas stretched between them that ran along each side of the aircraft. About every six feet a cross-brace, also made of pipe, held the two longitudinal pipes apart. If you had the rotten luck to get a seat above one of those cross braces and your canvas sagged into it, a long flight was hell for your assatabulum. One could lean back uncomfortably against the plane's overcurving aluminum ribs and skin; these not only over-taxed the spine but were pimpled with rivets and crossed by hydraulic and electric lines perversely aimed at critical points on one's anatomy.

It was not uncommon for military travelers to spend twelve or more groggy hours slumped amid a long line of guys who stared with unfocused eyes over a hump of tarpaulin-covered freight at an almost mirror image of the guys on the opposite side of the plane.

It was just as well to be exhausted when you reached the next island, because that made it easier to overlook the accommodations. On arriving at one island at two in the morning, a bleary-eyed, stubbly-faced duty steward in wrinkled khakis addressed the motley arrivals. "Y'all gonna stay in one big room with cots." He pointed to a dark silhouette. "See that building there? Just go on in and take any bunk. You're flying out again at 6 A.M. The light switch is just inside the door."

We stumbled sleepily toward the indicated building, and the first man, inexperienced in the islands, snapped on the lights. Instantly he fell back. "My God! The place is full of rats!"

"Naw," said the second man, an old hand in these parts, "them's only cockroaches. The ones on my ship could eat these." That triggered a series of giant-cockroach stories, but most of us didn't hear them. We just collapsed on the nearest cots.

After intensive study of the charts, the shape and layout of Enewetak Atoll had become fixed in our heads, but we did not clearly understand the scale. It is one thing to know in an abstract way that a reef is over a mile wide and a lagoon is over 25 miles long, but it is quite another to walk that mile both ways across the rough surface or buck the waves for hours in a small boat to cross the lagoon. What looked like a tiny ring on the chart of the western Pacific grew considerably in our estimation when we finally got to know the place.

Enewetak atoll is an irregular oval elongated on the N/S axis with a wide passage in the reef at the south end. The thin line of shallow reefs and low islands along its rim surrounds a 300-square mile lagoon that is rarely deeper than 180 feet and dotted with coral knolls that rise to within a few feet of the surface. The shallower knolls can be a hazard to ships, but each is surrounded by a blizzard of fish and thus are wonderful places for a diver to observe tropical sea life in the clear warm water. On the outer side of the atoll the coral reef slopes steeply into oceanic depths of as much as 16,000 feet. Low tide exposes hard flat reefs on the northern and eastern rim, but on the west side the reef is softer and a little deeper, so that it never dries completely. A dozen low islets and islands rise from the eastern side of the atoll, but the highest point of land is no more than 15 feet above mean sea level.

My immediate interest was in inspecting the layout of the islands and the support facilities. Enewetak Island, at the southern end of the atoll, next to the wide entrance, held the airstrip and some military quarters. Next up the chain was Parry Island, home for the civilian contractors like Scripps; it was a city of tents, mess halls, and administrative facilities, where our team would live until our ship arrived. Immediately to its north was Japtan, a relatively unspoiled islet used for recreation, which still had tall coconut trees, a pandanus jungle, and giant lizards. Japtan was also distinguished by the huge rusting wreck of the *Knickajack Trail,* a supply ship that had run high up on the outer reef; it had been entering Enewetak's East channel just as a warship decided to leave.

A little further up the chain of islands was Runit, a long skinny sand spit with a Japanese hulk on the lagoon side. And finally, all the way at the north end where the reef was very wide, there were five small islets, one of which was to be the site of the Mike test.

These northern islets had been cleared of their normal vegetation by bulldozers, and now they were dazzling white wastelands of broken coral rock and sand. Construction crews were busy building a long straight causeway between them, apparently intended to connect the site of ground zero with some measuring station. Elugelab, the island selected for the honor of vaporization, was identifiable by a drilling rig on the north shore with a half dozen men around it. Near it there was a small shack with the words "Basalt or bust" crudely painted on the side. The fellow in charge was Dr. Harry Ladd, a paleontologist from the U.S. Geological Survey, who was pleased to see some trace of interest by a visitor from the outside world. Harry was short and stocky, a fatherly fellow with a trim white mustache who had lived much of his life in the south seas, studying volcanoes and reefs. We talked of Charles Darwin's hypothesis of the origin of atolls and how a hundred years of subsequent geological investigation had proved him right.

Darwin was a naturalist, a very astute observer and inductive reasoner who sought logical natural explanations for the creation of the world around him. As he sailed through the Pacific islands, he observed recent volcanoes with narrow coral reefs growing against their sloping flanks, as at Tahiti. At Bora-Bora he found a high volcanic spine on a central island encircled by a narrow lagoon and ringed by a coral reef. Elsewhere he noted places where there was only a nub of volcanic rock in the center of a large circle of coral, and at atolls like Enewetak, where there was no volcanic center at all, only a ring of coral.

That, Darwin reasoned, must have been the sequence. The original volcanic cone had built slowly upward from the deep sea floor by repeated outpourings of basaltic lavas, until it broke the sea surface and grew to be a mountain. Then a ring of coral formed around it at sea level. Over millions of years the great mountain subsided, and as it did, the coral reef that ringed it grew vertically to keep even with sea level. As the volcano submerged, the central volcanic island grew ever smaller until finally only the coral ring remained. The question to be answered here at Enewetak atoll was this: How deep is that ancient volcanic surface?

In Harry Ladd's drill hole at a depth of 4,630 feet, drilling progress slowed markedly, indicating that hard rock, presumably basalt, had been reached. However, when the drillers sent a core barrel down to retrieve a sample, the hole walls collapsed. It was a problem hole anyway; sometimes chunks of coral would fall from the hole walls and jam the drill string, other times the drill bit would encounter open space and suddenly drop a dozen feet. In any case, it was too near to shot time to start another hole near ground zero, so the drill rig was moved to Parry Island where, eventually, Ladd and his drillers brought up a sample of basalt. Darwin's volcanic-subsidence theory of the origin of atolls was again confirmed.

That afternoon Harry and I went for a swim in the shallow waters of the nearby reef, and shortly afterward I recorded my first impression of life on the reef that now holds one of the largest man-made craters on earth:

There are few greater thrills than one's first intimate look at a pristine tropical coral reef. The water was less than five feet deep, and my entire outfit consisted of a face mask and tennis shoes. The water was warm and clear, sparkling as the breeze dimpled its surface. White and gold coral colonies branched gracefully upward and among their branches swam dozens of brilliant blue fishes only a few inches long; scalloped clams pulsed their blue-flecked mantles; hermit crabs retracted swiftly into whatever shells they had appropriated, and white sand produced by the coral-eating parrot fish tumbled slowly along the bottom, propelled by a gentle current.

Spellbound, I floated face down on the surface, sometimes forgetting to breathe. Then periodically, like a sea-lion who has overslept, I would rouse from this reverie with a start and a snort to raise my head and gulp frantically for air. I should have been content to look but no, I had to touch, and was thus reminded that most corals are both sharp and firmly secured against waves and currents. The most fragile-looking coral branches have surprising strength; probably one could shave with the natural edges of some of the shells.

When the time arrived to catch the boat back to Parry, it was with dozens of tiny cuts on my hands, a sunburned backside, and a firm resolve to return with a color movie camera.

The reef along the eastern rim was quite different. It was hard and flat like an irregular concrete floor and had no live coral. At low tide this reef was dry, and one could walk to the slightly raised outer edge, known as the Lithothamnion ridge because of a calcarious algae that lived there, and see slate pencil sea urchins clinging to cracks churned by breaking waves. There was no doubt that this kind of reef would support a bulldozer. In fact, the reef surface was harder than ordinary concrete; metal fasteners fired from a gun, used by contractors to hold objects to concrete floors, would not penetrate it, and we had to devise another way to secure cables to the reef.

My next stop was Bikini, a slightly smaller atoll 200 miles to the east, decidedly off the beaten path and uninhabited since the Crossroads nuclear tests five years before. For this trip the Navy provided a Martin Mariner seaplane, bulky and slow but reliable. All six of us aboard were as familiar with Bikini as it was possible to be from charts, photos, and stories about the Crossroads tests; although we had never been there before, we had the odd feeling that we were "going back" to revisit an old friend. As the plane circled the area where the test fleet had anchored in 1946, we could see that the anchorage was a very small part of the 240 square miles of lagoon. Our pilot circled to make sure there were no buoys or coral knolls or mast tops of sunken ships that might wreck the plane before setting it down in the unruffled lee of Bikini Island.

The plane taxied into water about 20 feet deep and anchored. The surrounding water was like green glass, and in the dark reflection of the wing we could see ripple marks on the white sand bottom. It was only about a hundred yards to the beach, but from the air we had seen dark sharklike shapes moving about. With some trepidation we eased naked into the clear water and swam for shore. Our main concern was that our vulnerable private parts were hanging down like bait within easy reach of the teeth of any passing shark who might be tempted. With no face masks and with pants strapped on top of our heads, we couldn't get a good look at whatever was below. When our feet touched down on the sand of Bikini beach and we could wade ashore, out of reach of imaginary predators, I made a policy decision to keep my pants on while swimming from then on.

A line of tall coconut palms along the gently concave white beach sheltered a thick growth of underbrush from which fairy terns with innocent black eyes rose to inspect us. Bikini was larger and lusher, its protective reef wider, its lee calmer than we expected. We began to understand why the native Bikinians would not willingly trade their special home for another atoll that looked the same on a chart to some bureaucrat in far-off Washington.

We struck off through the thick scrub that came right to the beach toward the old camera towers that rose well above the tops of the coconut palms. They were covered a hundred feet up with vines thick enough to prevent us from climbing the rusty ladders. Stumbling about in the brush, we became archaeologists, intent on examining the concrete slabs and wooden frames of the ancient civilization that had existed here five years before.

The priorities of those long-departed inhabitants were clear. The ancients had lavished their attention on one building, the officer's club, and it had survived, sort of. Some of its roof had fallen in, and the floor was knee deep in weeds and rubble, but the prize piece of art work, an atrocious nude painting, still hung, skewed, behind a decaying bar. The spirits of the departed tipplers, in the form of huge blue and red crabs armed with claws capable of opening a coconut, reared back to defend their rightful places. We did not accept their challenge.

Back on the beach we looked out across the lagoon; our hulking midnight-blue seaplane that occupied center stage on the green lagoon looked very small against the vast backdrop of blue sky and white clouds. The tops of the coconut palms that crowned the islets on the south side of the lagoon were out of sight below the horizon.

In our imaginations, the photos of the Crossroads underwater shot came to life; our minds superimposed the old test fleet onto the vacant panorama of sea and sky before us. We could easily visualize the 2,700-foot wide column of water rising into the huge mushroom cloud and the 600-foot-high base-surge rolling over a fleet of large ships as though they were bathtub toys. It was hard to imagine how the mushroom cloud and visible effects from the upcoming test could be much larger, even though we knew Mike would have hundreds of times more energy.

Our daydreams then turned to the wanderings of the former residents. The Bikinians had left gracefully for what they expected would be a short time. They were amiable people, and the Great White Fathers in America had explained to them that their atoll was needed to test explosions that would somehow bring the kind of peace they enjoyed to others in the world. They would return soon, and they would be compensated for any of their coconut palms that were damaged. In the last days before they departed, their leader, King Juda, had squatted on this beach close to where we were standing, his attention fixed on a tiny but very loud gasoline generator. He had shown no sign of awe at the huge battleships that had been parked in his front yard or the other evidence of civilized technology that surrounded him, but Juda was mystified by how or why that two-foot cube of machinery could sit on his beach all day making such a racket.

When the Bikinians left in 1946, they decorated the graves of their ancestors and took along the essential, movable parts of their church. Weeping and singing, 167 people boarded an LST for Rongerik, an atoll not far away that they themselves had chosen. Their biggest initial concern about the new location was that it was rumored to be inhabited by a witch who poisoned fish and fruits. Soon after landing, they discovered the real problem: Rongerik was too small to support so many people.

By 1948 they were near starvation, so the Navy picked them up again and moved them, after a brief intermediate stay at Kwajalein, to Kili, a large island 500 miles south of Bikini. Kili was not satisfactory either; although there were plenty of coconut palms, there was no lagoon. This meant fishing was poor, and for several months each year breakers on the reef cut off contact with the outside world. At Bikini atoll all the people had lived on Bikini Island, the largest island on the upwind side, but each of the thirty islets on the encircling reef had a specific ownership and purpose. Some were used to keep pigs and chickens, some were temporary fishing camps, some had family coconut groves, some were pandanus jungles, some were best for gathering turtle eggs, and some were sanctuaries for edible birds. These special places and food stocks, as well as the spirits of their ancestors, were irreplaceable, and the Bikinians wanted to go back.

In 1968, long after the last nuclear device had been exploded and shortly after the death of King Juda, the high commissioner of the Trust Territory sent word they could return. His ship called at Kili to pick up an advance guard of nine men, who would inspect Bikini and return to tell the others about the changes there. There were bunks aboard the ship for government officials, military men, radiation monitors, reporters, and photographers, but none for the Bikinians who had to sleep on deck. Nor would they be allowed on the first boat to land; it was reserved for photographers who were to take pictures of them splashing ashore.

After a farcical ceremony at the old native graveyard, the only vestige of their civilization remaining, the ex-natives were free to explore. Their initial joy at being home again soon faded. Their home island was now overgrown with jungle; only a few coconut palms remained of the thousands they had left behind. A deep crater replaced two former islets; others had been linked by causeways, their pandanus and coconuts replaced by brush and concrete foundations. Massive bunkers and iron wreckage littered some islands, but the Bikinians were assured that any still-radioactive objects would soon be removed. The only favorable thing they would have to report to the others at Kili was that there were plenty of fish in the lagoon. With great sad eyes they faced the Americans and asked, "What have you done to our island?"

In spite of the bad news about the condition of their atoll and uncertainty about the kind and amount of residual radioactivity, some of the Bikinians returned. Then in 1978, 139 of the group who were living on the island were found to have ingested more cesium 137 than acceptable health limits permit, presumably from eating crops grown on contaminated soil, and once more they were evacuated to Kili. There they still wait for isotopes to decay and for Mother Nature and the United States government to clean their atoll enough so they can return.

The next island to be inspected for wave-recorder sites was Guam, the large southern anchor of the Marianas group, a thousand miles west of Enewetak. There was a large Navy base there, and I had vague hopes of getting some support from it in the way of lodging and transportation. The plane to Guam was another DC-6, but this one had a civilian-type interior with real seats and windows; as it rolled to a stop at the Naval Air Station, the steward announced, "Everyone keep your seats, you will disembark in order of rank." Looking around I could see some captain's eagles, so it was with some surprise that I heard the next words: "Mr. Bascom, please step forward."

I did and, on looking out the door of the plane, saw a red carpet stretching away from the bottom of the stair with a line of sailors at attention on each side. Out of the side of my mouth, I asked the steward if he'd mixed up the names; he checked the list and shook his head. So I descended grandly and walked the red carpet to be greeted with a salute by the commander who waited at the end. "Mr. Bascom, sir," he began, "your car is right over here. Your bags will be along directly."

Once inside the official car, I said, "What the hell is going on here? I don't deserve VIP treatment. I just came to look around the island for a couple of days." The commander gave a sigh of relief, took off his tie, and dragged out a rumpled set of orders. They proclaimed me to be a Class A traveler, an honor usually reserved for cabinet ministers and ambassadors. That was a SNAFU easily fixed, and he was glad to negotiate. All I wanted was a jeep and a map; in return he could release the honor guard and scratch the rest of the formal reception. I could keep the VIP quarters but without a private steward. We shook hands on the deal and went to get a drink at the Top of the Mar, the officers' club high on the cliff above the beach where Magellan had landed in 1521.

The jeep freed me to roam, and while checking out wave recorder locations, I drove several hundred miles along roads through jungle so thick and formidable that Japanese soldiers were able to hide in them for years after the war ended. In 1951, during my visit, there were still dozens of them in the dense brush; in the mid 1960s nine emerged to surrender, and twenty-eight years after the end of the war, after all the others who shared his hideout had died, Shoichi Yokoi finally gave up.

Above Pati Point on the eastern side of the island, facing Enewetak, a limestone cap rock slopes gently upward to end in a steep cliff several hundred feet high. The Strategic Air Command had built an airport atop that natural foundation, from which it could launch bombers. Below, amid a coconut plantation on the flat coastal plain, there was a fenced circular opening in the limestone a hundred feet wide and fifty deep. This carefully guarded cenote was the water supply for the Air Force Base above.

With an Air Force sergeant for a guide, I descended a steel staircase to a rocky beach at its bottom. Beautifully clear water with a glass-calm surface sloped down into a dark, infinite cavern beneath the opposite wall. I guessed that this fresh ground water moved as the pressure of sea water changed on its outer edge and asked the sergeant if the water level in the well changed with the tide.

"Yes," he said, "the tide in here is about a quarter the height of that in the sea, and high tide comes about four hours later." Well aware of the muggy heat of Guam's surface just above, the two of us did not hurry to leave; we sat quietly in the cool stillness and reflected.

Midway, at the extreme western end of the Hawaiian group, the "unsinkable carrier" of the central Pacific in World War II, also turned out to be another good place for a wave recorder. It is a low mound of coral sand that rises only a few feet above the sea, these days noted principally for the flotsam that collects on its remote beaches. Strings of glistening green glass globes, four to 18 inches in diameter, fugitives from the fish nets they once supported for Japanese fishermen, accumulate there, emerald beads on desolate white ridges of sand.

In Japan I visited seismic recording laboratories; then, with the inspection tour complete and a much better picture of what had to be done, I returned to La Jolla to start work.

DEEP MOORED BUOYS

Once it was evident that I was not going to die of cancer on the original schedule, the final preparations for measuring the waves from Mike moved rapidly. There were four kinds of locations where instruments would be placed: in the open Pacific not far north of Enewetak where the water depth averaged nearly 16,000 feet, in the shallow lagoon at Enewetak, on the rim of Bikini, and on distant islands including Guam, Wake, and Midway. I would build and install the wave meters in the first three categories; Bill Van Dorn would do the same for the distant islands.

The first question was this: What kind of a device could be used to make wave measurements in water 16,000 feet deep? Earlier instruments used had been designed for water less than 100 feet deep. The choice was between trying to invent some new kind of instrument that would measure surface waves while resting on the deep sea bottom or inventing some kind of a steady platform that would come within about 100 feet of the surface and would hold the kind of wave meters we already knew how to make. I chose the latter, being influenced considerably by charts of the region north of Enewetak that revealed two small, unnamed sea mounts twenty-five miles and seventy-two miles distant. These basaltic ex-volcanoes rose steeply from the three-mile-deep abyss to within about a mile of the surface. A mile is still pretty deep water, but a pedestal nearly two miles high is a big help. All that remained was the problem of how to build an instrument platform on top of the seamount, steady enough to hold the pressure sensors that would come almost to the surface.

For this I devised the first deep-water taut-moored buoy, an invention that has since been put to many uses by others. The principle is simple: a submerged buoyant float pulls upward against a slender wire, whose lower end is attached to a heavy clump anchor, with enough force to keep the wire taut and almost vertical. The top of the buoy, over a hundred feet below the sea surface, would hold the wave pressure sensor. For this scheme to work it was necessary to select the proper combination of wire, anchor, and float, taking into account the fact that the sidewise drag of ocean currents on a mile of wire could be very large.

For wire I used a single strand of high tensile strength music wire only about one tenth of an inch in diameter, which could be obtained in long pieces. This minimized the drag, but if we were to take advantage of the wire's full strength, it could not be bent sharply where it was connected to the buoy or anchor. The Scripps machine shop devised conical steel wedges that could grip the straight wire so that it could support about a ton.

The anchor was a pair of steel railroad car wheels weighing about 600 pounds, and the underwater buoy was made from the steel casing of a 500-pound practice bomb, stripped of fins and fitted with a tire valve so it could be pressurized with air against external water pressure.

The fact that the buoy was to be over 100 feet beneath the surface meant that a surface float of some kind was needed to mark the spot and support the recording devices. For this we used a raft Isaacs designed made of two layers of plywood shaped something like a four-leaf clover with large truck inner tubes sandwiched between them for flotation. In the raft's center, there was a stout pipe, which would support the recorder case and batteries a couple of feet above the water and which could be used to lift the raft in and out of the water. This first deep-water mooring was rugged and cheap, and it could be assembled from easily available parts. There was no time to build anything fancier.

The pressure recorders we would use to measure Mike's special waves were direct descendants of those built at Berkeley, except that for this application they had to screen out both tides and ordinary waves. This filtering was achieved by means of three internal air chambers connected by short glass capillary tubes whose size was calculated by Dr. Walter Munk.

With the help of senior technicians Chuck Fleming and Warren Beckwith, a prototype of the system was installed in relatively calm seas off La

Jolla, where the water depth is about 2,000 feet. It went in easily, and we hoped we could do nearly as well installing one in trade-wind seas while our ship maintained position above a small hump a mile below. We shipped the moorings to the Pacific Proving Grounds on the Scripps ship *Horizon*. A couple of weeks later our party followed by air.

Just before leaving, I signed a contract for the construction of a new house on the beach near Scripps. While waiting for me to return, Rhoda and Anitra would spend their time supervising the work and playing on the beach.

LIFE ON ENEWETAK

At Enewetak the Scripps scientists and technicians lived with hundreds of other men in four-cot plank-floored tents that stretched in parallel lines raised slightly above the sand along the eastern side of Parry Island. Practically perfect weather made living in these an enjoyable experience, at least for a few weeks. The tent flaps remained open, except during unusually hard rains, and the trade winds steadily moved soft sea air across our bunks. In the middle of the day the temperature was pleasantly warm; at night a single sheet was sufficient cover. In the central closet that held our stateside clothes, mildew was prevented by an eternally burning 100-watt light bulb. It was a simple, all-male life in which everyone wore khaki shorts, T-shirts, sandals, visored caps, and a deep suntan. There were no privileges of rank; senior scientists and laborers shared alike and would have been hard to distinguish without color-coded plastic badges.

In the evening the principal entertainment was a movie in a theater that consisted of a screen with crude wooden benches open to the stars. John Isaacs, more to be one-up than to be dry, once brought a large black umbrella to the performance. When the inevitable rain began and the first cool drops hit our warm naked shoulders, John hoisted his bumbershoot with a triumphant grin, precipitating a chorus of gibes and jeers which he acknowledged with a sweeping bow. Having gotten a satisfactory amount of attention, he decided that it was less effort to get wet than to hold the umbrella and subsequently left it in the tent.

When our party was not burying cables in sand and coral, traveling about the lagoon, building brackets and shelters for the recorders, or working over the instruments for the umpteenth time, we had all the advantages of living in a tropical paradise. There were brilliant white sand beaches and coral knolls to explore populated with thousands of bright colored fish. There were a few disadvantages, the main one being that it is hard to have a tropical paradise without women, and there were some unexpected dangers. One morning a scientist was dragged from the water unconscious and hospitalized. His blood pressure dropped to about 40 and his pulse rate was even lower. He had been pricked by a spine of one of the deadly and extremely well camouflaged stone fish that inhabit the reefs. For a few hours his condition was listed as "critical," but a couple of shots of adrenalin revived him, and by 6 P.M. he was back at the dinner table.

The Scripps part of Operation Ivy went very smoothly, largely because Commander C. N. G. (Monk) Hendrix was assigned as our project officer and my liaison with Joint Task Force 7. Monk was stocky and blue-eyed, a fellow with a decided set to his chin who was determined to get everything "squared away" and "keep 'em informed" about what's going on. He was a graduate of both the U.S. Naval Academy and the Scripps Institution of Oceanography and had spent most of his naval career in submarines.

During World War II he had served twelve long war patrols in "S boats and Fleet boats" pursuing Japanese shipping throughout the western Pacific from the Java Sea to Yokahama Bay. In turn his submarines were also pursued, and somehow he had survived being depth-charged a great many times (I think it was over 140). Monk believed that a sub's best defense against an exploding can of TNT was to go deeper, and as executive officer, he controlled the diving planes. His skipper was of the opinion that shallow was best, and after hearing Monk's rather sparse account, I had a mental picture of his sub following a porpoiselike pattern while under attack. Later I learned he had won two Silver Stars and two Navy Commendation medals and some other decorations for action, none of which he ever mentioned.

On Operation Ivy Monk was not only able to get messages through the swamped communications system of the Joint Task Force and obtain permission for us to do things that were bizarre by Navy standards, but he could "cumshaw" powdered ice cream from the Navy for the *Horizon,* and our crew loved him for it. Today the oceanographic laboratory at Annapolis is named in honor of Captain Hendrix, its first professor of oceanography.

One of the wave-measuring stations was to be at Bikini atoll, about 200 miles to the east, and so the *Horizon* carrying Scripps' men and instruments made that its first stop. Warren Beckwith and I were able to meet it there by hitching a ride from Enewetak on a passing LST (Landing Ship Tank). It was a surprise to see how limber the 328-foot-long LSTs are in a seaway. When the ship was bucking into 12-foot trade-wind seas, persons standing aft could watch 1-foot waves move along the steel deck as the crests and troughs of the water waves passed under the ship.

The research vessel *Horizon* was an ex-Navy tug, 142 feet long with a diesel-electric drive and a huge propeller. Two 1,000 horsepower engines ran electric generators that produced direct current to drive the electric motor that turned the screw; with this system the ship's speed could be controlled precisely from the bridge. This was a great help for holding position against a current or maneuvering in tight places. Like most tugs it had a large deck aft (also called the fantail) that gave us plenty of working room. A husky A-frame had been built over the stern, so that long or heavy objects could be conveniently handled overside by means of a special deep-sea winch that carried about 18,000 feet of half-inch wire. Otherwise, except for a coat of white paint, the ship was little changed from its Navy days.

Best of all, it was captained by Fred Ferris, a real can-do skipper with a lot of experience in tugboat operation. He was a man after my own heart, willing to take whatever risks with the ship were necessary to get the job done, including operating in tight channels between reefs. After a too-close encounter with a coral knoll while I was laying a cable, he whimsically applied my name to a large nick in the propeller. The occasional risks he took were balanced by his great knowledge of the ship's machinery. On one occasion far at sea, he waded through a dark, partly flooded engine room to find and close a submerged valve after a seawater intake pipe had broken and threatened to sink the ship.

With the *Horizon* to live on and work from, and a couple of new hands, Ned Barr and Alan Jones, who would stay and maintain the wave station on Bikini, we set to work to get the instruments operating. The waves from Mike were expected to be only a few feet high at Bikini, so it was decided to install the pressure sensor in quiet water where the natural wave background would be low. The site chosen was a pass in the southwestern reef next to a small islet named Chieerete.

The *Horizon* had also brought along some Aqualungs, the first SCUBA gear to reach that part of the world, and we intended to use them to install the Bikini wave recorders. Considering that the water was warm, visibility at least 100 feet, and the currents were low, it looked like an easy job. But there was a special problem; none of us had worked in tropical waters before, where there were sharks of several species, some over 8 feet long. We were unable to judge whether the danger of sharks was real or only imaginary. There was no one to ask or books to consult; we would have to find out for ourselves, by experiment.

Warren Beckwith, Todd Carey, and I shared the diving work, assisted, arguably, by Isaacs who never went in the water but who was generous with his forecasts of doom for those who did. John was known as a calamitologist, and now he energetically scoured his memory for stories about man-eating sharks which, we had no doubt, he embellished in the telling. Even worse, during our first few dives, he insisted on standing guard when we were below with a rifle at the ready and a bandolier of cartridges crosswise on his chest. He, we, and possibly the sharks, all knew that a bullet was effectively stopped by a foot of water, but near the surface we were much more worried about Isaacs than the sharks, especially after he had referred darkly to some floating coconuts as "suspicious" objects and blown them to smithereens. Once below we did not show our heads without first raising some familiar object.

The *Horizon* had brought special knives, spears, and antishark cages that someone on the home front thought divers would need as a defense, but it was clear to me that if we spent much time worrying about sharks or trying to work inside a cage, the wave recorders would not be ready in time. I resolved to see how brave these creatures were at the first opportunity; that came when we borrowed a dukw from the Army and began in earnest to inspect Bikini's southern entrance channel. For the first dive Todd Carey, my diving buddy, joined me in the water while the other prospective divers and Isaacs stood in the dukw's cargo compartment peering downward. The water flowing out through the pass was clear, and the the bottom below was a mixture of white sand and hard coral from which large gorgonian fans sprouted.

As we reconnoitered the area, a gray shark about my size glided slowly toward us, doubtless wondering what manner of fish we were. Now was the time to establish my territorial right to this part of the channel. I opened the 3-inch-long blade of my Swiss pocket knife and swam toward him, and as I did, the image of Peter facing the wolf with a popgun flashed through my mind. He, or perhaps she, looked astonished at this development and turned away. I followed, without the slightest intention of overtaking the shark, until with an exasperated glance and one swift flick of the tail, he tripled his speed and disappeared into the blue gloom outside the reef. The issue was decided for the moment; at least one shark did not care to contest the area we needed.

The other divers got the point, and after that no one was much concerned about sharks, although almost always there were one or two circling just at the limits of visibility, as curious about us as we were about them. They seemed no more hazardous than large strange dogs. Anyhow, the spears and shark cage were never used, and in six months of diving daily with sharks, we never felt seriously threatened—nor did we ever see dark fins projecting above the surface as shown in comic books.

The location was just right for a wave recorder. It was sheltered from the northeast waves by Bikini atoll but open to the southwest to admit any of the long waves that Mike might produce. In two days the installation was complete. The pressure sensor, in 60 feet of water, was attached to a railroad wheel with half an axle projecting up from it; from there an armored cable took the signal to the islet, where the recorder, batteries, and two men were housed in a tent.

All that remained to do was calibrate it by making a record that looked like a series of 5-foot waves with a 3-minute period. Dr. Walter Munk, Scripps' resident authority on waves who had come to observe the operation, agreed to help. We swam down to the instrument, and while Walter held a rule vertically, I stood legs apart with stomach braced against the half axle, adjusting my leg span until the top of my head was exactly 5 feet above the bracket on the wheel. Then, while Walter timed the operation with a watch, I hoisted the pressure pickup to the top of my head and stood immobile while curious little blue fishes circled me and peered into my face mask. When Walter signed that a minute and a half had passed, I set it down on the bracket for a similar length of time. We repeated this several times and then swam ashore to find the recorder had dutifully traced a series of waves in red ink on the moving chart. It worked!

Later the two of us visited that site again, and this time I carried a movie camera to record the scene for our friends at home. As I finished a long pan of the cable weaving its way between the coral heads, a shark about 6 feet long appeared in the viewfinder. This was my first chance to get a close-up of a shark, and I moved toward this fellow until he filled the viewfinder. Warily he moved away and circled left. As I cut across to meet him head-on, he turned behind Walter, who was prying away at a coral colony. Walter looked up, pleased that I wanted his picture; then it occurred to him that there might be another reason. He turned to look over his shoulder as the shark approached his rear; both were spooked by what they saw. Walter did a neat backward flip, and the shark swerved hard and zoomed off.

The wave meters on the distant islands were under the control of Bill Van Dorn, a doctoral candidate at Scripps whose thesis would be on wave generation. Bill was a bright, athletic, ingenious fellow who who had been raised in Pasadena and trained at Stanford. He became one of the first employees of Aerojet at a time when that company's principal business was building Jet Assisted Take Off units. The object of JATO was to give a heavily loaded airplane an extra thousand pounds of thrust for ten seconds to help get it into the air. These units were steel bottles about a foot long into which a warm mixture of roofing asphalt and potassium perchloride was poured. When contained behind a nozzle, this material burns explosively and furnishes enough thrust to move an airplane; without the nozzle the units only smolder violently.

One of Van Dorn's jobs was to take rejected bottles without nozzles to an abandoned quarry on Sunday morning and destroy bad batches of propellant by igniting the mixture with an acetylene torch. Watching lines of bottles pouring black smoke into the sky eventually became boring so, either out of curiosity or just for the devil of it, Bill decided to put a nozzle back on a bottle to liven up the day. He was not disappointed; the result was much like the Fourth of July compressed into ten seconds. The instant the propellant was lighted, the bottle took off at speeds of as much as 1,000 feet a second. While Bill cowered behind a boulder in the quarry bottom, this unguided missile performed leaping arcs, spun like a pinwheel, caromed from quarry walls, and went high into the air. Somehow the JATO unit and Van Dorn both remained within the confines of the quarry for the full ten seconds. With such qualifications Bill Van Dorn was just the fellow to measure the waves from Mike.

At Midway the island's military staff was extremely busy with activities related to the ongoing Korean War, and they were reluctant to allow messages on other subjects to interfere with their work. After he had installed the wave meter there, Van Dorn realized that he was unlikely to receive any information about whether the test would go on schedule or even if Mike actually exploded. So he built his own seismograph in the thick concrete basement of a former Japanese communications bunker. This consisted of a vertical post from which a weighted arm about 3 feet long was supported horizontally by a slanting wire. The pointed base of the arm rested against the post on a jeweled bearing made from a fused Coke bottle; the free end of the arm held a pen that rested lightly on a recording cylinder borrowed from a barograph. Once this ingenious piece of instrumentation was installed, Bill was sure he would get better information about Mike directly through the rocks than via official channels.

Back at Enewetak I had installed one of the wave meters on the lagoon side of a long narrow islet named Runit, only a dozen miles from ground zero. This location was favored because there was a small concrete bunker at the remote north end that could shelter the recorder. Not far from there, the wreck of a Japanese cargo ship, a vivid reminder of the terrors of World War II, stirred my curiosity, and I would occasionally dive on it.

The ship's bow, with hawse pipes and anchor windlass awash, was almost intact, but toward the stern the shapelessness increased until finally the hull became a splatter of ragged steel chunks on the sandy bottom. Twisted plates, girders, and unrecognizable chunks of iron, some weighing 20 tons or more, spread across an area ten times the ship's beam. One of the king posts, a tube of steel over two feet in diameter, had been ripped loose and bent double like a soda straw. Belts of machine-gun cartridges were strewn across the bottom. One side of the black hull rose above the sea surface so that the scuppers, backlighted by the sun, became a chain of bright lights. As I peered through the openings trying to decipher the strange shapes in the well deck, my head came up smartly against a rusty piece of pipe drooping over the rail. The odd looking machine from which it sprouted seemed strangely familiar, and a closer look resolved it into a 30 millimeter gun. Bright green algae growing along the ragged waterline contrasted with the orange rust of the exposed deck; fish darted from the shadows of the guns to the shelter of the holds. In the midships section the hull plates resembled a giant cheese grater, full of holes, each ringed with jagged iron thorns pointing outward.

This had been an ammunition ship. The ammunition in the holds had somehow been touched off, sending hundreds of rounds outward in all directions. Probably some of these had set off high explosives in the after hold, and there had been one final tremendous explosion. As I swam along the rusting wreck that explosion played on the screen of my imagination: a small cargo ship—dark silhouettes of men running frantically along the deck—a blinding yellow flash followed instantly by the sharp, stunning thunderclap of the shock wave—dark objects arcing outward, trailing steam—a few final screams—a rising plume of water—hot metal falling into choppy water. Those terrible few seconds had happened here only seven years before. Now the water was cool and quiet, patrolled by a school of bright blue fish who would soon have the opportunity to witness a very much bigger explosion.

The wave recorder at Runit had been running on test and needed to be checked. So, about a week before Mike, I stopped to see it, diving from a passing boat clad only in khaki shorts, swimming down to look at the pressure sensor, following the cable ashore, and inspecting the recorder. All was well. But as I emerged from the bunker, I was confronted by a Marine guard pointing a .45 caliber revolver.

Young marines holding big guns with quavering hands in uncertain situations are not to be taken lightly. I had no identifying badge, no rank, no serial number—only a fishy story about measuring waves. That's crazy to a marine; he was pretty sure he had discovered an enemy agent infiltrating the operation, and he marched me back to the construction camp at the other end of the island to check me out. It was just about noon when he paraded me past the mess hall; there my friends jeered in such a familiar way that I was released, and we all went inside.

The mess hall consisted of a single long table down the center of the aluminum-roofed dining area where anyone on the island was welcome to eat, family style, from heaping platters. Eating was a serious business that certainly would not be interrupted by any of the ordinary rain showers that fell each day, but now began a rain to be remembered—a truly superior deluge of an intensity that could only come from the collapse of a thunderhead, the bursting of a cloud, or the onslaught of a hurricane. Huge drops drummed so intensely on the roof that it was impossible to hear anything less than a shout from the guy sitting next to you. The water on the concrete floor rose over our ankles, and the view through the window openings on all sides was of an impenetrable curtain of falling water. Water spurted through the ordinarily impervious roof and poured down on the table, the food, and the diners. But in spite of this extraordinary and unexplained performance by the rain god, the men at the table bravely kept on eating. In ten minutes the rain passed, and in an hour Runit was dry again.

MIKE AND THE FIRST FALLOUT

A few days before Mike the *Horizon* headed for the first seamount north of Enewetak. The plan was to install a wave recorder there—if the seamount existed, and we could find it. All we had to go on was a single depth shown on the chart of 860 fathoms (5,160 feet) in a region where 2,500 fathoms was the norm. Since the quality of the reporting ship's navigation was not known, it seemed likely that we would have to search for the undersea mountain. We were prepared for that; *Horizon* was equipped with a 12 kilohertz, high-powered Edo, the best echo sounder available. A few miles before the ship reached the presumed location, our attention turned to the record of pings echoing from the bottom nearly three miles below. At first the bottom was nearly level; then, stroke by stroke, the record of each echo began to climb the slowly moving chart paper. When the markings went off the top of the chart, the next stroke would automatically come on the bottom; the result was that the ocean bottom rose in a succession of saw-like traces beneath our keel. The slope became ever steeper until the trace was almost vertical, as though the rock was rushing up to meet us. Then the line of echoes became relatively flat and in a few minutes fell rapidly away again; *Horizon* had crossed a flat topped basaltic peak.

We congratulated Captain Ferris for having found it on the first pass; that was good navigation. Now it was up to me to make a local survey of the rocky spire, navigating from the ship's laboratory; this meant watching the echo sounder and giving course directions to the bridge over the intercom. As is usual with such surveys, after *Horizon* was well out on the flank of the mountain, I brought her left 120 degrees, ran for about 15 minutes, and then came left another 120 degrees to cross the center of the peak a second time; then we repeated that pattern once more. Now we had profiles of the shape and size of the crest of the seamount on three different courses that trisected its center. The area on top was small, less than 1,000 feet across, and it looked as if it might be very lumpy, but it was an acceptable foundation for a deep mooring. The profile of the seamount looked a bit like Devil's Tower in Wyoming, although it was very much larger and its steepness had been accentuated by the way the record was made.

Now all that remained was to lower the 600-pound trolley wheel anchor on the one-tenth-inch steel wire and hit this small area while the ship was being pushed westward by the current and bounced around by trade-wind seas about 8 feet high. The music wire was wound onto a spool attached to the axle of the ship's main winch; now we ran the end of the wire over a large-diameter pulley, added the special fitting to its lower end, and shackled it to the anchor. The plan was to touch bottom with the wheel, then hoist it back 100 feet, stop off the wire while the upper fitting was added, attach the underwater buoy and raft mooring, then lower the anchor to its final position.

A combination of luck, skill, and clean living allowed us to do just that. As *Horizon* released the raft, Chuck Fleming and I stepped aboard it to get the wave recorder running. While the ship waited nearby, he and I slid the pressure sensor down the raft mooring line until it was seated on a rigid pin atop the buoy. Then we secured the electric cable to the mooring line at several depths with a tarred sailor's yarn called marlin, and began to watch the recorder that was mounted on the center post of the raft. In a few minutes, when the internal air chambers in the sensor came to equilibrium, the recorder centered itself to await waves of the period it was tuned to measure.

Now it was John Isaacs' turn. He had invented a new kind of instrument that would be dropped to the top of the seamount to record Mike's waves in another way. It contained a clockwork mechanism that would turn it on an hour before shot time. Then, twenty-four hours later when the *Horizon* could return to the area, this clock would fire an explosive device that would release its anchor weight and allow the instrument to float back to the surface where we would pick it up.

As with all "free" instruments (those with no cable to the surface, which must return automatically) there were substantial risks of losing this one, so, quite logically, John would allow no one else to touch it for fear we would screw it up. When the moment for it to be launched arrived, the rest of us stood back in a respectful semicircle as Isaacs checked it one last time, made sure the ship was above the seamount, and shoved the package overboard. We were still cheering when the expression of horror on John's face told us something was wrong. He had forgotten to remove a safety device that prevented the explosive trigger from going off on deck; with this still in place, the instrument would be unable to return. There was nothing to be said, and the rest of us slipped away, leaving the poor fellow with his head in his hands contemplating months of wasted effort.

Then *Horizon* moved on to the other seamount, some seventy-two miles north of ground zero, where two more deep moors were installed the day before the shot. Our plan was to stand by one instrument until the waves from Mike came past, so we would know at once how large they were. If they were unusually high, we would alert the Joint Task Force, who would issue a tsunami warning to all points around the Pacific.

So it came about that at 7 A.M. on November 1, 1952, I was standing on a cloverleaf raft with seawater sloshing around my ankles, daintily using an insulated screwdriver to adjust a 137-volt circuit in the wave recorder. John Isaacs and Monk Hendrix stood by at the oars of a skiff nearby, waiting to transport me back to the ship as soon as the waves were recorded.

When Mike dawned, its initial bright burst was well below the horizon, and even this hugest of fireballs cast only a pink light in the southern sky, being screened from our eyes by layers of low clouds. Apparently the cryogenic machinery in the large black building on the northern reef had done what it was supposed to do.

This was a moment in history to be remembered—the birth of the superweapon. Probably most of the Scripps personnel involved in this test thought of themselves as curious but detached spectators watching other kids playing with the biggest chemistry set on the block. We just wanted to know if these special waves would turn out as we had forecast. Our thoughts were not of war but about the opportunity to learn a new kind of physics and to see how the ocean would react to a huge explosion. If someone had asked us foretell the future, we would have guessed that it would be many years before the contents of the huge black building that had just exploded could be reduced to a bomb small enough to be carried by an aircraft. We would also have ventured the view that there would be plenty of time before that happened for the United States to reach political accommodation with other nations, and there would be no need to build bigger weapons. On both counts we would have been wrong.

In a few minutes dozens of thunderlike noises, sounding something like gunfire, came at us from all directions as the air shock waves caromed off clouds. I suppose we had expected a single loud report, but as John and Monk and I talked and waited, riding our small craft up and down on the trade-wind seas, we understood why it was that when the sounds of the eruption of Krakatoa were heard at Singapore, 500 miles from the volcano, search ships were sent out to see if a naval battle was in progress or some ship in distress was firing its cannons.

Mike's sounds had barely died away when the *Horizon*'s whistle began to blow, rapidly and insistently, a signal for us to return to the ship at once. Reluctantly I resealed the recorder box and stepped into the skiff to be rowed back to the ship without having seen the wave record. That turned out to be a mixed blessing. Captain Ferris met us at the rail. "We've just received orders from the commander of the Joint Task Force to steam northeast at flank speed. The shot went very big, and a lot of fallout is expected downwind."

As ordered, *Horizon* drove eastward into the trade winds at its best speed with the intention of getting upwind of any fallout of radioactive particles, which were expected to land on the open ocean to the west of us, all the time listening to radio reports from the fleet, 100 miles to the south. There was no doubt that Mike had been larger than expected; more energy had been released, the fireball and crater were bigger; the cloud had risen higher and spread further. Later we would learn that this first thermonuclear explosion was 600 times larger than any previous explosion.

The *Horizon* was prepared to deal with any radioactivity that might fallout on it. It had aboard an Army captain named Rogers, who was trained to measure radioactivity, and a "wash down" system consisting of a half-dozen fire hoses with preaimed spray nozzles was ready on deck. The system had been tested, and we knew that when pumps below deck were started, the roof tops and decks would be sprayed with seawater and washed clean of fallen particles.

In the early afternoon when *Horizon* reached a position over 100 sea miles from ground zero, Captain Rogers informed us that the level of radioactivity indicated by his geiger counters and ion-chambers was rising. At that the skipper ordered everyone inside, with doors and portholes closed; the ventilators were stopped to prevent them from sucking in air containing radioactive particles, and the wash-down system was turned on. Most of the thirty men aboard lay sweating in their bunks, reading and napping.

When the radiation level on the ship reached 35 milliroentgens per hour Captain Rogers, showing signs of considerable nervousness, visited Isaacs and me in the cabin we shared. Only then did we learn he had been trained at a Bureau of Standards Laboratory where the maximum permissable dose for a week was 3 milliroentgens. With this ultraconservative background, he considered the level on the ship to be extremely high and was near panic. A similar reaction had been observed of the monitors at the Crossroads tests, who had been as worried about the radiation from the ordinary luminescent dials in aircraft cockpits as that from the explosions. Before long the radioactivity on the *Horizon* leveled off, the particles being prevented from accumulating on the decks above by the spray system. Then it started to drop, and five hours later it was down to 1.5 milliroentgens per hour. The worst was over.

What we guessed, but did not know for sure until several days later, was that for the first time in history the hot updraft carrying particles from a nuclear explosion had burst through the top of the lower atmosphere into the stratosphere above, where the high-velocity winds blew eastward. Then, as the radioactive particles fell back into the lower atmosphere, they were blown westward again by the trade winds where we encountered them. The *Horizon* had been ordered into a region where it would have avoided fallout from a small explosion and was accidentally sent to a place that received fallout from the first really large explosion. Everyone aboard the ship probably received a dose of about 100 milliroentgens or about half the average annual dosage of people living in the Rocky Mountains. For me, who had been paying for anticancer radiation in far higher doses, it was a trivial addition.

Mike did not create the first fallout; all surface nuclear explosions necessarily produce some fallout of radioactive particles. These are bomb components and bits of nearby objects in which radiation has been induced by neutrons from the burst. But Mike produced the first really significant fallout, covering thousands of square miles of ocean and a few islands. What made the fallout from this explosion special was not only the size of the explosion but the unusual amount of material involved. This included all the metals in the cryogenics plant (compressors, tanks, piping) that had been vaporized, plus the fragments of coral in a crater a mile across and several hundred feet deep. This huge mass of material had been reduced to tiny particles, all radioactive with various half-lifes, and spread by both stratospheric and lower winds over a thousand square miles of sea, unpopulated by humans except for those of us on the *Horizon.*

When we went back on deck, we discovered that all the painted metallic surfaces within the range of the wash-down system were quite clean, but the masts, antennae, and hardware above the reach of the spray needed more hosing and scrubbing. Rough-surfaced materials such as manila cordage, canvas lifeboat covers, and our collection of giant clam shells were deemed uncleanable and heaved overside. Soon the *Horizon* was clean enough for longtime habitation, but its background level of radioactivity was permanently raised, so that the ship could no longer be used for very low level scientific measurements of radioactivity.

Meanwhile, back at Bikini, Ned Barr and Alan Jones who were watching the wave recorder on Chieerete had seen Mike dawn in the western sky and were listening to the general chatter on a short-wave radio while waiting for the waves to arrive. They were startled to hear that the explosion had gone far beyond expectations and a large tidal wave was on the way. If that were true they correctly figured that they would be better off aboard a boat in the lagoon than on a low islet that could easily be overtopped. The only boat on Chieerete was a whaleboat that had been dragged up the beach with some difficulty by six husky men. But fear endows a man with great strength, and later Ned claimed he and Alan had picked up the heavy boat and launched it on a dead run. After an hour passed with no sign of the "tidal wave," they returned to see the trace of a small wave on their recorder.

On Midway Island Bill Van Dorn waited in the deep, cool bunker watching his jury-rigged seismograph to find out if Mike had gone off on schedule. A few seconds after the scheduled shot time, the instrument's weighted arm held steady while the island and the bunker moved beneath it, causing the pen to mark the arrival of Mike's seismic waves on the chart paper. About two hours later the water waves arrived.

Dr. Edward Teller, the principal driving force behind the development of hydrogen weapons, watched a similar but more elaborate seismograph at the University of California at Berkeley for the same reason. About a minute after Van Dorn had, he learned Mike was a success and sent the famous code message "It's a boy" to Washington.

Ted Folsom, who operated the wave recorder on Wake Island, also had the problem of finding out whether Mike went off on schedule. His solution was to devise a sensitive barograph from pieces of junk; with it he detected the faint air shock wave 450 miles from Mike. John Knauss, on Guam, finally learned from Navy scuttlebutt that Mike had exploded several days earlier. Although his instrument did not pick up any waves from the blast, it was working because a week later he recorded a natural tsunami.

The day after Mike, we returned to the deep moorings to pick up the wave record and retrieve the instruments. The raft looked about the same, but we took along a geiger counter to measure any radioactivity that might remain. The raft and instrument case were clean, any particles that landed having been washed off by rain; so were the electric cable and pressure head. But one item was not. At about 20-foot intervals, we had secured the instrument cable to the raft's mooring line with marlin, a tarred sailor's twine. It was just the sort of material that would catch any particles moving past in the current, and it was radioactively "hot." We checked each piece of marlin as we cut it off and saved the pieces in a glass jar to give to the radiation measuring group back on Parry Island. This subsurface radioactivity was additional evidence that there had been extensive fallout on the sea to the east, which had mixed with the upper layer of seawater and been carried past our instruments by the currents. However, this piece of data, as well as the information about fallout on the ship, was received with a ho-hum attitude by the authorities; a year and a half later the Atomic Energy Commission regretted not having paid more attention in planning subsequent tests.

As for the wave record, it showed that quite large long-period waves from Mike had passed the instrument. On the basis of what we knew then about wave theory, if we had seen the record immediately after the shot, we would probably have sent out a general warning that a large tsunami (tidal wave) was on the way. But since no such wave reached distant shores, it was lucky we had been ordered away when we were.

We recovered what we could of the wave measuring equipment and returned to Enewetak. As the only group with diving capability, we were asked by Admiral Wilkins, the task force commander, to search the atoll's "wide entrance" for an aircraft that had crash-landed short of the runway, after having sampled Mike's radioactive cloud. Warren and I did so for four days, wearing Aqualungs and riding on a chunk of iron towed at middepth from a landing craft, sometimes followed by sharks that seemed to think we were bait.

In that beautifully clear water, we covered about a square mile. We saw fish by the thousands (tuna, sharks, and rays), coral clumps by the hundreds, torpedo net panels by the dozen. There were submarine cables, sunken landing craft, and oil drums, all slowly rusting away. One pass took us along what had been a field of moored mines, methodically placed to guard the entrance during World War II, but we never saw the aircraft in question. Our reward for the effort was a letter of appreciation from the admiral.

Our last day was saved for King shot. It was to be a small air burst, and Dr. Russell Raitt, a Scripps seismologist, wanted to determine how much of the energy of its air shock wave would be transmitted into the ground. I tagged along to observe and be helpful when Russ lowered hydrophones into Harry Ladd's old drill hole on Parry Island. We talked as we waited, noting that the sky was unusually clear and that there were fewer than usual of the characteristic flat-bottomed cumulus clouds.

King was to be launched by rocket from Runit, and the broadcast countdown started just before noon. Russ and I turned away, squeezed our eyes shut, and waited for the jolt of the shock wave. Then, after a glance at the seismic record, we turned our attention to the spot in the sky where the burst had occurred. A small white cloud had formed, presumably made of water condensed on bomb particles. We watched for a minute as this cloud grew against the blue sky background and drifted slowly toward us. We headed back to the mess tent for lunch, and just as we reached it, about 20 minutes after the shot, the bomb cloud arrived overhead, and rain from it began to fall. By then nearly everyone in the camp was already eating, including the officially designated radiation monitors.

I was curious about how much radioactivity the rain might contain and suggested to their leader that samples be taken immediately and that the level of radiation in the tent be measured. Reluctantly the fellows with the radiation instruments agreed to make some measurements after lunch, but by then the rain had stopped, and probably the suggestion was forgotten. The point is that in those days there was very little interest in radiation levels, even among those whose specific job was to measure them.

With the Ivy series at an end, the Scripps people piled on the *Horizon* and headed for Kwajalein to begin a great new adventure to be known as Expedition Capricorn.

John D. Isaacs in 1947 (UC)

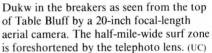

Me with Dick Wilson, Navy photographer in heavy weather clothes, relaxing between surf surveys at Table Bluff, California, in 1946. (UC)

Dukw in the breakers as seen from the top of Table Bluff by a 20-inch focal-length aerial camera. The half-mile-wide surf zone is foreshortened by the telephoto lens. (UC)

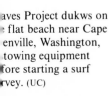

aves Project dukws on flat beach near Cape enville, Washington, towing equipment fore starting a surf rvey. (UC)

Dukw surfboarding o[n] 12-foot plunging breaker, considered small enough for train[ing] drivers. (UC)

A fleet of purse seiner[s] called Monterey harb[or] home in 1947. Their motions while at anch[or] made this an ideal pla[ce] to study "surging" caused by long-period waves. (USN)

Bud Hughes and me getting ready to lower [an] early wave recorder to[ward] the bottom off Canner[y] Row in Monterey Bay[.] (UC)

Point Sur, California, is a spectacular observation point. A sheltered cove on its south side is a good place for submarine cables to come ashore. (WB-USN)

Laying a submarine cable at Point Sur with a steel life raft towed by a dukw. This wave recorder was quite successful. (UC)

The great tidal save of April 1, 1946, washed over the pier at Hilo, Hawaii, tearing away the superstructure and killing the unlucky longshoreman. Photo was taken by an alert crew member on the *Brigham Victory* tied alongside. (World Wide Photo)

he April 1 tidal wave swept ross Hilo, Hawaii, moving many ocks of houses into a line of iftwood. (Ichii, Honolulu Advertiser)

The *Ioannis Kulukundis* hard aground on the out bar near Point Arguello, California, 1948. As wav slammed into the ship, driving it shoreward, crewmen hung over the r clamoring for transport ashore. (WB-UC)

Me with a photographer's mate in the blister of PBY Catalina seaplane with the cameras used to photograph much of the Oregon–Washington coast. (USN)

Marines practice an amphibious landing at Camp Pendleton. Photographed by the auth as part of his studies of amphibious operations for the Navy. (WB-USN)

wrecked landing craft on the soft sand
⸱h at Waianae, Oahu, during Operation
⸱i, 1949. (WB-USN)

The Scripps Institution of Oceanography in 1949
consisted of a pier, a seawall, and a few buildings,
all in need of repair. (WB-USN)

Supervising the placing of
dummy mines in San Diego
Harbor from a landing ship.
⸱(USN)

Working with the
Underwater Demolition
Teams from the Amphib-
ious Training Base at
Coronado, California. (USN)

Me on the cloverleaf buoy above the taut-moored bu just before Mike, the first thermonuclear explosion. The skiff is rowed by 'Mu Hendrix and John Isaacs. (Scripps)

Adjusting part of a wave meter at the tent on Chieeriete Islet, Bikini ato October 1951. (Scripps)

The Crossroads Baker sho 20-kiloton explosion fired a depth of 90 feet in Bikin lagoon July 25, 1946. The white base surge is 600 fee high, the column is 2,200 in diameter. The Bravo sh 8 years later, was 600 time powerful. (U.S. Navy)

...e sinking of the aircraft carrier
...ratoga after Baker shot. The
...ling column of water toppled
...r stack and swept the planes off
...r deck. (U.S. Navy)

...e signal bridge of the Saratoga about 50 feet
...derwater in early 1954. (WB)

The *Spencer F. Baird* on Expedition Capricorn.
The winch wire to the corer hangs low off the
stern; the hydrophone cables are held high. The
pointed object secured to the mast is the author's
Fijian outrigger canoe. (Scripps)

...iting on the stern of the *Baird*
... an instrument to surface.
...ipps)

Ready to go overside with a motion picture camera. (Scripps)

Me sitting on the edge of the abyss at Falcon Island, a very active volcano in the Tonga group. (Scripps)

Walter Munk floats in midwater above one of the huge coral vases of Alexa Bank, a sunken atoll. (Scripps)

V

Through the South Seas

FROM THE MARSHALLS TO THE FIJIS

Waiting at Kwajalein for the *Horizon* was its sister ship, the *Spencer F. Baird,* a newcomer to the Scripps fleet. In a few days these two ships would start on a major geophysical expedition that would follow a track over 12,000 miles long across the tropical Pacific. The scientists aboard would measure the elasticity and magnetism and temperature of the undersea rocks; they would sample air and water, mud and ooze; they would dive and photograph and visit remote islands; they would explore the Tonga Trench and the Andesite line. The ships would range as far south as the Tropic of Capricorn and as far east as the Easter Island Rise—where it is over 1,000 miles to the nearest land in any direction. This was to be the Capricorn Expedition of the Scripps Institution of Oceanography, led by Dr. Roger Revelle and utilizing the talents of scientists from Sweden, France, England, and Australia as well as the United States. We would be the first to examine this major part of the earth's crust; our findings would lay the foundations of modern geological theory.

Kwaj, as it is familiarly known, is some seventy-five miles long, the largest atoll in the world. Its flat coral rim is dotted by many islands, but only two of the southernmost ones were then inhabited. These were Kwajalein Island with the American military base and airport, and the island of Ebeye where the Marshallese natives lived. All the other islands were off limits, because they still contained the wreckage of war, including land mines and unexploded bombs. The most distinguishing feature along the low horizon

115

was the rusting hulk of the German battleship *Prinz Eugen,* which had been beached high on the shallow western reef after it survived the Crossroads explosions.

While waiting for alternate crew members and new scientists to arrive, I busied myself building a "shooting shelter" on the stern of the *Horizon.* Deep seismic work to look at the structure of the earth's crust requires large shock waves, and Russ Raitt, our expert on these matters, planned to set off twenty tons of high explosives in the course of this expedition. That meant a special working space was needed on the fantail, where blocks of TNT would be safely made up, fused, lighted, and heaved overboard. This shelter was only a rough frame of wood covered with plastic sheeting but it kept out most of the wind and spray. Since *Horizon* was tied up to a pier that was also used by the boats that transported native workers back to their island after working hours, this was a chance to observe rush hour, Kwajalein style.

The Navy provided twin-engined steel landing craft (LCMs) for transportation, and most of the natives used these for commuting to and from their home island. After work a procession of them would plod down the pier, crowd aboard, and stand during a noisy 3-mile ride home at 9 knots. But half a dozen of the young men would run down the pier and make a racing dive in the direction of three Marshallese-style outrigger canoes that were anchored a short way offshore. With powerful strokes these chaps would swim out to the boats and slither aboard them in a smooth continuous motion. On each boat one fellow would pull in the anchor line while the other raised the mast and unfurled the sail that had been rolled up on the deck; in less than a minute they were sailing. With the tradewinds on their beam, these graceful craft seemed to rise up and skim the surface, usually beating the powered landing craft home.

I was filled with admiration for these sailors and their canoes, and later had a chance to study both their sailing practices and boat construction. The keel and lower hull of these swift boats is hewn from a single log whose sides are then raised with sewed planking. A relatively small outrigger float is connected to the hull by a wide, lightly constructed platform on which crew and passengers ride. These craft sail well into the wind, using a radically different method for tacking than that known to the western world. The two ends of the hull are the same, so that either end can be the bow but the lee side is flat; this is necessary because the boats are always sailed with the outrigger to windward. In order to change tack, the mast and its sail of fine pandanus matting must be lifted from its step at one end, carried to the other and restepped.

Although this sounds cumbersome, the expert local sailors make it look easy. These craft are steered mainly by adjusting the sail but with occasional help from a broad steering paddle. To ensure that all the lashings holding the boat together remain sound, ancient taboos require that these be renewed every year. In these small boats, rarely over 30 feet long, Micronesians repeatedly voyaged hundreds of miles between islands, steering by the stars and by wave patterns resulting from refraction around unseen islands. These fast light boats are not unlike the flying proas that impressed the early Spanish voyagers by literally sailing circles around their clumsy caravels.

By the time Expedition Capricorn left Kwaj, the scientific ranks had been considerably enlarged by a dozen new men who had flown out to join the ships. John Isaacs, Russ Raitt, and I, all of whom had been at Enewetak, transferred to the *Baird,* which was to be Roger Revelle's flagship. Nearly everyone aboard was young, eager, and naively enthusiastic about our new ship and the scientific equipment it carried. This is probably the normal way for expeditions to start. By the time we got back to San Diego three months later, there were rust streaks on the white hulls, the shiny instruments were battered but finally working properly, and all of us had a much more realistic view of the ocean.

Roger liked to say, "We need a lot of strings for our bow," by which he meant that we'd better take along many kinds of measuring instruments so that, if some failed, or if we got some new ideas while at sea, or if unexpected opportunities presented themselves, we could make the most of this expedition. He also believed that there were two ways to go about oceanographic work at sea: The scientist could pose a specific question and devise instruments that would make appropriate measurements to answer that question. Or an expedition could make many kinds of measurements in many places to see if some previously unsuspected truth could be sifted from the data. Capricorn, because it made many new measurements in rarely explored waters, and because it had many scientists representing different disciplines and working on their own projects, was a mixture of the two.

Because space is tight on a small ship, the scientists necessarily worked closely with colleagues from other disciplines, whom they would rarely encounter elsewhere, so the voyage provided a good opportunity for each of us to learn about other branches of oceanography. All members of the scientific party put in a full day's work either on their own specialty or giving a helping hand to work that was going on at the time; in addition, everyone stood four-hour watches day and night, keeping an eye on the echo sounder and taking the temperature of the upper waters. Between these intermittent jobs, there was a virtually continuous discussion of every aspect of the scientific program and a constant readjustment of our views as new bits of evidence surfaced.

Aboard the *Baird*, Art Maxwell and I shared a tiny bunk room below the waterline in the bow. His job was to measure the flow of heat from the earth's interior through the sea bottom. Heat flow was an appropriate subject for a fellow on his first tropical voyage in a room at the far end of the ventilation system. Another problem with living in the bow is that, in a storm when the ship went over large steep oncoming waves, the bow would fall suddenly, and the bunks would drop out from under us; when it slammed into the next trough, we would crash into the bunk, and the collision would force air from our lungs. Clothes were stored in drawers under the bunks, and the only furniture was a canvas chair. These inelegant quarters had the advantage of privacy, and in my off-duty time there I kept the detailed set of notes about our daily adventures from which this account is taken.

The main reason for a two-ship expedition was so Dr. Russell Raitt, our seismologist, could explore the thickness and sound velocities of the layers of rock beneath the ocean. This was done by measuring the time it took explosion-generated sound waves to travel between the two ships by each of several pathways. *Horizon* would shoot, and *Baird* would listen.

The first trial of this method was in deep water west of the Ralik Chain of islands, and the conversation went something like this:

Horizon (about fifty miles away, on the radio loudspeaker): "We're ready to start firing."

Russ (in the lab of the *Baird*): "We're trailing our hydrophones and we're on 'quiet' status." (Most machinery was stopped to minimize local noise.) "You can shoot anytime."

Then *Horizon*'s countdown (on the radio): "Ten seconds. Five seconds. The fuse is lit. Mark."

At the word "mark," Russ Raitt would start the oscillograph, and in a few moments we would hear (and he would record) the sound of the explosion of 80 pounds of TNT on the radio. That was the zero moment, from which the velocity of the shock waves racing toward us would be measured. The first to arrive, as indicated by the movement of pens that recorded the sounds picked up by the hydrophones, came by the longest but fastest route through the dense rocks in the upper mantle of the earth below the "Moho." The next sound waves came via the shallower and less elastic crustal rocks; then came the waves that had traveled in the soft sediments, and finally the largest waves arrived that had come the shortest distance through the slowest medium, water. In those days it was something of a triumph to have "shot the

Moho"—measured the speed of sound using explosives in the rock beneath the Mohorovicic seismic discontinuity.

Anyone else might have shouted "Eureka," but Russ, inherently suspicious of initial luck, muttered to himself and fussed over the instruments. From then on the *Baird* waited as the *Horizon* moved toward it, dropping charges every few miles. Every record was a good one, and as the gap closed, smaller explosions could be used; by noon *Horizon* was close enough to the *Baird* so that the crews could stand along the rails exchanging banter: "Anyone who wants to change ships, swim across." Then *Horizon* continued on, shooting, until it was fifty miles away on the other side, and Russ Raitt grudgingly admitted the day's work was "OK." From the travel times of sound, he would later calculate the thicknesses of the rocky layers of crust beneath 3 miles of seawater. This was only the first of dozens of days of seismic measurements in regions whose geologic structure varied considerably.

While Russ Raitt's hydrophones sprawled lazily on the water behind the ship, waiting for the next explosion, other scientists were busy taking cores, lowering cameras, and collecting water samples. In order to do those things, a winch was needed that held enough wire rope to reach the bottom. The *Horizon*'s winch was a simple powered spool on which was wound 18,000 feet of wire—enough to reach the bottom of most of the Pacific basin with a bit to spare.

But one of the objectives of Expedition Capricorn was to investigate the bottom of the Tonga Trench, which, at about 35,000 feet, was believed to be the deepest known place in the southern hemisphere and perhaps in the world ocean. For that the *Baird* had been equipped with a new winch that held 40,000 feet of wire rope tapered in steps from 3/4-inch diameter at the upper end to 3/8-inch at the lower end. This winch was not only large and complicated, but it had never yet lowered a cable at sea.

No one with good sense would take a huge virgin winch thousands of miles out into the Pacific before testing it, but the combination of a steel strike and the Mike schedule left no other choice. With some apprehension we inspected this untamed creature, which would control a piece of our destiny, hoping that it would do everything it was designed to do but suspecting that many bugs were hidden in its mechanical innards.

The Great Winch had three main parts, the largest being a huge storage spool below deck that held nearly 8 miles of wire weighing some 15 tons. From it the wire ran up through fair leads to the main deck and three times around a winding gypsy (a pair of large powered pulleys in tandem) that powered the wire in and out. The deliberate separation of the storage spool from the motive power had the advantage that the wire was not stored under tension; this scheme had first been used by the Danish deep-sea research ship *Galathea* and worked well, so the Scripps engineers decided to try it. The third element of the winch was a control panel covered with lights, buttons, switches, knobs, and valves. It was located under a shelter on the boat deck that looked down on the *Baird*'s fantail and was controlled by Bud King, first engineer.

The day of testing the winch, and ourselves, was at hand. Not far west of Namorik atoll in the southern Marshalls we attempted to take a core in water about 15,000 feet deep, a modest depth relative to the planned capabilities of our new toy. With some apprehension, we observed that first unreeling, not knowing what to expect. The corer reached bottom and was on its way up when a loop of slack got into the wire between the winding sheaves and the storage spool and spun itself into a screweye; when tension was applied this loop pulled into a tight, twisted, knobby kink. There we were with a corer dangling 10,000 feet below on the wire, unable to move the winch in either direction. Luckily we had a very resourceful bos'n aboard named Brownie; he and the remarkable Bud King, who had lived with the winch during its construction, decided our best bet was to cut out the kink and make a splice in the wire.

Brownie got a grip on the overside cable with wire stoppers; then he excised the kink, and Bud cautiously winched 200 feet of the overside wire aboard so we had a long end to work with. The stoppers were reset, and we pulled another 200 feet of wire off the spool below deck by hand. Now all that remained was to splice the two ends together. It is not a simple matter to make a 200-foot-long splice in a five-part wire rope on the cluttered decks of a ship only 142 feet long. Each of the five strands at the end of each wire was coded with a color and one man was assigned to hold it. Then, at Brownie's order, the man on the green strand of the up wire would unravel 200 feet of his strand, and the green man on the down wire would replace it with his strand, patiently twisting the sticky strand to make it lay properly every inch of the way.

The line of men that extended from the fantail along the port passageway to the forward deck and back again were literally up to their knees in wire; they were also streaked with sweat, salt, and cable dressing. They joked sardonically that when the process was finished, we would probably find that some major part of the ship had been spliced into the middle of the wire. (Several years later on another expedition a finished splice was discovered to have an overhand knot in it; John Isaacs proposed they solve the problem by enlarging the loop and passing the ship through it.)

As we sweated through the afternoon, two tiny outrigger canoes, each filled to capacity by a single man, approached hesitantly. In answer to a friendly wave, these residents of the nearby island of Namorik paddled up. One, a handsome dark man named Tapaka wearing jeans and a yellow roll of cloth around his forehead, climbed over the rail. He thoughtfully inspected the snarl of wire on deck while we stood like dummies holding our ends of strands, but he gave no indication that he thought our behavior odd. Instead he asked us in calm, unaccented English what our business was in this part of the ocean. On hearing that we were examining the structure of the earth beneath the sea, he seemed to understand the value of such research at least as well as some of the admirals we had talked to. Somehow we got the feeling that these two fellows were only scouts for a larger group. Presently they said good-bye and paddled off in the light rain while we continued with the splice.

Our instincts were correct. Late in the afternoon, when the decks had been cleared and the newly spliced wire was cautiously being wound back aboard, Tapaka returned in a whaleboat with a dozen friends. The new arrivals scrambled over the rail and then backed against it in a self-conscious line; our crowd formed an equally self-conscious opposing line and gawked back as they presented us with great bunches of green bananas. In return Roger presented them with some bright yard goods, old white shirts, writing tablets, and combs. Solemnly he set these down and, with Tapaka translating, made a speech to the visitors that embodied all the best features of an old Western movie in which the hero gives presents to the Indian chief as a token of friendship. Gone was Roger's easy style and smile. In accents grave and words stilted, he spoke of friendship between lands across the sea and the exchange of cultures.

By the time our visitors mumbled good-byes and cast off, it was dark. A slim moon backlighted the tops of the distant coconut palms and glinted on the copra boat's wake as it headed for three tiny orange fires that pricked the black background. When our core barrel finally came aboard, it was empty, but we were too tired to care and flopped into our bunks. By midnight *Baird* was underway again.

The ocean's story, at least for the last few millions of years, is written in the soft sediments on its bottom. These sediments are composed of ash from volcanoes, meteors from space, and dust from the Gobi Desert, all borne by the wind. Some of the bottom material is fine silt from land carried far by currents, and some comes from minuscule skeletons of marine animals that live in astronomical numbers in near-surface water. Slowly, very slowly, these tiny particles settle to the bottom, where they become clays or muds or oozes.

In some places the buildup in the deep ocean is as little as 1 millimeter in a thousand years; in others, a hundred times that much material may accumulate in the same length of time. Sometimes the age of the sedimentary layers can be determined from the fossils present or the decay of natural radioisotopes. In any case, these marine muds are a valuable source of information for marine geologists, who want cylindrical samples of them, called cores, that go as deep into the bottom and as far back in time as possible.

On the leg of the expedition between Kwaj and Fiji, I (who had never before seen a deep ocean core tool) was in charge of the coring. The *Baird* carried a new type of piston corer developed by a Professor Kullenberg of Sweden, and our overside crew studied the pieces of this device. The core barrel itself was a piece of steel pipe 3 inches in diameter and 30 feet long, with a sharp cutting edge at the bottom; inside its lower end was a "core catcher," a circle of thin stainless steel fingers that opened to admit the sediment and bent shut behind it. Because the slim barrel was weighted at the top with 500 pounds of cast iron, one would expect such a huge hollow needle to bury itself in a soft bottom and take a long core. But it would not, because after a few feet of sediment entered the barrel, the friction of the sediment against the inside of the pipe was greater than the resistance of the mud ahead, and the open end was effectively plugged. The result was that many long corers had been hauled up that had gone deep into the bottom but contained only a few feet of core inside.

Professor Kullenberg solved that problem by providing an external trigger holding a slack loop of wire that would allow the core barrel to fall freely through the water for its own length. Inside the barrel, there was a tight-fitting piston attached to the lowering wire. When the corer fell, the wire to the piston would come taut just at the moment the tip of the core barrel touched bottom. As the piston touched on the mud surface, the core barrel would slide down around it, and the mud would be sucked into the core barrel, filling it. The principles were clear, but rigging the cable properly was not so simple.

A day or so after the splice, we tried for another core in the region west of the Gilbert Islands, confident this time that the cable would not snarl again. Again we lowered the core tool, and as far as we could tell, the core barrel went into the bottom, but on a heaving ship, when the weight of the corer is only a small fraction of that of the wire overside, one is never sure. If the ship puts out an excess of wire to be certain the bottom has been reached, the cable can come up snarled. On this occasion when the corer returned to the surface after a round trip of four hours, the mechanism seemed to have worked, and we hoisted the long pipe clear of the water. As we did, the weight of the water in the upper part of the pipe washed out the sample of calcareous ooze in its lower end, and our "core" trailed off in a white stream behind the ship.

With true scientific objectivity we discussed the problems of coring in oozes, adjusted the core catcher, rerigged, and sent the core tool down again. Four hours later we repeated that performance, except that this time our discussion had less objectivity and more cussing. We were still not sure whether the bottom was uncorable or whether we were not rigging the tool properly, so we sent the corer down for a third time. It was late at night when the corer returned to the surface; by now our morale was at stake. If this core washed out, we would have wasted a long day. I was determined to prevent that by diving down and driving a bung into the bottom end of the core barrel before we tried to lift it out of the water. Conditions were less than ideal; it was almost midnight, and the sky was black as sin; the pitching and rolling of the ship jerked the pipe about, and as usual, there were several sharks circling the ship looking for garbage. Some of the crew expressed doubts about my sanity.

I was lowered over the stern on a bos'n's chair wearing the usual swim trunks, tennis shoes, and Aqualung, and carrying a bucket on a line, a wooden plug, and a hammer. At the last moment Roger passed me a packet of shark repellant, a chemical so repugnant that sharks leave it alone. Swinging in the harsh glare of a floodlight with my white feet dangling in the dark water, I had momentary misgivings, but once below the surface my confidence returned. Sharks and shark repellant were forgotten; I clung to the core barrel just below the weights and hung on, waiting for my eyes to adapt to the dim light. In a couple of minutes, the ship's screw and rudder became barely distinguishable, and I could see the orange pipe disappearing into blackness beneath. The corer was suspended from a block 30 feet above the deck, and even though the rigging crew had snubbed it at deck level, the barrel jerked erratically.

Feeling something like a man on a flagpole in a windstorm I descended slowly, knowing that if the pipe broke from my grasp, it could knock me out. Now I could see why it seemed to buck. From the first it appeared to be bent, but I had attributed this to underwater optics; now I found it really was bent and that this caused the bottom end to rotate and jerk with every roll of the ship instead of swinging like a proper pendulum.

When I reached the bottom end, only 35 feet down, it was at once apparent that three hands were required to hold pipe, bung, and hammer simultaneously. Being one hand short I found it necessary to cling upside down, with the barrel clamped between my legs, while I tried to drive the plug home. That was not possible so I followed the backup plan, which was to put the bucket over the end of the tube and signal the line tender above to hold it tight against the bottom of the core barrel. That done, I surfaced and was picked up by the bos'n's chair. Barely was I back aboard when one of the watchers on the fantail said, "Look, your friends are back." Two sleek sharks cruised in the water I had left moments before reminding me that the packet of shark repellant was still unopened.

When we finally landed the core barrel, there was a core inside—not because of my effort but because the tip of the barrel had penetrated a layer of red clay that was readily retained by the core catcher.

Soon the ship was "making easting" again, headed for a legendary boundary. Between the Santa Cruz Islands and Samoa is a great undersea ridge; it extends in an east-west direction across hundreds of miles of open ocean, breaking the surface with only one small island, Rotuma. To the north, this ridge slopes away into the great depths of the Pacific but to its south the water is relatively shallow and speckled with islands; here, some scientists believed, was a great drowned continent (now known as the Australian plate)—a continent that would include Australia and New Guinea and encompass the Fiji, Hebrides, Solomon, and Tonga island groups. The concept of a partly underwater continent was based solely on geophysical and geochemical evidence, for, to a scientist's way of thinking, continents are not determined by the shoreline of the ever-changing sea but by the kind of rock of which the earth's crust is composed. By this standard the ridge had been determined to be a continental boundary called the Andestite line; to its north the crustal rocks are oceanic basalts and to the south they are continental andesites.

Astride the ridge, 300 miles from the nearest island group, are dozens of shallow areas called banks, including Penguin Bank, Alexa Bank, and Pandora Bank. These have the proportions of normal coral atolls but their highest parts are below the present ocean surface. The scientists of Capricorn wanted to know if these were really sunken atolls and, if their tops were made of coral, why had the coral been unable to build fast enough to keep up with the rising level of the sea as it had at other atolls? Had there been some oceanic change that had caused the corals to lag in growth (water too cold, inadequate nutrients, dust blotting out the sun) or geologic changes (rapid subsidence of the crust)? About all we had to go on was a chart made by a British vessel eighty years before, showing the depth of water on the banks.

There's a lot of satisfaction in going to a place where no man has ever been before. But few such virginal places remain on earth, especially in the tropics where every island was reached by native sailors hundreds of years before European explorers arrived. When the *Baird* reached Alexa Bank, a few of us had the opportunity to visit a place where it was virtually certain no human foot had ever trod. For Alexa Bank is a "drowned" atoll in a rarely traveled part of the ocean; a ring of reefs and submerged islands surrounding a sunken lagoon whose highest point is a hundred feet below present sea level. We would be the first to explore it.

As soon as the *Baird* dropped anchor on Alexa's highest point, six of us entered the water: Dr. Robert Livingston, the ship's doctor; Walter Munk, our all-purpose oceanographer; expedition photographer John MacFall; two technicians; and me. The water was clear and calm; the sun was bright. A swimmer on the surface wearing a face mask could just make out the mottled blue and brown bottom directly below. Filled with enthusiasm for this new kind of exploration, we started downward as a famous line from "The Ancient Mariner" flicked across my mind: "We were the first that ever burst into that silent sea." At least we were the first men, for a pair of sharks circled below; as we descended they widened their circle to watch the strange intruders. Visitors to the drowned lagoon were a first for them, too.

The bottom looked like a desert with a line of low rounded hills perhaps 8 feet high; to their west the bottom flattened, and alternating rows of light and dark material could be distinguished. Atop the hills, which would have been islets crowned with vegetation if they had been a hundred feet higher, there was an occasional Midas Cup (gold colored) coral, but the scene was drab by comparison with the white corals and bright fishes of a normal lagoon. I decided to take back a Midas Cup nearly two feet across to show those on the ship, but the one I chose preferred to remain where it was. In the wrestling match that ensued, I clasped the entire cup in a bear hug and heaved. As it came loose, the thousands of tiny animals that made up the living colony demonstrated they were not without defenses; each contributed a tiny poison dart that left my chest and inner arms burning for several hours. Eventually this trophy was carefully packed and brought home to the Scripps aquarium.

That night, in a scene oft repeated, Bob Livingston set up his tape recorder on a green-covered dining table, and, half naked in the equatorial heat, the divers clustered around describing what we had seen. These sessions always began with Livingston intoning the date, latitude and longitude, and the circumstances of the dive, and ended with Roger Revelle asking detailed questions. On this occasion the consensus was that we had examined the "lagoon side" of the reef below and that we should make some more dives along the outer edge of the reef, a quarter mile to the east.

Down we went the following morning, armed with sampling gear and cameras. The first samples we brought up of what appeared to be the dull brown bottom turned out to be a glistening emerald green at the surface. If released from the light-filtering effect of seawater, the somber landscape below would look like a verdant lawn of algae.

On the next dive I collected a chunk of rock about a foot across with a dozen species of colorful plants and animals attached to it and started up the descending line. My air supply was almost gone, and before I reached the top, it gave out completely. However, I was able to hoist myself up the line to the surface where, coughing and spluttering but still clutching the rock, I spat out the useless mouthpiece and gulped air. Revelle, watching from the deck, thought this was very funny and called down, "I believe you'd rather drown than drop that damn rock."

Our next stop was Suva, capital of the Fiji Islands and hub of the South Seas. There had been more problems with the winch and more kinks in the cable; every time we lowered a tool overside, it was with trepidation. So it was decided to have professional riggers unspool and resplice the wire while the scientific party went ashore for a few days. Scientists always have excuses for sightseeing; geologists claim they need more rock samples, biologists need to examine reef life, etc. But Dr. Ted Folsom, our mustached, bespectacled isotope physicist, had the most original project in mind; he proposed digging in old cannibal middens to see if they contained human bones. I opted for that one, and Ted and I went off to visit the old capital, Serua, once the home of cannibal chiefs. Serua is a very small island, reached by wading a quarter of a mile at low tide. At each end of the island there is a steep conical hill about a hundred feet high, one of which is occupied by Government House, the other by the tribal cemetery.

Fijians are distinctive-looking people, dark as other Melanesians but larger on the average and crowned with a basketball-shaped mass of hair. Like almost everyone else we met in the Pacific Islands, they were very friendly, and we did not want to change that attitude by grubbing among their ancestors' garbage. When they heard of our interest in the sea, the Seruans were eager to show some of the Scripps scientists a reef-enclosed pond area that provided seafood for the village.

A crowd of scientists and locals started out swimming and riding in pole-propelled punts. The natives carried spears tipped by a circle of sharp, unbarbed bamboo points that splayed outward. From time to time one would spear a fish and hold it up for the close inspection. Some of these creatures were pretty exotic, one being a large lion fish, whose dorsal fin was arrayed with long venomous spines that could inflict a very painful but rarely lethal wound. Our guides also warned us about various cone shells and sea urchins by pointing and repeating the Fijian word for "dangerous." But it took Ted Folsom to find one of the venomous cobralike sea snakes that lived in the area.

In the process of picking up a couple of rare white cowrie shells from the shallow bottom, Ted disturbed a snake, and it brushed against him as it swam to the surface. When the snake wriggled by, one of the boatmen was able to lift it out of the water with a paddle for a few seconds, so we could see that it was about three feet long and dark green with small black and yellow dots. Although their venom can be deadly, these snakes are not much feared by the locals, because their small mouths with fangs at the rear make it inconvenient for them to bite humans. Not being eager to test this allegation, we allowed this one to go on about his business.

Then Ted and I went to Korolevu Bay for a couple of days to live in a Fijian grass shack called a bora, a few feet from the water's edge. Our bora was a wonderful piece of woodworking and weaving; a circle of fernwood posts supported the sill timbers to which were lashed a circle of sloping poles that came together above the center and served as rafters. The grass roof they supported was over a foot thick and was said to weigh five tons. When I looked doubtful about that statistic, our host pointed out that in the recent hurricane these "grass" houses survived when many European-style wooden houses had not.

With real beds and solid ground beneath us for a change, we had planned to sleep late, but at 6 A.M. a smiling Fijian with a silver tray insisted we have "wake-up tea." Then, somewhat groggily, we went out to inspect the sandy shore, which had been washed clean by the night tide, and to watch a glorious sunrise. An outrigger canoe high on the beach seemed to have been put there for our convenience, so we launched it and started across the bay. While other guests and natives watched, we put on what they later described as a hilarious performance as we tried to balance the darn thing and paddle in a straight line at the same time. This one did not respond at all like those we had seen in the hands of natives so, naturally, we blamed the canoe. To our surprise that turned out to be a sensible reaction; our outrigger was not authentic, but had been made as a movie prop for *His Majesty O'Keefe* with Burt Lancaster, which had been shot there the previous year.

From then on I looked more carefully at real canoes, and on our last day in Suva while walking along the street, I saw a beautiful little hand-carved boat in someone's backyard. It was 12 feet long and looked something like a wooden kayak; above the dugout lower hull, there were seven other pieces of wood that had been sewed together in the ancient fashion with sennet fiber. The owner had made it for his young son, who had since grown up and gone away; he was willing to sell it for three Fijian pounds. The whole canoe only weighed about 20 kilos, so I hoisted it on my shoulder and, with the outrigger barely clearing the ground, carried it through the main streets of Suva to the astonishment of passersby, and lashed it to the mast of the *Baird*. Thirty-odd years later it became part of the Kelton Foundation's collection of Oceanic art.

TONGA, SAMOA, AND THE COOK ISLANDS

The region around the Tonga Islands is one of the most fascinating places on earth to a marine geologist. Along the islands' eastern flank, there is a very deep trench that extends for nearly 1,500 miles from New Zealand to Samoa; their western side is relatively shoal with numerous active volcanoes. According to the theory of plate tectonics, the Tongas are the leading edge of a great crustal plate that is riding up and over the plate to the east; the trench between the two is a "subduction," zone where the rock of the ocean bottom is slowly being forced back into the subcrust and perhaps the mantle of the earth. These ideas were not formalized until fifteen years after our expedition, but the principles on which plate theory was based were discussed in many variations by the Capricorn scientists led by Revelle in our evening talks.

We began by making a preliminary survey of the Tonga Trench, crossing and recrossing it with both ships, charting its profile at several latitudes. The depth was so great that the bottom was beyond the reach of our deep echo sounders (that is, their sonic pings were not loud enough to reflect back), and it was necessary to toss blocks of explosive overboard and listen for the echoes of their shock waves. *Baird* passed over a deep spot that showed an apparent depth of 5,300 fathoms, and *Horizon* found one 5,440 fathoms deep (32,600 feet). But the echo sounders measure only the time between outgoing and returning pings and use an assumed sound velocity in seawater; when these indicated depths were corrected for the increase in the speed of sound due to the compression of seawater at great depth, another few thousand feet were added to those records.

The Tonga Trench is 15 to 30 miles wide and 7 miles deep, with a steep-walled gorge (discovered by Bob Fisher of the *Horizon*) in its bottom. As the seismic soundings produced evidence of this great chasm beneath us, we sought words to express it properly: "A mile deeper than Mount Everest is high." "As deep as seven Grand Canyons but with much steeper sides." "Higher than a stack of thirty Empire State Buildings." None of those seemed adequate to explain this heroic crack in the crust that was only about 200 feet shallower than the deepest place ever found in the ocean—the Challenger Deep in the Mariana Trench.

Curious about what the bottom was made of, we lowered a simple gravity corer, basically a 6-foot long core tube weighted with lead. After a very long time, during which we speculated about whether there would be sediments or coral debris or other things at the bottom of the abyss, the tool returned coreless and badly battered. Imbedded in the soft lead weight was a small chip of basalt.

The initial look at the trench completed, we headed for Nukuʻalofa, the "land of love," capital of the Kingdom of Tonga. Arriving on Christmas Eve, the expedition was royally received by Prince Tungi, heir apparent and prime minister. He was a huge man in his early thirties, over 6 feet tall and weighing 23 stone (322 pounds) whose family had ruled Tonga since before the Norman conquest of England. Tungi was a very affable fellow, and he listened attentively to Roger's account of volcanoes, earthquakes, and trenches that beset and surrounded his kingdom. Then he graciously entertained our crowd with a cocktail party at the palace and an authentic native feast, where we sat cross-legged on mats near the surf and used our fingers to eat local delicacies from huge leaves. In return Roger invited Tungi and other Tongan dignitaries to a party on the broad afterdeck of the *Baird*.

Tungi was most curious about the SCUBA diving equipment, the first that had ever reached his country, so Phil Jackson and I made a "command dive" for the prince. We felt a little immodest dressed only in swim trunks amid the elegance of the high-buttoned white cotton jackets and fine pandanus skirts worn by our Tongan visitors. While Bob Livingston explained, we put on the Aqualungs and stepped off the stern of the *Baird* into the harbor. The water was reasonably clear, but the bottom was cluttered with bottles, tin cans, tires, and batteries. We knew the watchers on deck would expect some tangible evidence that we had reached the bottom, even though that would not have been hard to do even without diving gear; finally I found a chunk of coral among the junk that could be shown to the audience on the fantail without causing embarrassment. Once out of the water and into a clean shirt, I talked with the prince, who had been educated in Australia and was reported to have been quite an athlete. "Well", he said modestly, for a man weighing over 300 pounds, "I was pretty good at pole vaulting." Later, thinking back on that afternoon, I saw myself as Huck Finn, bragging to his friends along the Mississippi, "Aw, it warn't really nuthin', but Phil and me dove before crowned heads."

While *Baird* was at Nukuʻalofa, John Isaacs decided to leave the expedition, and I replaced him as senior scientist on the ship, a job that required me to pay much closer attention than I had before to navigation, record keeping, chart making, and watch standing. At the same time, Russ Raitt's wife, Helen, who had traveled to Tonga to be with him on Christmas, was invited to become the first woman to join an American deep-sea scientific expedition. At first the idea of a woman aboard was received with mixed enthusiasm, but in time Helen's sunny personality won over the dissenters.

The first stop at sea was late in the afternoon on a broad shallow bank with an average depth of about 120 feet. The objective was to inspect the nature of the bottom and the amount of sea life. A weight was lowered about 80 feet on a line from the stern of the *Baird* to help orient the divers. On descending, Walter Munk, Phil Jackson, and I could see the ship was drifting with the surface current, moving about a knot and a half relative to the bottom. After experimentally diving to the bottom, we decided it would be easier to let the weight tow us while we watched and save our deeper efforts for some specially interesting object. The light was beginning to fade when I saw some special piece of coral and dived to get it, forgetting that I had on a single bottle of air and the other fellows had doubles. As I wrestled the coral free of the bottom, my air supply ceased. I started up at once, this time leaving the rock behind, but I could not overtake the weighted line that was towing the other fellows off into the gloom.

A hundred and twenty feet is a long way to go without air, but I hoped that as the pressure decreased, more air would become available from the tank. In the interim there was no choice but to swim upward with all my might, allowing air to escape from my lips as the pressure lessened, so as not to get an embolism or burst a lung. With 50 feet still to go and my legs aching from the oxygen deficiency, I slipped out of the Aqualung harness but still hung on to the darn thing (it was university property worth $50) and swam even more desperately upward. It's hard to remember to exhale when you really want to inhale and your senses are reminding you that you're not far from the edge of consciousness. A long time later, maybe 20 or 30 seconds, I broke surface and gulped free air again. That's a wonderful feeling, when oxygen begins to spread through your body again after you've been through the worst part of drowning.

After a few delicious seconds, it occurred to me that the ship might drift away in the fading light and leave me, so even though the *Baird* was only a few hundred feet away I hollered, "Help!" Almost at once Bob Livingston appeared on the stern of the *Baird,* naked except for a line around his waist. One long racing dive, and he was on his way to give me a hand, but by the time he arrived most of my strength was back.

A little west of the Tonga Islands there is a long chain of intermittently active volcanoes that the geologists wanted to visit and sample. One in particular took our fancy, partly because of its intriguing history as recorded in the *Sailing Directions for the Pacific Islands.*

Falcon reef was first upgraded to an island on the charts in 1885, but two years later smoke rising from the sea was all that was visible of this uncertain volcano. By 1889 a new island was surveyed by the British vessel *Egeria,* but a few years later it was gone again, and only breakers and discolored water marked the position for the next thirty years. By 1928 it was 2 miles long and sported a cinder cone 600 feet high. In that year Harry Ladd of the U.S. Geological Survey swam ashore with Prince Tungi's father and climbed warm cinders to plant the Tongan flag on its summit. (We weren't aware that kings did that sort of thing in person.) By 1936 Falcon was a reef again, but volcanic eruptions every 15 minutes were reported, and in 1938 the island was 30 feet high and a mile and a half long. Not surprisingly Falcon became known as the Yo-yo Island; those on the *Baird* wanted to see if this unreliable piece of property was above or below water at the end of 1952.

Captain Larry Davis eased the *Baird* toward the position shown on the chart for the island, and since navigational charts in such an area are often unreliable, he slowed to 3 knots. The horizon was unbroken by land, and the sea was calm, almost glassy, as we studied its surface for any indication of Falcon reef. Then three low swells passed under us, and a faint diamond-shaped pattern showed on the water surface just ahead. That could only mean a shoal, and although the echo sounder still indicated there was 80 feet of water under the keel, the skipper called "hard left rudder" and took the ship a half a mile away.

This meant we would have to row at least a mile and probably have to search for the high point, so I said, "Cap'n, couldn't you drop us off a little closer?"

He turned with a half smile. "Did you know that Captain Bligh was put off the *Bounty* just a little north of here and had to travel over 3,000 miles in an open boat? I don't want to take a chance on being shipwrecked and making that trip again."

Neither did I, so we divers launched a skiff and started rowing toward a patch of turbid greenish water. Phil Jackson hung over the stern with a glass-bottomed view box and described the changing bottom as rippled volcanic sand about 40 feet deep, then rounded yellowish knobs somewhat shallower with a shark cruising around, then deep water again. When we reached the discolored water, our 200-foot-long sounding line wouldn't reach bottom, so for 30 minutes more, we rowed back and forth looking for the high point that the ship had visited an hour or so earlier. When we finally found it, Falcon reef turned out to be about 30 feet wide and 100 long with almost vertical sides; its highest point was only about 20 feet below the surface, just shoal enough to threaten the transducer of the *Baird*'s prized echo sounder. The skipper had been right to play it safe.

Over the side we went, Walter Munk carrying a thermometer that he inserted in the bottom of a sand-filled gulley to see if he could detect any excess temperature. The top of the reef was rough, dark volcanic rock, largely covered with coral colonies 1 to 2 feet in diameter. We were privileged to see the beginning of a coral reef—new corals growing on a new basalt foundation. This was the way that great atolls had started. We were directly observing geological change at breakneck speed; this new coral reef had been an island a mile long only 15 years before.

Around the pink and white coral colonies, black and yellow striped Moorish Idols swarmed, trailing their long pennantlike dorsal fins. How could these small shallow-water fish find and populate a new reef isolated by relatively deep water for many miles in all directions? I perched on the edge of the vertical cliff peering into the dark depths and, even in this weightless world, wondered about falling off.

When we returned to the ship, Roger seemed more eager than usual to get underway. "Don't you know that only three months ago a Japanese survey ship, doing just what we're doing over an undersea volcano, was caught by a sudden eruption and lost with all hands?" That was a pretty good reason not to loiter longer.

Baird's next stop, at Vava'u in the northern Tongas, was ostensibly to determine the proportion of coral and volcanic rocks there, but also to give us a look at an island rarely visited by outsiders. In the cool beginning of a very bright morning, when the tall palms still threw long dark shadows across the green bay, *Baird* dropped anchor. In ten minutes we were in the skiff headed for shore, stopping only to visit some submarine gardens that featured huge black pearl shells and echinoids with slender needlelike spines a foot long that waved in frantic defense as we approached.

We beached the skiff and followed a trail inland lined with tiny pineapple plants and patches of taro and tapioca until, as we crested a hill, we found ourselves in a village made largely of plaited palm fronds. Ambling down the main street, we were hailed by a group of men sitting cross-legged in a circle under an open shed. It was morning kava time, and the chief, who spoke English well, invited us to join. All those present were naked except for the usual skirtlike mats of the Tongans and our swim shorts, but the words and ceremony were as stately as a cabinet meeting.

Bob Livingston, M.D., a *real* doctor (Ph.D.s didn't count with this crowd), had the place of honor on the chief's right at the head of the mat, and the rest of us took places in the circle between the ordinary citizens. At the other end of the mat, a handsome woman of about thirty knelt, making kava. Legend has it that in the old days kava root was chewed by a young virgin, who spat juice into the bowl. But times had changed, and our hostess pounded the root to a pulp between two rocks; then she put the stringy result in the bowl with some water and coaxed the flavor from it by kneading, wringing, and squeezing. When the drink was ready, she set it before the chief with a graceful gesture and the talking-chief took over. With graceful and dramatic gestures he told of the glories of Tonga, the virtues of Vava'u, and the beauty of the village.

Then this fellow, also the kava passer, scooped up a half-coconut-shell full of the brownish liquid and passed it to the person being honored, kneeling before that person as he did so. As the lucky recipient drank, the talking-chief led the rest of the participants in a slow, measured clapping of hands. The kava was quaffed with a single raising of the cup; at the moment it was finished, the others would moan in sorrow that it was gone, and the kava passer would go back to refill for the next man. With this procedure, it takes a while to get a drink, although after you've had a cup, you realize there was no reason to hurry.

The chief was served first, followed by Livingston, the honored guest, and then by us lesser folk. As each person's turn came to drink, the ceremony permitted a detailed inspection, that would otherwise have required an impolite stare. In the course of this inspection, I observed that several of those around the mat had filariasis (a gross swelling of the limbs caused by a mosquito-carried disease) and the chap next to me, with whom I literally rubbed elbows, had blotchy depigmentation of his arms and was missing a couple of fingers. Everyone drank from the same cup, and when it is passed to you in such a gracious manner before such a distinguished gathering, there is no option but to accept and toss the contents down.

Kava has a distinctive but mild taste and leaves a not unpleasant tingling sensation in the mouth. Mainly it helps create a social atmosphere and stimulates conversation; as we sat in the shade in this pleasant village, an ever-increasing audience of children, pigs, and dogs gathered around us. We chatted about rhinoceros beetles (a coconut pest), the chief's bad arm (filariasis), the value of aspirin in rural medicine, and the price of copra. By the time we left, we felt like citizens of the village.

As we walked back to the boat, I asked Bob Livingston, "What was wrong with that fellow who sat next to me?"

His answer was matter of fact: "He's got leprosy."

I gulped. "Why didn't you say something?"

"Aw, you won't catch it."

And I didn't.

When Stanley met Livingston, the scene was as I had always pictured it: a circle of curious, dark-skinned natives in a tropical village of grass huts. But it was not in central Africa; it was on the island of Niuatoputapu in the northern Tongas. The line of coral limestone islands and the line of basaltic volcanoes that formed two separated chains of undersea mountains in the southern Tongas seemed to converge to the north. Therefore we wanted to obtain rock samples from the northern islands to support magnetometer data that was sometimes ambiguous. Although we were anxious to get to Samoa, only 200 miles beyond, Roger decided we could spare a few hours to look at Niuatoputapu and Tafahi. As we approached from the southeast, it was obvious that Tafahi was a volcano, and there was no need to sample it. However, Niuatoputapu was much larger and quite flat except for a rounded hump in the middle that made it look a little like a sombrero. It could be either terraced limestone like Vava'u or that hump could be of volcanic origin so, about about half a mile offshore, we put a boat overside and started rowing ashore to find out. As we did, a small native boat with eight persons aboard came by, sailing briskly in the choppy channel. It slackened sail so we could come alongside and talk.

Their first words were "Do you have a doctor aboard?"

"I am a doctor," answered Bob Livingston, who was rowing at the time and wearing only swim trunks.

"We have a very sick baby here that we're taking to the hospital at Niuatoputapu."

"Let me take a look," said Bob, passing the oars to one of us and stepping into the other boat. It quickly resumed speed and sailed well ahead to lead us through a break in the reef to a small copra-loading pier. En route Dr. Livingston diagnosed the baby (and its mother) as having typhoid fever; the mother had not been able to supply milk, so the baby was badly dehydrated. Before they reached shore, he had unplugged the baby's bottle and gotten some water into it.

While we rowed toward shore, the *Baird* had made radio contact with the local radio station, whose tower rose above the hill in the center of the island. When the local operator heard that scientists led by Dr. Livingston would soon be landing, he put on his best suit, jumped on his bicycle, and started for the waterfront.

Thus it happened that as our group walked down the center of this cluster of "grass" houses, a man in a white suit on a bicycle rode toward us, dismounted, and said to Bob, "Dr. Livingston, I presume? My name is Stanley."

Pago Pago in American Samoa is quietly content to be the most beautiful and dramatic harbor in the South Seas. Its great green bowl was once a volcano that exploded in the distant past and blasted one side of its crater completely away, allowing the sea to rush in and fill the depression. It is a city of wooden bungalows and oval Samoan houses sprawled along the water's edge. Across the bay, a great tower of rock, the hard spine of the old volcano rises 1,700 feet almost vertically; this is The Rainmaker, and everyone keeps a weather eye on it, knowing that when a cloud passes over the peak, rain will fall on the city in a few minutes. Indeed, this was the setting for the play *Rain* in which Sadie Thompson seduces the minister, and sex triumphs. We rather hoped these colorful days were not entirely gone.

Baird tied up to a concrete pier whose center section looked as though it had been bombed. As usual, *Horizon* was already there, and we braced for the inevitable taunts. (Their crew insisted it was because they were more efficient; ours maintained that we worked longer and harder at sea.) Above and behind the pier on a rocky prominence was the governor's mansion.

The ships had put in mainly to load and exchange explosives; *Horizon* was running low on TNT; *Baird* carried ten tons that needed to be transferred. The air was hot and humid, and everyone wanted to do a little sightseeing, but the combined scientific and ship's crews of both vessels formed a brigade to make the exchange. Hundreds of 50-pound boxes of TNT were passed hand to hand up from the lockers and lazarets of the *Baird* to be carried to the *Horizon* and passed downward again for restowage. When the job was finished a few hours later, we were drenched in our own sweat, and after a quick swim we headed for town. The center of Pago Pago consisted largely of a grassy oval playing field surrounded by white-painted boulders and a row of stately white houses. Hiding behind these were much less pretentious buildings, one of which housed the Moonlight Café, formally known as a sailor's dive.

That evening at a reception at the mansion, Governor Ewing told us about the wrecked pier. One of the last Navy ships to call at Pago, only three years before, was the *Chehalis,* an oiler over 300 feet long. While transferring fuel, it caught fire, and the ensuing explosion shattered the pier. The fire could not be controlled so, to prevent a possibly larger explosion, the bilge cocks were opened and the mooring lines cut. The blazing ship drifted only a short way from the pier before it capsized and sank. The governor wanted to know the exact location of its highest point, if the hulk was a menace to navigation, and whether it might be salvaged. Would we use the new kind of diving equipment to take a look?

Would we ever!

The following morning with the governor waiting on the pier, five of us descended in the clear water at the place he indicated and saw no sign of the ship. We surfaced, moved over a few hundred feet in accordance with a new set of instructions, and dived again. This time I landed on a six-pointed white asterisk, the Plimsoll mark; away from me in all directions stretched a gray steel floor. This was the center of the ship's port side at a depth of 110 feet; I swam over to the rail and peered vertically downward across the deck at a twin machine gun pointed surfaceward. Then I descended through the rigging, being careful not to become fouled with any of the booms and pipes and wires. A cable that once had stayed the mast broke under my hand and fell away, leaving a trail of rust, but most of the heavier metal parts seemed to be in good condition, with the gray paint still intact. Finally I reached the bottom 160 feet down and swam along it from the bow to the stern.

Somewhere along the length of the hull, a feeling of unconcern and detachment came over me. Later I vaguely remembered feeling the broad bronze blade of the propeller, swimming between it and the rudder, and seeing bubbles rise from one of the deckhouses (meaning other divers were inside), but the latter part of that dive never registered on my brain. Somehow I returned safely to the surface but it had been a long deep dive, foolishly made without a buddy, a watch, or a depth gauge. Later we decided that I had probably experienced a touch of nitrogen narcosis. We had read that other divers in similar circumstances had experienced this careless euphoria known as rapture of the deep and had swum off into oblivion. In any case, none of us saw any evidence of either the fire or the explosion; the effects, if any, were on the ship's downward side, against the mud.

But science was our main concern. The objective of a geophysical expedition is to assign numbers to the properties of rocks along the route. What were the seismic velocities? At what rate was heat flowing out of the sea bottom? What variations in the earth's magnetic field gave evidence of crustal structure? The first two questions could be answered when the ship was stopped for a "station," but the magnetic properties were only measured when the ship was underway. By the time *Baird* returned to San Diego, it had dragged a magnetometer 7,800 nautical miles over a part of the crust never before examined in that way.

The earth's molten core moves like a dynamo to produce lines of magnetic force whose horizontal components of direction and strength can be sensed with an ordinary compass. These lines would be uniform (at equal magnetic latitudes) except that volcanic rocks containing magnetic particles have the ability to modify them, a quality called susceptibility. When a magnetometer measures changes in the earth's field caused by variations in the rock of the crust, this gives scientists a clue about the nature of the rocks below.

Dr. Ronald Mason, a slight, mild-mannered Englishman who usually peered quizzically over the top of his horn-rimmed glasses, was in charge of the magnetic measurements. Ordinarily this consisted of his getting the "fish" containing the magnetometer launched and running, and then marking time and course changes on the records. But as we approached Palmerston atoll, an isolated dot of land far to the northwest of the other Cook Islands, it was decided that we would make a magnetic survey around it and try to determine if a slight embayment on the western side was caused by a landslide. As the ship approached Palmerston, the magnetometer needed some last-minute repairs, so we hove to in the lee of the only inhabited island. Almost at once a couple of boats put out from shore to greet us.

It seemed a little strange that all our seventeen visitors were named Masters until we heard their story; then it seemed even stranger. An Englishman named William Masters arrived at this remote spot in 1862 with his wife, her sister, and a dark-skinned female friend. This chap had a certain talent for reproduction and, with the help of the three women, begot twenty surviving children, and they in turn produced an average of ten children each. Thus, in the space of two generations, one man peopled a country; now the descendants of each of the three original mothers keep their lineage straight by maintaining three tribes and three chiefs.

We gave our new friends some small presents, and in turn they entertained us. One song and dance, rendered in the lingo of the Cooks, was considered to be hilarious by those who understood the words, so we asked for a translation. A large dark fellow (third wife), who was the island's radio operator, explained: "We sang last a love song. A boy and a girl make love on a mat in the dark. Suddenly someone strikes a match and there is light." He spread his hands expressivelly. "The love is spoiled. The flame below his belly went out." We were sorry we had found it necessary to ask; the dance had described the situation adequately.

The only available chart of the atoll, issued by the British Admiralty, was based on a survey by Captain James Cook himself in 1774 when he discovered the place. (It was later named after Lord Palmerston.) It was my job to plot the ship's position and depth on the Admiralty chart as the magnetic survey progressed. But soon, as *Baird* ran a zigzag course around the atoll, I discovered that its position, based on angles to the ends of the atoll on Cook's chart, fell far behind my dead reckoning positions, which were based on the ship's speed, time, and direction. By the time we were halfway around the atoll, it was quite clear that the Admiralty chart showed the atoll smaller than it really is. Instead of the 4-mile length given there, I made it 5.5.

Probably Cook sent someone up the mast to make a sketch and estimate distance as the *Resolution* sailed past. I could appreciate that mapper's problem, but this required me to go back over all the data and make a new chart, readjusting our ship's position until all the angles and distances to the enlarged atoll came out properly, a time-consuming job. Having done that and drawn bottom contours down to 1,500 fathoms, I could plot magnetic isolines. These showed two magnetic highs, one under the south end of the atoll and another about two miles off its west side, where the water was 1,200 fathoms (7,200 feet) deep on a smooth slope. Probably these represented the tops of two volcanic peaks, long buried beneath coral rubble that had been accumulating for tens of millions of years. By the time that chart was completed we were halfway to Tahiti in the Society Islands.

TAHITI AND THE EASTERN ISLES

Tahiti's reputation made it every sailor's dream, and as the green slopes of Mount Orohena rose ahead, blood pressures rose with it. In early 1952, before the days of cruise ships and airplanes, Tahiti had relatively few contacts with the outside world, and this made it seem even more remote and desirable. All hands were eager to hit the beach, but because *Baird* and *Horizon* had visited islands infested by the rhinoceros beetle, a pest that attacks coconut trees, the ships were required to drop anchor offshore in a quarantine area. The reason for this was that the beetles fly only at night and never as far as 400 meters. So, we hung over the rail waiting for transport ashore, "breathing the balms of the fronded palms" and gazing dreamily at Papeete Harbor. White yachts, moored stern to a low sea wall, trading schooners crammed with colorful humanity and festooned with green bananas, low buildings behind with red tin roofs, and the steep green mountain behind were all reflected by the glassy harbor surface. We had no doubts that, somewhere in the green forest beyond, tall waterfalls splashed into dark pools and that naked brown girls would laugh and swim toward us when we approached. In ways we did not expect, that dream was close to the truth. There were girls available everywhere; it was the time-honored custom of the island. Pretty unattached young girls who asked no reward except fun swarmed over the ships during our three-day stay.

From a famous bistro named Quinn's Tahitian Hut to Tahiti-iti at the far end of the island we rogued and ranged. On the first night we were drawn to a local bar by the loud beat of insistent Tahitian rhythms; on entering we found half-clad patrons involved in a mad lascivious dance—a raucous, sweating, hip-shaking, pelvic-thrusting hula. That was only a normal evening's relaxation in an ordinary bar. The next day, while wandering up a small valley, I came on one of the local beauties standing in the midst of a small brook wearing Eve's traditional attire. She saw me too and modestly pulled a leafy twig in front of herself. In spite of this mildly inhospitable gesture, I continued to advance. Then a mango splashed hard in the path, and I looked up to see a husband in the mango tree above gathering more ammunition. My luck was indeed too good to be true.

On the last day in Papeete, I was ashore after lunch, patiently waiting for a shop named Établissement Donald to open so I could ask them to speed up repair on one of the ship's refrigerators. It was a solid old building on the main street along the waterfront, with large doors set deep in thick stone walls from which jutted a wide wooden awning. As I waited, it started to rain, then in moments this turned into a torrent and finally a gale. The harbor was obscured; two small trading schooners across the street broke their outboard mooring lines and swung around to batter against the quay; doorways filled with huddled people. When the gusts reached 50 knots, some of the corrugated roofing blew off the Yacht Club, where our meteorologists, as Walter Munk put it, "in their usual prognostic frame of mind had chosen to have lunch." An elderly native and I flattened ourselves in the corners of the deep stone doorway, furtively watching sheet metal skim down the main street, until one extraordinary gust blew open the door and projected us inside. Quite plainly the gods had intended for the two of us to remain safe and dry until the worst was over, so we closed the door again and accepted our fate in comfortable chairs.

When we sailed from Tahiti, it was with the resolve to return again for a longer stay; indeed, since we headed directly into a northeast gale, I would have been happy to have had that stay begin at once. As the *Baird* bucked through short steep waves, its bow would fall, and every few seconds my bunk would drop out from under me. When the bunk and I came together again, the shock of the impact and the sound of the air being squeezed from my lungs prevented comfortable sleeping. Eventually I bounced and wheezed into dreamland, thinking of the marvelous line based on a remark by Samuel Johnson: No man need go to sea that hath the wit to get himself in jail.

For a couple of days and nights we pitched, heaved, rolled, and took enough green water over the fantail that the captain declared our main working space out of bounds. When the storm abated, it was discovered that a large drum of fish samples in formalin had been swept off the afterdeck, but this was a small setback compared with two lost working days. By now we were amid the Dangerous Islands, as the Tuamotu Archipelago is known to mariners. Seventy-two atolls, interspersed with reefs and shoals, are strewn across a thousand miles of ocean. Emerald lagoons, partitioned from the ultramarine blue of the ocean by rings of white coral, create a beautiful but dangerous setting. These low islands are hard to see on dark and stormy nights, and many a sailing ship left its skeleton ribs high on the reef. Instead of rising from water 3 miles deep, like those in the Marshalls, these atolls are based on a broad arch in the crust only about 1 mile deep—a fact whose geological significance was not then clear. With the help of radar, the Scripps ships passed safely through the maze and reached Takaroa, one of the northernmost islands. Takaroa is famed for naked pearl divers and for the hurricanes that sweep across it every 50 years or so, leveling much of the village and many coconut palms.

Henri Rotschi, insouciant French chemist, Helen Raitt, our durable chronicler, Bob Livingston, and some others including me were the expedition's emissaries. By the time our skiff was in the channel through the reef that leads to the village, half of the populace was lined up along the quay, as curious as we to see what manner of people they would meet.

After a short tour of the village, the crowd collected on the quay to talk about diving, their main livelihood. We learned that the divers were after pearl shell, used for buttons and ornaments, not pearls, which were rarely found. The pearl shell, each as large as a man's open hand, had nearly been wiped out at the turn of the century when diving equipment had been permitted. But for some years conservation rules had required that only primitive native equipment be used, with the exception of tiny diving goggles that fit inside the eye sockets and minimize the air space between the glass and the eyeballs. They told us that a good diver could make over fifty dives a day to 20 fathoms (120 feet). The diver descends rapidly, feet first, with the help of a 4-kilo weight. At the bottom he gathers the few large shells that are nearby, puts them in a net, and starts climbing the line. When the alternate diver in the outrigger above feels the tug, he hauls in on the line to speed his buddy's return to the surface. When he's safely aboard, they haul in the net with the shell. Mapuhia, their oldest diver who retired at seventy, had dived to 168 feet on one occasion and could hold his breath for 3 minutes.

When we asked for a demonstration of the breathing technique that enabled them to dive so deep and long, a couple of the local boys volunteered, jumping with me into the clear water alongside the quay. They wore the tiny goggles; I had on a face mask and fins. As they sucked in air and whistled it out, I imitated them; then we all heaved up for one last deep breath and dived together, turning over in the first few feet and swimming down head-first. First we went to about 15 feet; then we returned to the surface, breathed again, and dived a little deeper. This time, in order to find out how much their breathing exercize had helped, I stayed down when the others returned, but I stayed so long the villagers asked their boys to check and see if I was all right.

The next dive we went deeper, but below 30 feet my face plate jammed hard against my nose, while the boys with small goggles were not bothered and easily went deeper. By the time we all surfaced, the current in the channel had carried us some distance from the quay, and as we started to swim back, the crowd perceived this to be a race. In Tuamotan they shrieked, "Swim faster," at the local boys. Although I did not understand this was a competition, I won easily. Then came the realization that these people had no knowledge of swim fins and did not understand their value so I slipped mine off and let the locals try. They were delighted by the increase in swimming efficiency and began debating whether the conservation rules would permit their use.

When our skiff pulled into the channel, I swore to return, but in a very short time our minds were fixed on a different kind of island that loomed ahead.

The Marquesas are high jungle-covered volcanic islands with no coral rim, intersected by deep valleys and rimmed with vertical cliffs. For Polynesians headed east in voyaging canoes, many hundreds of years before the first Europeans arrived, these islands were a jumping-off place but they did not know it. For over a thousand years their ancestors had moved steadily eastward across the central Pacific in open canoes, stepping from one island group to the next to reach here. Their stepping-stones had rarely been more than a hundred miles apart—a few weeks of sailing. They could not anticipate that the next step would be 2,200 miles to the South American mainland—unless they chanced on tiny Easter Island. Even so, a few of the hundreds of canoes that sailed east actually did make it across the Pacific, the evidence being Polynesian spearheads and adzes found in Chilean graves.

Cannibalism, at least in ritual form, had been a way of life until the early missionaries persuaded the Marquesans to get along without "long pig." In turn, the missionary influence was overwhelmed by that of slavers, whalers, and traders, many of whom brought along vices ranging from alcohol to opium, as well as diseases against which the natives had no immunity. By the year 1800, the population of this group of islands had been reduced to

100,000; when we arrived, there were said to be 3,000.

When *Horizon* and *Baird* pulled into Taiohae Bay on the south side of the island of Nuku Hiva, Dr. Gustaf Arrhenius, our affable Swedish geochemist, stood beside me at the ship's rail and talked of the things he had learned five years earlier when he had come as a scientist on the Swedish deep-sea expedition aboard the sailing ship *Albatross*. As our ship dropped anchor and the inevitable outriggers showed up alongside, one of the occupants, a stocky well-built fellow with a wide smile, hailed us in French. His first words were, "Gustaf, my friend. You must come and stay with me while you are here. I am sorry my daughters have gone to Papeete, or you could sleep with them."

"Tunui! I am glad to see you again," Gus answered. "Thank you for the generous invitation. This is my friend Bill Bascom. Can you take us ashore?" As we climbed down into the canoe, Gus gave a superfluous explanation: "I am considered to be a friend of the family." So it seemed.

As we glided ashore in style, the two of them chattered away in French about events in the years they had been separated. Then we walked through the village and up the hill to Tunui's modest houses; there was a sleeping house, cooking house, eating house, and storage house, plus a roofless shower and latrine partly enclosed by woven palm fronds. Because heavy rain is not unusual, these houses are built on stilts with walls of woven bamboo and roofs of palm fronds. Gustaf gravely inspected the mats on the sleeping house floor, counted the dozen inhabitants, and marked off the floor spaces where everyone would sleep. There was a little extra space, and he graciously accepted the invitation to stay.

Tunui beamed and wagged his finger, "For such honored guests I have another house with a real bed in it. It is used by me and my eldest son, but you and Bill must sleep in it tonight." He led us to a new building containing a large brass bedstead from which sagged an old mesh spring covered with a thin padding. We protested that he and the boy deserved their normal rest and that we were unworthy, but there was no turning back. To have refused would have been a great insult, so we smiled squarely and acknowledged the dubious honor. That night when Gus and I, both over 6 feet tall, eventually got into bed, the spring instantly sagged into a steep-sided hammocklike bag, and we spent a miserable night trying to keep our sweaty skins apart. Years later, in going through some of the private papers of the artist Paul Gauguin in the Tahiti museum, I came across a newspaper account of his landing at Nuku Hiva in 1901: "Four stalwart natives waded ashore carrying his great brass bed held high." On an island where beds are not generally used, it seemed possible that Gus and I had occupied one originally belonging to the great artist himself.

The eating house was an extraordinarily practical structure, whose floor was high enough so the pigs could live below and whose encircling window was used for garbage disposal. One could eat the best part of a melon, throw the rind over his shoulder, and be rewarded by the sound of a scramble and grateful oinks below.

After breakfast we mounted two horses named Maybe and He Won't Run, and followed jungle trails to ancient marae (temples) built of great slabs of basalt. There pagan practices intermingled sex and religion to the rhythm of throbbing drums before huge glowering tikis. We did not follow Herman Melville's tracks to the highest ridge and look down his escape route into the valley of Typee, but then we didn't intend to jump ship and remain here.

In the course of our stay, we became acquainted with Bob McKittrick, a person of some repute in the islands, who had operated a tin-roofed trading post on the shore of Taiohae Bay for 40 years. He was full of strange pieces of information about the islands. For example, he said that when he first came to the islands, the average man's shoe size was 11; now it was 8. This "racial shrinkage," as Roger called it, may have been the result of seminal immigration by the few Chinese who came to these islands.

The old Scotsman was responsible for the operation of an ancient generator that supplied electricity for the village's few dozen light bulbs. He was exceptionally friendly to me for a reason that was not immediately clear. But long after returning home, I got a series of letters from him asking me to find spare parts for his generator. With considerable effort, I located piston rings, a cylinder head, valves, etc., and shipped them off, all the time wondering how he had chosen me from the eighty or more members of the expedition who visited his store.

THE OPEN SOUTHERN OCEAN

Next the ships drove southeastward into the largest area of open water on earth. Behind were the myriad of islands of the central Pacific; ahead there was only sea and sky for 2,000 miles. At last we became true ocean people for a few weeks in an idyllic interlude, a dreamlike and timeless world, a space warp bounded by the ship's rail and populated only by the ship's company. Our interest in the other world, somewhere beyond the waters, was suspended; we neither thought about nor wanted it.

In these windless doldrums where the sea was often glassy, the *Baird,* low on fuel, supplies, and water, rolled heavily even when low swell passed beneath. We were used to the motion and noticed it only at mealtime when 30-degree rolls would cause food to slop off one side of the plate and then the other, and the portholes vertically scanned water and sky. The water ration was cut to 12 gallons per man per day, which barely covered cooking and face washing. These minor disadvantages were recompensed by some special rewards, one of which was a clear unbroken horizon that permitted us to see a green flash nearly every evening as the sun set and the Southern Cross a bit later.

It is hard to describe the immensity of the Pacific Ocean to anyone who has not voyaged on it in a small ship. Flying across it on a modern aircraft takes many hours at hundreds of miles per hour, but the result is merely boredom, not real understanding or appreciation of size. The Pacific is larger than all Asia and North America combined; it holds enough water to flood a smooth sphere the size of the earth to the depth of a mile. But facts like those are not as impressive as a few weeks on its broad bosom, far from the nearest land. No one should be permitted to call himself an oceanographer who has not experienced this vastness. It gives much needed perspective to a land-based species.

Everyone stood watches around the clock, worked at geophysics all day long, and slept nights while the ship churned toward the next station. But now the operations were much smoother; we had developed a lot of skills and debugged the equipment in the previous two months. The coring crew could take a full core (28 feet) almost every day; the seismic measurements had become routine, and the magnetometer appeared to work (although we discovered later that somewhere along the way a lost brass bolt at the two point had been replaced with a steel one; this may have influenced some of the readings but we were not sure which ones).

Now we fixed a new string to our bow; a Bottom Temperature-Gradient Recorder, commonly called the heat probe, that could measure the amount of heat flowing outward from the interior of the earth. The principles on which this instrument was based were the brain children of Sir Edward Bullard of Cambridge, England, Arthur Maxwell, and Roger Revelle, but its operation on the *Baird* was in Maxwell's hands. His earlier bad luck with leaks and faulty circuits turned good just at the right moment.

The heat probe consisted of a slender hollow spear about 2 meters long with thermisters (for precisely measuring temperature) near each end, topped by a heavy pressure case that contained a device for recording the difference in temperature between the two thermisters. This was lowered on the main cable until its spear plunged into the soft bottom; after a 30-minute time for the temperature to stabilize, it was withdrawn. By combining temperature gradient with measurements of the thermal conductivity of the mud in cores taken at the same station, it was possible to calculate the amount of heat flowing outward through the earth's crust at that point. This was a new kind of measurement and one that was essential to understanding the nature of crustal rocks.

There are two sources of the earth's heat. One is the disintegration of various radioactive isotopes that are present in all rocks in tiny amounts. The other is original heat from the formation of the earth that is stored in the molten core and the plastically moving mantle of the earth. Because continental granites contain about three times as much radioactive material as oceanic basalts, it was once believed that the average heat flow from the open ocean floor would turn out to be much lower than that from a continent, especially because the continental crust is also much thicker. But when actual measurements showed that the two kinds of crust had about the same amount of heat flowing outward (about 1 millionth of a calorie per square centimeter per second) scientists recognized that the heat coming from the earth's interior must be carried upward under the ocean by hot, upward-moving rock in the underlying mantle.

As *Baird* moved across the vast southern ocean, the water shoaled from 3 miles to about 2. We were approaching the Easter Island Rise, also called the Albatross Plateau. One of the expedition's main objectives was to learn more about this region, because there was some evidence that the crust below this undersea plateau was much thicker than the average ocean crust and seismically more active.

Atop this broad rise the heat flow was more than three times greater than the norm, but on its western flank the rate was less than half the average value. This suggested that hot, plastically flowing rock of the mantle was upwelling beneath the hot spot and downwelling in the cool spot. In the same region, seismic velocities showed significant differences in density distribution in the deep crustal rocks. Although we were unable to explain these new data at the time, the relationships would become clear 15 years later when the theories of plate tectonics and sea floor spreading developed. By the time we turned north again, we were sure that Expidition Capricorn had contributed something new and important to the understanding of the oceanic crust.

The equatorial sediments were Gustaf Arrhenius' home ground; he had cored them from the *Albatross* and done his doctoral thesis on their chemical composition and history. We therefore expected him to be able to read the age and history of a lump of brown clay or a gray ooze at a glance, and often he would do that, saying with dignified nonchalance, "I believe that is a Miocene globigerina ooze," or whatever. The deposits that really excited him, however, were the sediments within a few degrees of the equator that are high in calcium carbonate. Gustaf was especially interested in the relationship between productivity (growth of small marine life) and water temperature and bottom sediments; he had a theory about the origin of these calcareous sediments.

A hundred meters beneath the surface in this part of the Pacific, which is unbroken by islands, a narrow, high-velocity jet of water with the volume of 500 Mississippi Rivers flows eastward. This is now known as the Cromwell current, and it moves along the equator because the Coriolis force there is zero. This special current is cooler and better supplied with nutrients than normal surface water, so small sea life is exceptionally abundant. As these creatures die and fall to the bottom, their shells deposit a white line of calcium carbonate, literally marking the earth's equator on a bottom that otherwise consists mainly of darker colored sediments.

Even in this remote and peaceful region, our thoughts occasionally turned to military matters. One defense of a warship at anchor against attack by an enemy's underwater swimmers is to drop explosive charges, but when we had tried to find out what size charges were used and the distance at which they were effective, no one seemed to know. So, far at sea with a supply of explosives and trained divers, we decided to experiment on ourselves while we waited for the core tool to go down and return with a core of the equatorial bottom.

Bob Livingston, who had a medical interest in the mechanism of concussion, was in charge. He arranged a system of colored flags so that the skiff accompanying the divers could signal the ship when it was ready for a shot, and a marked cord that could be stretched from ship to skiff to measure the distance between the divers and the explosions. When the red flag waved, Russ Raitt would drop a half-pound block of TNT overboard that would explode at a depth of 100 feet. Starting a thousand feet away, pairs of divers alongside the skiff would descend about 20 feet and wait to experience the shock wave; when it passed, they would surface and report what they felt.

The first shot was a bit of a surprise, not to the distant divers who barely felt it, but to the crew below deck on the *Baird* immediately above the explosion. They surged on deck thinking the ship had struck a rock, or the hydraulic system had exploded, or that their ship had rammed the *Horizon;* Gustaf even worried about damage to his core that was on the way up.

When such explosions occur, the solid TNT instantly converts to a gas; a bubble forms that expands outward until the pressure of the gas is equaled by the pressure of the surrounding water. At that instant a compression wave (the main shock wave) breaks away from the rim of the bubble and moves outward. Then the bubble collapses, sending out a rarefaction wave, which lasts until the gas is overcompressed into small bubbles. This expansion-collapse then repeats, sending out a secondary pulse. All of these waves reflect off the sea surface and the bottom, so an explosion has more complex effects than one might expect.

Dr. Livingston was of the opinion that the combination of a steep-fronted shock wave followed by the negative pressure (rarefaction) wave could cause cavitation in the nervous system and thus bring about a brief period of unconsciousness. The question was: How much additional pressure is required to do that? We moved a hundred feet closer after each shot, especially listening for the echo from the bottom some three miles below, which arrived about 8 seconds after the original blast.

At the longer distances the divers (Walter Munk, Bob Dill, Bob Livingston, and I) heard sharp "double bangs" but felt little sensation of pressure, and when the range closed to 500 feet, we began to give an involuntary reflex jerk in response to the explosion; these reflex reactions became stronger as the distance decreased to 100 feet, but we never felt pain or distress. As far as a defense against swimmers was concerned, we decided it would take much larger explosions to drive off determined attackers, and if this was done on a regular basis, no one on the ship being defended would get any sleep.

A few days later Walter Munk, Bob Livingston, and I were again in the water on another experiment, when a blue shark about 8 feet long began pacing up and down on the other side of the ship. We could see him through the opening between the propeller and the rudder, but we were accustomed to having sharks around. Unfortunately, someone on deck also saw the creature, and the captain gave the order "Get those men out of the water!" Since our heads were below water, they could not call to us, and the frustration on deck grew; sharks always look more dangerous to observers from the surface. Walter Munk was first to break the surface, and when he heard all the shouting on deck about sharks, he caught the panic and started up the rope ladder with his lung and flippers on.

It is not easy to climb a rope ladder anyway, and in his haste Walter slipped through, ending up straddling a rung. In trying to get out of the way so we could follow, he churned up enough water to cause the shark to come around to our side to see what the commotion was. By then John McFall had the movie camera running; the watchers on the rail were cheering; Livingston and I were laughing. At that moment the curious shark brushed between us to get a better look; when he did, I could not resist tweaking his right fin, just to get in the last word. But he showed no sign of either annoyance or interest and continued on his way, leaving us wondering how such a large animal could make a living so far from where we supposed his food supply lived.

The ships had gone far enough east that they had to steer a northwesterly course to make San Diego. On the voyage home, some lymph nodes along my left collarbone became enlarged, and I asked Bob Livingston what he thought caused the swelling. He looked me straight in the eye and replied in a kindly tone, "Why do you ask? You know very well what they are." He was right; I did.

Capricorn was a success in several ways, one of which was that nearly all of the scientists on the two ships (some of whom have not been mentioned here) became leaders in the rapidly expanding field of oceanography. Within a few years, they were directors of laboratories, deans of colleges, presidents of corporations, heads of government agencies, and well-known scientists.

It was a long way home with the winds and waves of the North Pacific dead ahead, but by the end of February the ships had bucked their way back to San Diego. The fact that we had not sighted another ship in the open sea from the time we left Kwajalein until we were three days out of San Diego was indicative of the vastness of the Pacific. We were glad to be home, but the coastal waters of Southern California, which had always seemed so clear and blue, now seemed a bit off-color by comparison with our memories of the super clear water and brilliant color of the open central Pacific.

VI

Living with Nuclear Explosions

ENYU

Once the explosion of Mike demonstrated that massive thermonuclear reactions were possible, there was intense pressure to make a deliverable thermonuclear bomb. So the competing University of California laboratories at Los Alamos and Livermore each proposed testing weapons of several sizes and configurations to find out which were the most efficient. Since large weapons were detonated in the Pacific on even-numbered years, by the time the Scripps scientists arrived home from Capricorn, less than a year was available to build a new set of instruments for the next series of shots. This time the new tests would be conducted at Bikini atoll in a series code-named Castle.

For these shots it was decided my group would measure waves only in Bikini lagoon; this was a much simpler problem than working in deep water, but several kinds of wave meters would be used to ensure that some would get a record. After I spent another 6 weeks getting radiation treatments in Los Angeles to knock down the cancerous lumps on my shoulder, work on the new instruments began.

This time there was no doubt that the explosions would be really big ones, and so an extra effort had to be made to prevent the nearest pressure sensors from being crushed by the underwater shock wave and the recorders being destroyed by the blast. Again the basic scheme was to run a cable from an underwater pressure pickup ashore to a recorder in a box strapped to the backside of a coconut palm. These trees are very resilient in hurricane-force

149

winds, and we reasoned they could withstand winds caused by explosions.

Another version of that method used a small fiberglass boat to hold the recorder anchored directly above the pressure sensor. This made it possible to get several miles closer to the explosions without laying a long cable to an island. The third kind of instrument was a self-contained "turtle," designed and built by Professor Dick Folsom for the Crossroads tests 8 years before.

Without a Scripps ship to use for a base and storage depot, all the tools, instruments, small boats, and diving gear were loaded into a truck trailer that had been converted to a workshop. The sealed trailer was then secured to the forward deck of a Navy supply ship headed for Bikini, where it served as our headquarters throughout the operation. About 2 weeks later, our small but experienced crew flew out to meet it. Besides Isaacs, Beckwith, and myself who were involved in wave measurements, Ted Folsom and Feenan Jennings were there to collect radiochemical data and measure the fallout outside the lagoon.

On arriving at Bikini, we were assigned space in a tent city on the island of Enyu, also called Nan, at the southeast corner of the lagoon, over twenty miles from the northwest reef where most detonations would take place. We would live under canvas for a few months in tents on wooden platforms that lined the sandy main street. As at Enewetak, the tents had two bunks along each side, a central writing table, and a closet heated by a light bulb to prevent mildew, plus supporting ropes festooned with drying clothes. This time our tents were not on the windward beach, and although tent sides were always rolled up so the breeze could blow through, we were partly screened from the trade winds by a jungle of coconut palms and underbrush. Many of these seemingly minor details would become important to us in a way we did not expect.

Enyu was a no-frills camp for working men; its offices, kitchens, mess halls, and latrines were temporary wooden shacks suitable for a short construction job. Only one structure on the island was solidly built; the firing bunker was made of concrete several feet thick so that the men who set off the explosions and started the measuring instruments would survive whatever happened. That turned out to be a good thing.

Three weeks before the first of six scheduled events, we set about getting the wave recorders installed on the islands and in the small boats offshore. The new ones were Turtles, made of a pair of half-inch-thick steel boiler-ends fitted together to make an oblate spheroid about 18 inches in diameter and 10 inches thick. Four bolts at the corner flanges used to hold the two pieces together made them look a bit like turtles. Most important, these cases could withstand the very high pressure that the shock wave in water might create. There were two tiny openings in the shell, one for a bourdon tube that straightened out when pressure was applied and moved a pen across the surface of a clockwork-driven circular chart. The other hole was for a wire to the triggering device at the surface. These marvelously simple instruments gave a better record if the chart ran fast enough to leave a little space between successive wave crests. But high speed rapidly used up the chart so the device could not be started until just before the waves arrived.

To solve this I jury-rigged a starting mechanism that consisted of two small strips of copper lightly soldered together but connected to small springs that tended to pull them apart. When they separated, an electrical circuit holding a relay open would be broken, and the chart would start moving. The copper strips were then painted black to absorb heat and put in a test tube, and this in turn was mounted on a piece of wood about 1 foot square, which floated on the surface. This tiny buoy was anchored by its own slender wires to the Turtle on the bottom. The great heat at the instant of explosion would melt the solder and open the circuit, starting the clockwork. This hastily contrived trigger system worked every time on several Turtles and several shots.

We installed the Turtles by swimming down to place them on the bottom in water 40 to 60 feet deep, securing them to adjacent coral heads with a line to make sure that a really big wave wouldn't roll them away.

We also planned to take motion pictures of the waves approaching the beach at Enyu and brought along an electric camera that could be started remotely. At the north end of Enyu a platform had been built about 20 feet above the ground on huge wooden stilts; on it various kinds of instruments were mounted, each intended to measure some aspect of the upcoming explosions. My camera, aimed at the beach in the foreground where waves might break, joined the line of equipment on the wooden bench.

Once the wave meters were in place and ready to go, we had a few days free to turn our attention to more frivolous matters such as collecting giant clams and exploring the outer reef. Equipped with our own small boat, diving equipment, and an underwater movie camera, we had a super vacation. First we explored the coral knolls that rose abruptly from the smooth light-colored bottom of the lagoon almost to its surface. Each was made of dozens of varieties of intergrown corals and equipped with a resident animal population. Blue-green parrot fish chomped on coral branches, extracting the living polyps and excreting grains of clean white sand; tiny bright blue fish sought protection between the finger-sized protuberances of soft brown coral; and dozens of species of other bright colored fishes filled any remaining openings.

The rippling, pulsing blue mantles of the giant clams with iridescent speckles were another attraction. Some of these had shells nearly 3 feet long with breathing ports some 4 inches in diameter. When disturbed, the clam shell would start to close, jetting water out that orifice as its adductor muscle squeezed the interlocking shells together. It has been alleged these can close on a man's foot, holding him under until he drowns, but that seems very unlikely because they do not snap shut but close slowly as the water inside is expelled. Large turtles would dart from the shadows of a knoll when we approached and swim frantically away. Above on the soft, white mud bottom between knolls, one species of small fish would hover; as we approached, they would dive directly downward, disappearing into the bottom without leaving a trace. And often sharks would approach us, sniff like curious dogs, and then continue on about their business.

It was a beautiful world, a diver's heaven, but after a week the lagoon began to seem tame, and we started exploring the outer reef on the windward side of the atoll. Seen from underwater, the rim of the reef is a vertical wall of gray coral about 20 feet high intersected by narrow, devious canyons called surge channels. Below the cliff a rubble slope slipped away into the deep blue of the abyss. Each day we worked our way a little further along the outer reef until we reached the most exposed region, where the heaving swell abruptly converted to breakers a dozen feet high and churned the surface to froth. We became increasingly bolder and finally learned how to move in along the bottom of a surge channel, holding on to keep from being hurled against the sides until the breaker above crashed; then we would swim up, scramble onto the reef, and get out of the way before the next wave broke.

A person standing on the northeast corner of Bikini's reef, watching large breakers smash down every 10 seconds or so and seeing the water race in and out the narrow surge channels, would find it hard to imagine that a man could make it ashore there without grievous injury. Throughout the war it had not seemed worthwhile to post sentries on the windward side of many Pacific islands to guard against subversive landings, and such landings had not seemed possible to us at first. Nevertheless, after a few days of experimentation, we were able to go on or offshore at will amid violent breakers while wearing only masks, trunks, and tennis shoes. This required careful timing, for anyone caught by a breaker while crossing the edge would have been badly mauled.

One day, dressed only in swim trunks and tennis shoes, I waded alone across the knee-deep water between two sandy islets a half-mile apart. The water was clear, the sky blue, and the sun warm; the coral sand was soft and white beneath my feet—there was a wondrous feeling of being in harmony with nature. Then an approaching disturbance of the water surface attracted my attention. A dark, amorphous shape over a hundred feet across was moving toward me. Curiosity gave way to amazement and then consternation. The nearest islet was too far to make a run for it, so I stood, rigid as a clump of coral, while hundreds of small sharks, not much over 3 feet long but armed with layers of needle-sharp teeth raced by, close around my bare legs. Happily they were just passing.

The construction of an airstrip across several of the small southern islets began with cleaning all vegetation from their surfaces. This must have seemed necessary but whoever planned it apparently did not know that the pandanus jungle on these islets was one of the principal nesting grounds of the fairy tern in this part of the world. From the day the bulldozers arrived, clouds of these snowy white birds with black eyes soared above the newly made wilderness looking for a place to set down again. Not many survived.

One day on returning to camp, we heard of the death of a scientist, the man who operated the instrument adjacent to my camera on the wooden tower. Shirtless and sweaty, he had been working over an open rack of electronic equipment and leaned forward, touching a condenser with his chest and discharging its store of very high voltage electricity directly into his heart region. This made a considerable impression on me, but I did an equally dumb thing a few days later and almost came to a similar end.

John Isaacs, who had no interest in swimming with the rest of us and spent much of his time on the flagship talking with senior Task Force officers about weapons effects, had convinced himself that the explosion of Bravo would cause a tidal wave that would wash over the islands. If that were to happen, our wave recorder on Enyu would be drowned out, and the wave measuring group would look silly. So, although this seemed unlikely to me because Mike had not created such a wave, I decided to hedge the bet by moving one recorder inside the big concrete bunker to protect it. The only opening into that bunker, when it was sealed for the shot, was an open-ended 2-inch pipe above the main entrance through concrete a couple of feet thick.

I stood on a ladder outside in bright sunlight and stuffed the end of the cable through. Then I went from the brilliant outside light to the dark passage inside, where I set up the ladder again and climbed up it to pull the cable through. My eyes were not yet dark-adapted, and I could barely see in the cavernous high-ceilinged passage. But, as I began to pull on the cable, the hair on the back of my neck began to rise, an unaccustomed sensation. So I stepped down to wait until my eyes adjusted to the low light level. When they did, I could see that immediately above were three large bare copper wires, almost as big as bus bars, that brought high voltage power into the bunker. The back of my neck had come within an inch or two of making contact and certain electrocution. Sweat leaped out on my forehead as I realized that the difference in luck and life between me and my fellow scientist who had discharged the condenser was only a couple of inches.

A number of the AEC's so-called scientific measurements at Bikini were known as "no moss" experiments because any darn fool could see in advance how they would turn out. The phrase came from comedian Bob Burns who claimed he had a scientist uncle named Schnazi. Once a large boulder was dislodged from the side of a mountain and rolled down the main street of a small town, narrowly missing various inhabitants. It finally smashed through the door of the bank and came to rest in front of an astonished teller. In close pursuit came Uncle Schnazi, who carefully inspected it and then recorded in his notebook: No moss!

BRAVO

The first shot, code-named Bravo, was expected to be somewhat smaller than Mike, so the original plan was to leave personnel ashore on Enyu and another island when Bravo was fired. After all, the distance from its ground zero to Enyu was about the same aş from Mike to Enewetak Island, and people had ridden out the shot there without harm. But Isaacs' opinion that a tidal wave might sweep over some of the islands sufficiently impressed the Task Force commander that, to be extra-safe, he ordered the camps cleared during the shot. Thus, in the afternoon before the long-scheduled firing date, the inhabitants of our tent city temporarily went to sea on a transport ship, the *Ainsworth,* wearing the usual shorts, sandals, and T-shirts.

As soon as Bravo was out of the way, we would return to pick up the scientific records of its effects and resume island life. Because everyone expected to be back ashore before noon the next day, few bothered to take shaving kits, a change of clothes, or even their wallets and watches. The tents were left open, with personal possessions strewn across tables and bunks. What could happen to them overnight on a deserted island? We were more concerned about the ship's dinner menu and accommodations. The latter, on the *Ainsworth,* consisted of four bunks in a tiny cabin with minimal ventilation; the ship was not intended for vacation cruising.

On March 1, 1954, well before dawn, an announcement came over the loudspeaker system: "Anyone with protective glasses who wants to see the shot can watch from the deck. All others remain below."

It was dark on deck when we put on eye shields so dense that they could be used to look directly at the midday sun. But we knew that even those were not adequate protection against the initial flash which would be "brighter than a thousand suns," capable of blinding a man in a hundredth of a second. That scary thought hung in our minds as we stood in the dark with a breeze across our cheeks and the throb of the ship's engines under our feet. The countdown crackled from a loudspeaker, and at the zero instant, our eyeballs could detect the flash, even through tightly squinted lids and dark shields.

A few seconds later when we opened our eyes, Bravo's orange yellow fireball was over a mile in diameter and growing at an alarming rate. That glowing sphere expanded outward and toward us faster than the speed of sound, seemingly vaporizing everything it touched. No aircraft could outrace it, and no bunker could protect against it; the speed and reach of the weapon were beyond human experience. Then, just as panic was about to seize the watchers on deck, the fireball stopped growing; it cast off the air shock wave and began to collapse. Its boiling extremities curled under, converting the incandescent globe to a rising toroidal (doughnut-shaped) vortex. As the glowing mass rose, trailed by a column of vaporized reef material, it briefly took on the characteristic mushroom shape before disappearing behind high clouds.

We knew at once that Bravo had "gone big"; later it was officially specified to have had the energy equivalent of 15 megatons of TNT, the largest man-made explosion ever. From a distance of 72 miles Mike, the previous champion, had seemed relatively subdued, but Bravo, seen in the dark at less than half that distance, had made a knockout debut.

As the pinks and purples of nuclear dawn faded back to blackness, we went below for breakfast, and by the time we returned to the deck the sun was up, the sky was blue, and the world seemed serene. We relaxed, expecting this to be a brief cruise; fool's paradise would have been a better description.

As we lazed on deck waiting for the ship to return to the lagoon so we could go ashore, a frantic, breaking voice came over the ship's loudspeaker system. "Don't be alarmed," it reassured us hysterically. "High levels of radioactivity are being measured on the bridge. Everyone on deck must go below." No radiation levels were specified, but the tone of the announcement indicated that they had scared the socks off the guy who was watching the meter.

We were more amused than frightened or surprised. The ship was getting some fallout, probably not unlike that the Scripps group had experienced on the *Horizon* 16 months before. As everyone filed below, the *Ainsworth* speeded up and changed course. Then, a short time later the hoarse disembodied voice came over the speakers again with only a trace less panic: "Cover all the ventilators."

That was OK; it made sense to avoid pumping radioactive particles into the ship.

Then: "Throw the deck chairs overboard."

The refugees crowding into the lounge looked at each other. "The deck chairs? What the devil does that mean?"

The next order was "Everyone into their bunks to conserve oxygen."

Anyone who has spent time below deck on a ship in the tropics knows that, even when the ventilators are running full blast, rooms get hot and breathing is hard. Your buddies up-duct from you try to divert the incoming air into their bunk by rigging cardboard deflectors or by stuffing old socks into your outlet. But on the *Ainsworth,* with the ventilators stopped entirely, the temperature in the cabins rose rapidly, and the bunks became sweat baths; the temperature in our stateroom was soon over 100 degrees, and the humidity was just short of rain. Nuclear testing was supposed to be fun, but after lying gasping and naked in semidarkness on soggy sheets for many hours, we reconsidered.

Eventually the ship left the area of fallout, and the ventilators resumed pumping air; the decks were washed down, and by evening all hands were permitted to go topside again. Bravo's huge blast had changed our world—and everyone else's too, but they didn't know it yet. Rumors began to circulate that men were trapped in the bunker at Enyu by very high levels of radiation and that Rongerik atoll, 120 miles to the east of Enyu, was really "hot," and the people there wanted to be evacuated.

It was quite clear that we would not be going ashore again for a while. One reason was that radioactivity decays very rapidly for the first few days, and the place had to be allowed to cool down. Another was that although radioactive fallout had been expected and the equipment to deal with it had been brought to the test site, most of it was stored on a barge that was still moored in Bikini lagoon. It was ironic that the first casualty of fallout had been the things needed to measure and protect against radiation: ion chambers, geiger counters, film badges, protective clothing, and such. A frantic call went back to the mainland for replacement items, and it took several days for them to arrive. While we waited, nonessential personnel were shipped out, and those who were to stay set about adjusting to their new homes on the *Ainsworth* by obtaining shaving gear and minimal clothing from the ship's store.

After a couple of days passed with no direct information about the level of radioactivity on the islands, it was plain that someone would have to go ashore to see what had happened and make some measurements. Isaacs volunteered the two of us for this singular honor.

So, on March 4, three days after Bravo, a strange party landed on Enyu in a dukw: Isaacs and I, the driver, and three military observers. All of us wore protective clothing consisting of white jump suits, canvas booties over our sandals, white cotton gloves, surgical caps and breathing masks to prevent inhalation of dust. These clothes were protection only against alpha- and beta-emitting particles but were of no value against the X-raylike gamma emitters, as will be explained presently.

As soon as the dukw was ashore, I dismounted and, carrying an ion chamber at waist height, began a solemn walk down the sandy ruts toward our tent city. Every few seconds I would call the meter reading over my shoulder to those in the dukw some ten paces behind, so that the levels of gamma radiation could be recorded. If really high levels were encountered, the dukw was to back out the way it came in.

An observer from a distant land might well have mistaken our strange procession for some religious rite. A solitary figure swathed in white plodding down a palm-sheltered avenue while gazing intently at a wavering needle that marked the invisible power of the newest earth-god, and chanting the intensity of that power to other shrouded figures in the juggernaut behind. In a way the observer would have been correct, for this first meaningful fallout of radioactivity on an American community was an augury of things to come.

The island's only road led from the landing beach to the main street with its lines of tents. Here and there a tent pole was broken or a cabinet overturned, but, in the main, the tents twenty-two miles from ground zero had survived the blast very well, probably because their sides had been rolled up, minimizing their resistance to the shock wave. The winds generated by the blast had no more effect on these flexible structures than a tropical squall would have had, but the wooden latrine buildings, cookhouses, and offices that had rigidly resisted their onslaught had not fared so well.

On entering some of the buildings to investigate, I found that the power plant was still running, and some lights were on; water still flowed from the faucets. In a once-busy office, telephones that had jangled incessantly on my last visit now sat silently on dusty desks. As I moved about, the needle of my ion chamber would climb and then fall again as it responded to variations in radiation from different sources. Sometimes I would have to shift to higher or lower scale to keep the needle visible on this device, which could be used to predict the length of time a man might survive.

Finally I came to my own tent, whose entrance was marked by a pair of giant clam shells. Except for a coating of gray dust, everything appeared to be quite normal. Shoes were under the bunks; bathing suits were draped over the tent ropes: wallets, watches, and half-read books lay on the table. But these things did not seem normal to my instrument, which showed a radiation level of about 1 roentgen per hour. I did not touch anything, including my own possessions, for we were under strict orders not to remove any object; there was official concern about bringing contaminated things back to the ship. The gray dust had settled on everything—a silent, formidable enemy. The clothes we had left were now unwearable and our bunks unusable.

The dust was mainly tiny bits of coral, or of the test structure, or of the bomb itself. These particles had lived for an instant in the center of a man-made sun and had carried off traces of its radioactive fire. As with Mike, a huge cloud of this dust had reached the stratosphere, been carried eastward by the upper atmospheric winds, and then settled back onto thousands of square miles of sea and a few islands. That which fell in the sea was safely shielded from man, but the islets that ringed Bikini lagoon would remain "hot" for a long time.

In an introspective mood, I gazed at the objects in our tent for a full minute and a line from Macbeth crossed my mind: ". . . have lighted fools the way to dusty death." I thought: I am the first person, ever, to walk through an American city made uninhabitable by radioactivity from a thermonuclear weapon. Perhaps this once we can afford it, but it must never happen again! Here the only casualties are minor physical assets, just enough to remind us that a nuclear war must never be fought. I thought of Pompeii whose citizens fled before the explosion of Vesuvius, leaving a complete city to be studied 1,800 years later. Would our civilization leave a thousand cities for future archaeologists to inspect? The fun of participating in these massive experiments in physics and oceanography was now shadowed by the thought that our toys really were weapons that could be used against cities, or as threats to hold nations hostage.

Shouts from the others aboard the dukw brought me back to the real world. Dripping with sweat inside the protective clothing, I reported that the level of radioactivity measured inside the tent was only a third of that outside. Then we went to recover charts from some of the wave meters. This required going up a road through a stand of tall palms; under the trees the radiation levels were twice that in the open because of dust radiating downward from the palm fronds above. A pair of fairy terns, blinded by the explosion, stumbled about in a clearing. Finally we reached the plywood box on the coconut palm that held my recorder, removed the chart paper, and found a satisfactory wave record. By now this was an anticlimax.

Once back on shipboard I set down the thoughts and experiences described above in a set of notes entitled "Dust and Death," being careful not to include anything that might be considered a security breach. There was no doubt in my mind that, if John Isaacs had not warned of a possible tidal wave, the islands would not have been evacuated and a lot of men would have received large, possibly fatal, doses of radiation. There had been no large wave, but on some of the islets nearer to the shot, there had been a minor run-up of lagoon water that stranded a few fish.

The reports we had submitted about Mike's fallout on the ocean to the east a year and a half before had been forgotten or ignored by those who planned the Castle series. The semipermanent camps on three of Bikini's islets had been intended for use throughout the shot series; such plans would not have been made if fallout had been given any consideration.

After Bravo everyone lived on shipboard. Most subsequent shots were fired from barges anchored in the lagoon or the Bravo crater to minimize the amount of solids involved in the explosion, which would be carried upward to create fallout.

Having measured the level of radioactivity ashore, we gave more thought to what could have happened if anyone had stayed on the islands. On the basis of a median lethal dose of 450 roentgens, anyone who had remained on Enyu and *who took no precautions* probably would have accumulated that much radiation during the first week after the shot, mostly on the first day.

About ten days later, we learned that fallout from Bravo had landed on some Japanese fishermen at Ailinginae, the next atoll to the east. The following year I was able to obtain some detailed information about their experiences from Dr. Ralph Lapp, an expert on nuclear radiation.

The *Lucky Dragon* or *Fukuryu Maru no.* 5 was a 100-ton wooden sampan-type fishing boat out of Yaizu, Japan, whose standard speed was 7 knots. It was a "long liner," whose twenty-three-man crew fished by setting out a buoyed fish line 25 miles long, which had 1,500 hooks dangling from it. Later the crew remembered that this voyage had begun with several bad omens.

Early in the morning of March 1, a seaman on watch named Suzuki saw the dark sky to the west suddenly fill with a great bright light that rose ever higher above the horizon. He ducked below and shouted to his sleeping companions, "The sun rises in the west."

The men who piled on deck to see this remarkable event knew they were in the region where nuclear weapons were tested and guessed at once what it was. "It's a Pika-don!" they agreed, using a contraction of the words for "thunder" and "lightning," which was the Japanese idiom for a nuclear explosion. The seamen thought at first the source was at Enewetak where the previous tests had been conducted, but as they stood on deck discussing this marvel the "sound of many thunders rolled into one" arrived. Radioman Kubayama, the best educated man aboard, estimated that 5 or 6 minutes had elapsed between the flash and the sonic arrival; from that he calculated that the *Lucky Dragon* was 85 miles from the explosion. This placed it at Bikini.

The fishermen were afraid they would be accused by the United States of spying, so they decided to pull in their fishing line at once and leave the area. However, it takes at least six hours to bring in a 25-mile-long fishing line and remove the fish from it. As they were doing so, about 2 hours after the explosion, a fog formed over the sea, and a light rain fell. The rain brought with it ashlike bits of coral they likened to rice powder. This powder got on everything, including hair, clothing, and the newly caught tuna that were being gutted on deck. It stung their eyes, and it looked like salt so they tasted it. Two men collected and saved small samples, one for luck, which he kept under his pillow for the rest of the voyage, the other because something was "wrong" with the ash. The fallout continued for 4 hours, but eventually they got their long line back aboard, washed down the decks in the usual way, and headed for home.

From the first evening on the men were sick; they lost their appetite, were nauseated, and occasionally vomited. Their skin and scalp itched, there were sores on their heads, and some had elevated temperatures. By March 14, when they reached their home port, the men's skins had greatly darkened, and their hair was falling out in patches. Blood tests showed their leucocyte levels were only one third the normal range, and their sperm counts were zero. Eventually all were hospitalized for over a year; radioman Kubayama died.

Over a month after the fallout landed on the ship, Japanese scientists measured the radioactivity of some items aboard it. The fish line still in the tubs stirred a Geiger counter to 15,100 counts a minute, a short piece of steel cable went 66,500, a white cotton glove caused 14,200 clicks. Tuna, taken by other fishing vessels and found in the Osaka central market, "cried" at 2,000 counts, and the sales of all fish came to a stop.

Calculating backward from the data that eventually was assembled, Dr. Lapp estimated the fishermen's total dosage of gamma radiation at 250–330 roentgens (450 is generally given as the dose that will kill half the people exposed) plus beta burns on their skins and possibly some damage from internal alpha emitters. On March 4, the day we went ashore on Enyu, the level on the un-*Lucky Dragon* (after the washdown and radioactive decay) must still have been half a roentgen per hour. Dr. Lapp concluded that the reason that most of the fishermen survived was because the deck area of the ship was so small that relatively few fallout particles could remain there to radiate gamma rays at them.

Enyu, where I had monitored fallout radiation at 1 to 3 roentgens per hour was only one-fourth as far from the burst as the fishing boat so the dosage there must have been considerably higher. Luckily both Enyu and the *Lucky Dragon* were south of the axis of worst fallout. Later the AEC announced that after these incidents, twenty-eight American personnel and 236 natives had been moved from other atolls in the Marshalls as a precautionary measure.

Isaacs and I were curious about the distribution of radioactivity from the shot, especially on the downwind islands to the east, and on about March 17 he arranged for the two of us to visit Ailinginae on a Navy PC (Patrol Craft). This class of ship looks something like a destroyer but is only a quarter the size. It is slim, made for speed not comfort, and rolls a little too readily. This tiny warship also had tiny bunks; the one I used was ordinarily occupied by the executive officer, who was on leave. It was not possible to roll over in bed because of piping that paralleled the bunk a foot above the mattress; one could only enter or leave it by sliding sidewise. After a couple of nights there, I decided that the absent officer was probably being treated for claustrophobia.

By noon the next day, the PC tried to enter a south channel into Ailinginae lagoon but encountered strong currents flowing out through a thicket of coral heads and turned back. So the captain decided to lay off a half mile and wait while our party went ashore in the ship's dinghy. This tiny boat was intended to carry two men, so when four of us, including Isaacs and some radiation detection equipment, piled in and started for shore, there was only about 4 inches of freeboard. For a while John and I bailed, one man ran the outboard and the other one prayed, but we had no great confidence in any of those endeavors so I agreed to swim alongside to prevent a disaster. It is not easy to get out of a tiny, overloaded boat without tipping it over, but I dived up and out from its precise center in an almost vertical leap, barely clearing the side as I came down and not quite swamping it.

Ailinginae, where the Japanese fishermen had seen the Pika-don, is a small atoll with one islet on the south rim that supported a stand of coconut palms and a few pandanus. When we landed there, the ground was still radioactive, but the level was down to 50–100 milliroentgens per hour. The evidence of higher levels in the recent past came from birds too sick to fly that staggered around on the sand.

Probably very few people have ever been to this remote atoll, and its high-tide line glistened with green glass balls up to 16 inches in diameter. These lost fishing floats were easily washed souvenirs, and we collected all the dinghy would hold—one for everyone aboard the PC.

Meanwhile, Ted Folsom and Feenan Jennings were at sea aboard a large Navy tug directly measuring radioactivity in the sea for the first time. They had scrounged up several Geiger counters and some odd lengths of pipe, and without the benefit of circuit diagrams or a proper place to work, they put together a remote detector that could be towed behind the ship to record the level of radioactivity in the water.

Eventually the tug covered 900 miles, crisscrossing the path of the cigar-shaped fallout zone east of Bikini, commonly called the upwind sector. Their work reconfirmed and expanded upon what we had learned at the Mike shot, namely that fallout landing on the sea immediately mixes with the near-surface water and is borne along by the wind-driven currents. It also showed that the band of really dangerous, very high-level radiation was only about 50 miles wide; a little to either side of that, the levels dropped quickly. If the Japanese fishermen or the people of Rongelap had been another 20 miles further north, they would have been exposed to very much higher levels of radioactivity. Twenty miles further south they would have had a much lower exposure.

At the time military men seemed pleased by the unexpected bonus that fallout gave to their new weapon. They saw the huge area downwind of a burst, where an enemy would be too sick to fight for long, as "more bang for a buck." Some, however, noted that fallout could be a two-edged sword, because it enlarged battlefields to include civilian populations, and it required that wind direction be taken into account. Publically, senior officials tended to minimize fallout and its effects until on February 15, 1955, almost a year later, a press handout by Chairman Lewis Strauss of the Atomic Energy Commission admitted that fallout was a serious matter.

Permit me a simplistic description of radioactivity and how it acts; it may be useful. Radiation, although invisible, is not magic, and a knowledge of how it causes damage can reduce fear of the unknown and help one make sensible choices if circumstances require them. Fallout particles, at least in the most seriously affected regions, probably will be visible, as they were to the Japanese fishermen.

When atoms of any material are exposed to the huge initial burst of neutrons from a nuclear explosion, they are likely to be converted into unstable isotopes, which decay in much the same fashion as a clock running down. The time for each isotope to lose half of its remaining radioactivity is its half-life. Half-lives vary widely; that of radon 219 is 3.9 seconds, that of carbon 14 is 5,600 years, that of uranium 238 is 4.6 billion years.

A great many kinds of isotopes with a wide range of half-lives are generated by a nuclear explosion, but as a useful first approximation, their emanations decrease in accordance with a simple formula derived from the average of their decay curves. That is, the radioactivity of mixed fission products (fallout) decreases to one tenth as the time increases sevenfold. Thus, at the end of 7 hours (or 7 days) there is only one tenth as much radioactivity as there was at the end of 1 hour (or 1 day) after an explosion. Therefore it is most important to seek shelter during the first few days after a nuclear explosion; as time goes by, the need for shelter becomes much less important.

Radioactive emanations from the nuclei of radioactive isotopes consist of three different kinds of particles or "rays" called alpha, beta, and gamma. *Alpha* particles are so light they are easily stopped by a layer of clothing or a sheet of paper. Consequently they are of little importance as an external radiation hazard, but if they are taken into the body with food and water, they can collect in bone marrow and eventually cause leukemia.

Beta particles are electrons, easily deflected but able to penetrate a little farther than alpha particles before their energy is absorbed. A few feet of air or a jumpsuit stops most of their effect, but they are a hazard when they come in direct contact with the skin. Beta radiation was responsible for most of the skin problems suffered by the Japanese fishermen; if they had washed thoroughly, they would have avoided most of those injuries.

Gamma rays from explosion-made isotopes act much like X-rays or cosmic radiation; one can think of them as a special kind of light to which all substances are partly transparent. These rays can penetrate all kinds of material, but their capacity for damage is reduced in proportion to the mass of the material through which they pass as they are deflected and back-scattered. Thick or dense materials prevent much gamma radiation from shining through; lead is better than concrete, which is better than water.

The physiological damage caused by gamma radiation is due to the ionization of body tissues, measured in rads (this has supplanted the dosage in roentgens per hour, for the ionization of air, which was in use during the tests described). There is a big difference in the response of various individuals to radiation, but half the people exposed are unlikely to suffer any serious damage from the receipt of 50 rads over the whole body at one time. An average person gets a lifetime dose of about 5 rads from natural radioactivity, cosmic rays, and diagnostic X-rays. (My own whole-body exposure over the years is around 50 rads, and so far no detrimental effects are visible.)

ROMEO

For the first week or so after the Bravo disaster, the Task Force seemed to be in a state of shock. The ranks were thinned as nonessential personnel were sent elsewhere; those of us with jobs in the lagoon or on the islands waited for the worst of the radiation to decay a bit more before resuming work. When we returned, it was in an LCM with a crew of three sailors to replace our contaminated wooden boat, which had been scuttled.

The film in the camera on the beach tower that was supposed to photograph arriving waves had been exposed by fallout and the contaminated camera was junked. One of the anchored boats holding a wave recorder had been flipped over by the blast, another had torn free and drifted off. However, the one that did survive got a good record as did the recorders ashore; all showed that the waves produced by Bravo were less than phenomenal.

Recovering the Turtles was a bit of a problem. In the midlagoon area where they were located, the radiation from the water surface was 20 to 50 milliroentgens per hour, and we spent about 6 hours a day for a couple of weeks in the area. We were not sure what that level meant to a diver in terms of radiation, and we had no instruments for measuring radiation underwater. Not knowing the risk, my solution was to make all the dives myself in areas where the water was "hot" and charge it to cancer therapy.

The Turtles had done their job well; they had made a precise record of the initial shock wave and the train of water waves that followed it. We reset them and replaced them in readiness for the next shot, Romeo, which was supposed to be in a few days.

Then began a routine that lasted much longer than anyone anticipated. The "shoot on a scheduled date" program gave way to a careful review of the winds aloft each day; their strength and direction would dictate the date. Since all islands likely to be hit by fallout had already been evacuated, this was locking the barn door a bit late, but it was a step in the right direction.

Every day in the early afternoon, there would be a meeting in the wardroom of thirty or so of the senior scientists. We would hear a forecast of the movements of upper and lower winds; then the test director and his top advisers would huddle for a few minutes and announce that Romeo would be fired early the next morning. At that several group leaders, including me, would start for the boats to get their instruments on various islands ready for the morrow's event.

The Scripps group now had five widely scattered wave meters that had to have the chart paper changed, the batteries checked, and the timers reset. With luck our party would get back to the ship just in time for dinner, after which there would be an evening weather briefing, and the announcement would be made that tomorrow's shot had been canceled. We would groan and fall into bed. Then, the next morning early we would be out on the lagoon again, stopping the instruments to conserve paper and batteries.

After lunch the cycle would begin again. What had once been fun became drudgery, and after a couple of weeks of starting and stopping instruments daily, many of those involved were heard quoting Shakespeare: "O Romeo, Romeo! Wherefore art thou, Romeo?"

When the evening announcement finally changed to "The shot is definitely on for tomorrow," there was a very loud cheer. The wind briefings seemed to indicate the atmosphere was behaving exactly as it had before, so we assumed the authorities had decided not to wait any longer; there was little to lose anyway.

When Romeo burst, the sun was below the horizon, but the eastern sky was already light. Isaacs and I were standing on a forward hatch cover of the *Ainsworth* as it circled southeast of the atoll, 40 miles or so from ground zero. As the fireball changed into a toroidal vortex that rolled upward, we saw a new phenomenon.

As the fiery ring rose, parts of it spread outward along a series of previously invisible atmospheric layers. In half a minute these growths from the main stem resulted in a fantastic tower of clouds with five or six distinct horizontal layers and a mushroom top rising 60,000 feet or more above ground zero. The tower was strangely beautiful, as it boiled pink, violet, and orange with a weird internal nuclear light, something like a lighted fountain in a public park. Despite its deadly purpose, Romeo was an artistic masterpiece.

When the clouds turned gray, John and I walked through the ship to haul in a pair of electrodes that he was trailing behind the ship to measure any underwater electrical signals that might come from the burst. As usual we had forgotten about the air shock wave, and just as we stepped out of the main passageway onto the fantail, it struck. Two Filipino mess boys who had been leaning on the taffrail some 30 feet away were thrown backward by its impact and tumbled down the deck in a series of somersaults to land at our feet, surprised but unhurt.

After Romeo the other shots in the series became a blur. Vaguely I recall living on several ships, recovering instruments, and diving daily in radioactive water.

Sometimes John Isaacs would come along on the LCM and talk about all sorts of subjects, including radiation. On one of those occasions, within earshot of the Navy boat crew, John revealed that film badges, the little white squares of radiation-sensitive material that everyone wore to keep track of how big a dose of radiation he had received, were also pressure-sensitive. In other words, when these badges were "developed" each week and doses recorded, the effect of pressure (squeezing) on the badges was indistinguishable from the effect of gamma rays. The rule in the Proving Grounds was that anyone who received more than 3.9 roentgens of radiation was sent stateside.

The Navy crew, which had shown no previous interest in radiation, had listened to John's story more closely than we realized. As with most other sailors, their highest priority was to get home to the girls; now they saw their chance. About two weeks later, as the LCM was about to head out on the lagoon one morning, the Shore Patrol arrested our three-man crew. Their film badges had been developed, and they were found to have received the highest doses of radiation ever recorded (85 to 96 roentgens) in the test program. They were ordered to a stateside hospital that day. Since we had been together every day, and all the other badges showed low dosage, something was funny about those high readings, so I took one of the boys aside to find out what it was.

"Well," he said, pleased to be going home, "after we heard Mister Isaacs say the badges were pressure-sensitive, we put 'em in our shoes for the rest of the week."

That accounted for the pseudoradiation, but these fellows were not as smart as they thought. In the end they spent more time in Navy hospitals under observation for radiation effects than if they'd served out the tests.

One of an oceanographer's tricks for salvaging instruments that have been flooded with saltwater is to immerse them in alcohol. By fortuitous circumstance, we had chosen to bring alcohol of a drinkable quality, a rare commodity on the *Ainsworth,* which was a dry ship. After weeks of frustrations and delays, not to mention the loss of several instruments, we felt that an evening of relaxation was in order and declared a gallon of instrument alcohol to be surplus. For a mixer we were able to borrow some oranges from the ship's galley and combined the two ingredients equally: one orange to one quart of alcohol. The result was a very stimulating punch that caused a nonnuclear explosion. Soon after the party began, I remember Isaacs and myself alternately reciting Rudyard Kipling's poetry to our assembled guests. None of those present remembered how it ended.

During one of the tests, I was aboard a small aircraft carrier, and about 5 A.M. a call came over the speaker system: "All those who want to see the shot report to the flight deck." So, half-asleep, half-dressed, and half-wishing they'd stayed in their bunks, a thousand men stumbled up flights of steel ladders to the deck. It was dark, the only light being that from the stars plus a little from the deckside ladders and the ship's island. One could barely make out men only an arm's length away. Then came the next order: "Everyone sit in rows across the deck facing aft." Gradually we achieved a ragged semblance of that formation while unfunny jokes and wild speculations about the incipient explosion circulated among the crew. All the while the ship moved slowly in the direction of ground zero, about 30 miles away. Eventually the hollow metallic voice said, "One minute to zero. Everyone close your eyes tightly and put your hands over them."

We did. The chatter subsided, and the deck became quiet as the final countdown began. It was broadcast to all the fleet from the bunker on Enyu, and since it had been prerecorded several days before, there was no nervousness in the voice: "Nine, eight, seven, six, five, four, three, two, one, zero." Then—nothing—for another few seconds. Until the voice on the speaker said: "You can look now."

We opened our eyes to broad daylight. The clouds, the sea, and the ship looked about as they always had, except, of course, for the thousand men sitting on the flight deck. There was no bright glare or hard shadows; the sky had simply filled with diffuse daylight during the seconds our eyes had been closed.

We rose, turned to squint at the source, and once again felt a moment of panic watching the glowing fireball expanding much faster than a jet at that distance could fly across the sky. Then the incandescent center began to rise into the familiar colored mushroom clouds, and a minute later twilight began to descend on the ocean off Bikini. Steadily the world grew darker, and when we could no longer identify our friends at close range, we groped our way to the ladders and went back to our bunks. Then, just as our minds had turned to other things, the shock wave arrived with a crack of sound as though some giant had struck the ship with a huge hammer.

Slowly the carrier moved closer to the southern rim of the atoll to reduce the range of the helicopters that would overfly the lagoon to inspect the effects of the shot and monitor the levels of radioactivity. About 11 A.M. I went out as a monitor on the first flight over the crater. The flight plan called for a course north across the center of the atoll, then west over the shot point, and southward to rejoin the ship. Our 'copter flew at 1,000 feet and about 50 knots, trailed by another chopper called an angel that flew 500 feet higher and half a mile behind. Its job was to pick us up if ours got into trouble and had to ditch (helicopters were not terribly reliable in the early 1950s).

Above the underwater crater, which was the source of a mass of muddy water that streamed southward and overflowed the rim of the lagoon, the ion chamber I was using read 40 roentgens per hour. As we moved, details of our position, altitude, and radioactivity readings were reported by radio to someone on the ship, who presumably recorded them. Since we spent 15 minutes or so over the discolored, highly radioactive water, the pilot and I must have gotten a dosage of at least 10 roentgens from that one trip. If our chopper had gone down in the "hot" part of the lagoon, even if we had been picked up in a few minutes, it is not likely that the pilot or I or our rescuers would have survived the radiation. But all of us understood the risks, and I flew a similar mission over another new crater a week later.

THE SARATOGA

The aircraft carrier *Saratoga,* famous for its exploits in World War II, had been one of the test ships at Bikini in the Crossroads tests of 1946. It had been moored within a mile of the underwater shot and suffered severe damage from it. The aircraft on her deck had been swept overboard by the huge column of falling water, and she sank 7 hours after the explosion, her bottom rent by the underwater shock wave. We knew the *Sara*'s position within a couple of miles, and many a day on our way to tend the wave recorders, we had crossed that region slowly, peering downward, looking for some sign of her presence. She did not reveal herself, but the more we thought about the possibility of visiting the great flattop on the bottom and walking on its sunken flight deck, the more we wanted to.

Finally we obtained a portable echo sounder for the LCM; with it we expected to find the broad flight deck, which should produce a record something like a table, 80 feet or so above the lagoon bottom. Or perhaps we would accidentally come upon one of the other ships that sank in the same area, such as the U.S. battleship *Arkansas,* the Japanese battleship *Nagata,* two troop transports, and several destroyers. It would be hard *not* to find one of these wonderful targets on the bottom. The problem was that we had so little free time. With the echo sounder operating we began taking a slightly different route across the old test area every day, intently watching the recorded echos but seeing nothing. Late one afternoon a few pings came back from some large object at middepth, and almost simultaneously a crewman shouted, "There's something!" and pointed to a vague greenish shape a few feet below the surface.

In a moment I was in the water wearing a facemask and called back, "It's the top of the *Sara*'s mast. The bridge is right below us." There was no time for a dive that day; a small marker buoy was secured to the mast so it could be found easily again, and we returned to the *Ainsworth,* jubilant. The next day, equipped with a movie camera and an ample air supply, we returned, moored the LCM to the top of *Sara*'s mast, and four of us went in the water to see the famous carrier.

A cross spar near the top of the mast had become the nucleus for the mass of algae and corals that we had seen from the surface. The outlines of the upper mast and that blob of life were covered with an unfamiliar fuzzy, diaphanous white growth. When we innocently touched and sampled the white fuzz, we found it was the well-protected larval form of some sea animal. To be more specific, it felt like a cluster of hot needles, pricking or burning our skins. After that we detoured around the mast top to a balconylike platform that was the uppermost observation point on the old carrier's island. Climbing into the tub that once had protected the lookout from wind, we pretended to shout into a long-drowned intercom system.

Descending further we could see what appeared to be a crumpled section of the deck through the bluish haze of water. As we continued down, this confusing shape resolved itself into the ship's smokestack, which had fallen partly across the caved-in aircraft elevator. We recalled that the *Sara* had a most unusual stack just behind the island; it was very thin, at least 60 feet long and high, and it stood next to the aircraft elevator in the middle of the flight deck. Apparently the mass of falling water from the explosion had stomped in the elevator, and subsequently the stack had fallen across the opening.

Visibility at deck depth was reasonably good, and we could see about 60 feet, but we soon realized that our dream of seeing a large part of the flight deck at one time was not to be. Fee Jennings hung over the edge of the elevator opening and looked deeper into the ship but saw only blackness; Warren Beckwith pried some beautifully sculpted clams from the side of the bridge; they were not only flat on one side but had gray paint stuck to them. Chuck Fleming toured the old command center, and I filmed the operation, knowing the light was inadequate but hoping that some images would come through. Most important, we walked, or at least touched down, on the flight deck, stirring up wisps of dust.

Back aboard our LCM the four divers were unusually pensive, our minds still communicating with the spirits of the *Saratoga*'s long-gone pilots and crew. Having visited their old haunts, our minds reconstructed the ship as it had been in its glory days. We could see the uniformed figures on the rail of the bridge and A-5s on the deck, as the ghost ship steamed through the fourth dimension, running into the wind like the Flying Dutchman to launch phantom aircraft.

During the Castle series a large tumor sprouted on the inside of my spine at stomach level, despite all the free radiation I had received courtesy of the AEC. One last time I returned home for another six weeks of hospital-style radiation. I was tired of this cancer foolishness by then and decided to quit, cold turkey, having had enough of this silly commuting between two kinds of treatments at Los Angeles and Bikini. I accepted a job with the National Academy of Sciences in Washington, and the cancer disappeared for good.

VII

Pausing to Smell the Flowers

THE NATIONAL ACADEMY OF SCIENCES

In the summer of 1954 Washington, D.C., was a gracious southern city with uncrowded tree-lined avenues and white buildings set in broad green lawns; trolley cars provided efficient public transportation; the federal budget was balanced; a good dinner cost two dollars. The worst problem in those days before air-conditioning was the heat and humidity; residents sweated and suffered, marveling that the founding fathers could have accomplished anything while wearing tight satin breeches.

Although Washington at first appeared outside the mainstream of activity for a working oceanographer, it turned out to be an excellent place to get an overview of the changing national attitude on the subject of ocean affairs. The Scripps Institution of Oceanography lent me to the National Research Council for a year to be executive secretary of a new Committee on Civil Defense, which needed someone who had "hands on" experience with nuclear explosions.

The National Research Council is the operating arm of the National Academy of Sciences, a private institution of distinguished scientists chartered by Congress in 1863 and dedicated to the furtherance of science for human welfare. It occupies a white marble building on Constitution Avenue, whose entablature is inscribed in Greek, whose windows and roof are weathered green bronze, and whose entrance steps encompassed cascading waters representing the flood of knowledge pouring forth.

The academy was a good operating base, because one could learn some-

171

thing about the policies and financing of science, get to know scientific celebrities, and work on a wide spectrum of scientific problems. The big boss was Dr. Detlev Bronk, who was not only president of the academy and the Johns Hopkins University but also chairman of the National Science Board, a government organization. His easygoing style of management permitted me to range beyond my specific job into other aspects of science and oceanography. Even so, I felt a little guilty not to be working at sea, and at the end of each year I would approach him to discuss returning to California. But Det always beat me to the punch by saying, "Why, Bill, I'm so glad you came in. I've got a wonderful new job for you." And I would agree to stay another year.

One of my jobs was executive secretary of a Committee on Atmospheric Research for the Weather Bureau, which was cochaired by Dr. Carl Gustaf Rossby and included such members as John Von Neumann, a famous mathematician who contributed much to the development of modern computers and the use of them to solve such complex problems as atmospheric motions. Rossby was an energetic, enthusiastic Swedish expatriate, who discovered what are now known as Rossby waves, which control upper-air circulation and was the first to forecast weather for airline pilots.

The committee was set up at the request of the Weather Bureau to take a fresh and deeper look at the application of mathematical modeling to worldwide weather forecasting and the use of radioactive tracers (mostly tritium) and very tiny fallout particles to follow global movements of the upper air. Immediately after a meeting on those subjects at Utrecht, the Netherlands, another international meeting on the International Geophysical Year took place in a palacelike structure in Göteborg, Sweden. The latter attracted many of the world's leading oceanographers, including the big three of the United States: Columbus Iselin of Woods Hole, Maurice Ewing of Lamont, and Roger Revelle of Scripps.

At that time only a handful of nations, mostly northern European countries and the United States, were actively working in the ocean. The official language of the meeting was English, and everyone used it in a good-humored way—everyone, that is, except two Russian oceanographers, who insisted they could not speak a word of it and waited for everything to be translated by a handsome woman who sat between them named Madame Trotskaya.

One day when about fifty of us were sitting around a huge oval table, the subject turned to the disposal of radioactive wastes in the sea. The meeting was informal, and various delegates expressed their personal views on what was then a subject for the more distant future. All was calm until Dr. Robert Dietz of the United States pointed out that the Black Sea is the one large body of water on earth whose deep waters are without oxygen or life and do not mix with surface water. "The Black Sea," he said, "would be a good place to dispose of the world's radioactive wastes."

No translation was needed this time; all three Russians were instantly on their feet denouncing that suggestion. Although they didn't actually say they considered the Black Sea to be a Russian lake and that Dietz was leading a western plot against them, the rest of us got the idea.

One evening Madame Trotskaya joined me for dinner, and although our conversation about ocean matters was quite innocent of politics, she seemed to be a bit nervous about being watched or overheard. Later that evening, as we crossed a very wide cobblestone square in the freezing moonlight, she abruptly stopped, grasped both my arms, and looked at me directly. I wondered what was coming.

"Tell me truthfully," she said with great earnestness. "What is the automobile production of the U.S.?"

Dumbfounded, my memory groped for some reasonable approximation. "I think it's about six million cars a year."

She thought for a moment. "Why do you need so many?"

That was even tougher. I thought of our traffic jams, of the difficulty of parking in major cities, of the crowded used-car lots. I answered that the time of gridlock was not far off but this must have sounded like a cover-up, for after that we lost some of the camaraderie. Each of us thought: There's some clue here about how the American/Russian mind works.

The Göteborg meeting ended in a fantastic banquet that climaxed when a long line of waiters in formal attire, each bearing a flaming brandy pudding, wove in and out between the candlelit tables while an orchestra played "Swedish Rhapsody."

The next day, while Jacques Piccard and I sped across Sweden on the train past frozen lakes and dark forests, he told me of his famous father's inventions. After setting the world altitude records in a balloon of his own design, his father, Auguste Piccard, promised his wary wife he would not risk his life by going up again. Instead he built several "underwater balloons," which he called bathyscaphes, to descend to the bottom of the sea. I was fascinated by the description of these strange craft, which consisted of a pressure-proof gondola slung beneath a balloonlike tank containing a lighter-than-seawater liquid. When we parted in Stockholm, Jacques extended an invitation no oceanographer could refuse: "Why don't you stop and visit our home on the mountainside above Lausanne?"

A few days later, on Professor Piccard's seventieth birthday, I arrived at his house with its panoramic view of Lac Leman and the snowy Alps beyond. August Piccard was slender and straight, exactly 2 meters tall, with flowing gray hair and well-defined features. Although he looked a bit fragile, the inner fire that glows in explorers of all ages showed through. That morning he had broken the thin layer of ice on his swimming pool to take a dip; now he wanted to go for a walk on the snowy mountain roads. So we did, conversing comfortably in an odd mixture of French and English.

"This is an important day for me," he said. "This morning I saw the green flash at sunrise after looking for it for many years. Just over there," and he pointed at the crest of a dark ridge in the snowy Jura mountains. In that unlikely Alpine setting, we talked about exploring the bottom of the sea. Two years before, he had cruised along the bottom of the Mediterranean at a depth of 3,150 meters (10,033 feet). At that time he was quoted as saying, "I am a professor of physics, not an adventurer. I want to show that the bathyscaphe is a dependable scientific device in which the father of a family may entrust himself without anxiety."

A few months later, a brief news item revealed that the *FNRS,* one of Professor Piccard's earlier submarines, while on the bottom at a depth of 4,000 meters off Dakar, West Africa, had seen a large shark. That was new and unexpected. Oceanographers did not doubt the existence of sea life on the deep bottom (the Danish deep sea expedition had brought up a tiny hydra from over 10,000 meters in the Philippine Trench), but few of us had expected to find large animals so deep. This brought home to me how little was really known about life in the deep, and I decided to see what could be done about getting United States oceanographers into deep water.

Captain Frank Lynch, an ex-submariner and special assistant to the Secretary of the Navy, met me at the Army-Navy Club and listened to my thoughts about the need for a deep diving submarine. It was time for oceanographers to go deep and see for themselves whatever wonders might be there; I mentioned that Professor Piccard had since built the *Trieste,* a much better boat than *FNRS;* the easiest way for the United States to get started was to borrow it. Frank listened with growing enthusiasm and finally said, "I'll see what I can do."

A few days later he called back. "I've made a start on the problem."

"What did you do?"

"I wrote a memo from the Secretary to the Office of Naval Research that said: 'Do something about the Bathyscaphe!' and slipped it into his mail. He signed it, and it went off."

ONR's opening move was to invite Professor Piccard to come to the United States and lecture, but he declined, sending his son Jacques instead. Within a couple of months Maurice Ewing, director of the Lamont Geophysical Observatory, served as chairman of a session on Deep Sea Research at the

National Academy of Sciences. There Jacques Piccard gave a lecture on the bathyscaphe *Trieste* to about a hundred assembled oceanographers and showed a short movie of its operation in the Mediterranean. During the discussion period, I proposed that those present should declare themselves in favor of obtaining a bathyscaphe for the use of American oceanographers, quoting Al Vine's earlier remark that the best instrument to have sent on the *Beagle* was Charles Darwin and suggesting that a live observer should be sent to the bottom.

On the spot, Maurice Ewing appointed me chairman of a committee that included Al Vine, John Isaacs, Admiral (Iceberg) Smith, Dick Fleming, and R. Montgomery to draft a resolution. By the next session it was ready. It said, in part, "The scientific implications of having the technical capability to operate manned vehicles at great depths in the ocean are far reaching. We wish to go on record as favoring the immediate initiation of a national program aimed at obtaining such vehicles for the United States." The assembly concurred enthusiastically.

Within the next few years, the *Trieste* came to the United States where, after being fitted with a deeper sphere, it was piloted by Jacques Piccard and Lt. Don Walsh, U.S.N., to the bottom of the Philippine Trench, the deepest part of the ocean. Although the academy's Undersea Warfare Committee had repeatedly suggested that the Navy build an experimental deep submarine, as distinguished from an underwater balloon, it was this odd chain of events that seems to have triggered the deep submarine program in the United States.

WOODS HOLE

During the summer of 1956, the Undersea Warfare Committee convened a study group something like the mine-countermeasures study of 1950, at the Whitney estate, Little Harbor, at the base of Point Nobska in Woods Hole, Massachusetts. The estate consists of a beautiful, old gray house with many huge rooms, and acres of grass that slope down to the water. One of the advantages of this location was that it was close to the Woods Hole Oceanographic Institution; Columbus Iselin, its famous long-time director was made chairman of the study known as Project Nobska. John Coleman of the National Academy of Sciences was the organizer and driving force behind the scene.

Columbus was a big guy, 6 feet 4, raised around the water by a wealthy family, a sailor from the age of ten. The names of his boats give some clue to his personality: the *Sponge* (because it leaked), the *Chance*, and the *Risk*. No illusions there about the sea. He went to Harvard to study mathematics, thinking that it had something to do with banking, the family business, and came under the influence of Professor Henry Bigelow, a zoologist cum oceanographer. On graduation Columbus took a crew of college students with him on the *Chance,* a 72-foot schooner, to study the Laborador current. When he returned with useful data, Professor Bigelow was pleased to learn that good oceanography could be done by such a small, inexpensive ship.

Partly as a result of Bigelow's report to the National Academy of Sciences, the Rockefeller Foundation gave him $3 million to outfit and endow the Woods Hole Oceanographic Institution (WHOI) in 1930. One of the first steps was to have a research ship, the *Atlantis,* a 142-foot ketch (foremast taller) constructed in Copenhagen. Columbus sailed her home across the Atlantic; then, for the next 10 years skippered her on dozens of cruises, as its scientists studied fundamentals of the ocean's temperature-salinity structure and underwater sound.

WHOI's few scientists of the 1930s, mostly shared with MIT and Harvard, comprise a *Who's Who* of the great men in oceanography at the beginnings of its golden age. They included Norris Rakestraw, chemist, Alfred Redfield, biologist, Carl G. Rossby, oceanographer, Henry Stetson, geologist, Selman Waksman, bacteriologist, and Captain Sir Hubert Wilkins, under-ice explorer, all of whom had substantial influence on those who followed.

Columbus Iselin became director of WHOI in 1940, and with Commander Roger Revelle, U.S.N., had much to do with interesting the Navy in using oceanography to develop its SONAR capability by understanding how sound travels in the ocean, a very complicated subject. He was a solid, common-sense director. On one occasion one of the institution's power boats was "borrowed" by an unknown person, who ran it without oil and burned up the engine. Columbus said, "If it was somebody from the Institution, he ought to be fired. But if it was somebody from outside who figured out how to start that engine, he should be hired."

The subject of the summer study was antisubmarine warfare, bringing university scientists together with Navy scientists and Laboratory directors to talk about new ideas for finding and destroying enemy submarines. Our European supply lines had been seriously threatened by them in two major wars; no one wanted to risk that a third time. Submarines had created the need for convoys in World War II, but the nuclear weapons that ended that war appeared to have made convoys obsolete. At least a convoy would be a clear military target for a nuclear weapon if civilians were to be spared. But no one could be sure.

This was a time of transition; underwater acoustics was developing rapidly as oceanographers learned more about how sound pathways in the sea were influenced by ocean conditions. Hydrophone arrays and computer processing of signals were coming into use as the SOSUS system developed; its intent was to track submarines in spite of the steps being taken to reduce submarine noise. Nuclear weapons made large ships and convoys more vulnerable, at the same time that the missions of ships and aircraft were changing. Intercontinental bombing was getting the attention and the money; ballistic missiles would soon be available. The Navy was groping for new directions and new missions.

There were two new and as yet unevaluated developments in submarines. The first nuclear-powered submarine, *Nautilus,* was now operational. It was noisy, and it used an already outdated type of reactor, but it demonstrated that a boat with nuclear power could range far from its base, remain below for a long time, and run under the ice. Once, while some of our group was at sea on a destroyer during a submarine-hunting exercise, the *Nautilus* ran under us without being detected and insultingly released a flare to show we could have been torpedoed.

The second was the *Albacore,* the first submarine built in the shape of a figure of revolution (symmetrical about a center line) with a single screw. This had the great advantage that the water velocity past the propeller was the same on all sides, thus reducing its noise output. The *Albacore* was the first submarine not built to operate on the surface; it had no diving planes and had to be "flown" like an airplane. The result was a 35-knot boat, twice as fast as any predecessor. Even more remarkable, it was classed as an experimental boat; it carried no armament and was not attached to a fleet.

If a warship couldn't detect the noisy *Nautilus,* what chance would it have against a really quiet boat like *Albacore?* The scuttlebutt was that a boat that would "marry" the capabilities of the long-range, long staying power of the *Nautilus* with the quiet, efficient *Albacore* was already on the drawing boards. Its mission still would be to fire torpedoes at other ships or submarines.

The office I used, once a bedroom in the Whitney mansion, was shared with a colorful admiral named Red Ramage. As a submarine skipper Red had won the Congressional Medal of Honor for having surfaced the *Parche* in the midst of a Japanese convoy at night and fired torpedoes in all directions, sinking at least four ships. The Japanese, not knowing what or where the enemy was in those days before they had radar, fired at each other and damaged several of their own ships. Vice Admiral Ramage, a sort of naval counterpart of Columbus Iselin, who later went on to command the Pacific fleet, held a very pragmatic view of how the new developments could be applied to naval operations.

On warm summer days, Professor Isador Rabi of Columbia University and I often met for lunch; we would sit on the pier eating sandwiches, our legs dangling over the water, and talk about practically every subject known to man. Rabi was a Nobel Laureate in physics who had been a senior adviser to presidents on nuclear weapon development, and he still retained the imagination, enthusiasm, and curiosity of an eighteen-year-old kid. His hero was Howard Hughes who, he said, had solved man's basic needs for money, girls, and travel by inheriting Hughes Tool Company and then buying a movie studio and an airline; he recommended that I do the same.

One of the many secret briefings related to the Navy's proposed strategic missile program. We were told about a plan to install Jupiter-type rockets on 20-knot, Mariner-class cargo ships. These would be stationed in the Arabian Sea, where they would be able to fire missiles over Iran and Afghanistan at targets in the Soviet Union. The idea that the Navy would consider using unreliable missiles on unprotectable ships in a remote sea was an indication of the uncertainties about where new technology would lead. When asked why the Navy did not plan to use submarines as missile platforms, the answer was that no one in his right mind would consider taking liquid rocket fuels aboard a submarine and that no sub had yet been designed with a large enough diameter to hold a rocket in a vertical position.

The Navy has long used Mercator projection charts of the earth for strategic planning. They are fine for low latitudes, but when they flatten the spherical earth onto a cylinder the size of the equator, the north pole becomes a line across the top. When your principal enemy is a northern power like the USSR those charts make it appear much larger than it really is, and they hopelessly distort the Arctic region. So I obtained a 6-foot-square polar projection of the northern hemisphere (north pole in the center) and nailed it to the office wall. This was regarded by some as a subversive act because it had been made by the Navy's real enemy, the Air Force, but it attracted a number of scientists into our office to recheck the distances between our northern shores and the Soviet Union, and reconsider the Arctic approaches.

A polar view of the earth helped emphasize the possibilities of submarines as strategic weapons. In our regular afternoon seminars I quoted Waldo Lyons, the Navy's technical expert on ice matters, on the thickness of the ice in the Arctic Ocean and the White Sea (it averages about 6 feet) and pointed out that a submarine could surface many places in those ice-covered seas.

That summer's work resulted in a series of recommendations to the Navy, but the most important one was "Go for missile-firing submarines. Even if present solid-rocket propellants will only throw a missile a few hundred miles, better ones with much longer range will soon be along. Don't worry about the details of missiles because they will continue to change. Start building larger submarines with standard-size missile tubes (as you standard-ized torpedo tubes at 21 inches in 1912) and push ahead." That basic idea was accepted by the Navy, and the result, before long, was the Polaris program, which added missile firing subs to strategic forces.

Navy matters had been taking up a large part of my time; in addition to the summer study, I had spent months writing position papers on the future of the Navy for Nelson Rockefeller and his chief of staff for special studies, Dr. Henry Kissinger. Military applications of new findings in the sea are an important part of oceanography, but they are often classified secret; I felt this restricted my independence. Having just reached the advanced age of forty and being unsure of how many more years of life were left, I became restless again and decided to take time off to smell the flowers, accompanied by Rhoda and Anitra.

We sold our possessions in Washington to finance the holiday and left for Tahiti, where Scripps gave me a three-month job helping install wave recorders for the International Geophysical Year (1957). As we were leaving I received a letter that said: ". . . Hope we stay in touch from now on. If you ever need an assistant in the South Sea Islands, please consider me. Henry Kissinger." I would fly out with Bill Van Dorn; Rhoda and Anitra would follow in a couple of weeks on a New Zealand freighter.

RURUTU AND TAKAROA

The best way of getting to Tahiti in 1957 was via a Solent flying boat that left Fiji every other week for a 2,000-mile mail run to the east. These huge planes carried passengers in style on two decks. They would take off from Suva in a roaring cloud of spray and once aloft would cruise leisurely at a low altitude above green and white atoll rings, gems in the midnight blue ocean. Plane and passengers would rest overnight at Apia in Western Samoa, and early in the morning would continue on to Aitutaki in the Cook Islands to refuel from drums on a raft while all aboard paddled ashore for breakfast on a deserted beach.

Early in the afternoon the plane would fly low along the shoreline at Papeete to announce its arrival before landing in the harbor. At that signal the islanders piled onto their bicycles and headed for the old green post office on the waterfront. By the time the plane had come to anchor and lighters had brought the new arrivals ashore, a crowd of young vahines gowned in bright pareau cloth were waiting on the quay to welcome these strangers with wiggling hips, girlish giggles, and flowers behind the left ear. Two weeks later, when the no-longer-strangers left on the next plane for Fiji, the girls would cry with desolation for a moment or two, put another flower behind their left ear, and begin again with the new arrivals. Neither love nor work were treated seriously in Tahiti; somehow one's needs would be met.

Bill Van Dorn was the scientist responsible for installing special tide gauges and long-period-wave recorders on several islands, and I was one of his three assistants. After he and the others returned stateside, I would concentrate on relaxing and writing. We had just arrived and begun adjusting to the easygoing pace of the Tahitians when Bill learned that a French warship would be leaving for the Austral Islands in a few days and would drop us off on Rurutu, a small island 300 miles to the south.

The *Dumont d' Urville* was a corvette, something like a large destroyer; it had served its time in Vietnam during the French Colonial period and had a cluster of battle flags painted on the bridge, but it had been stationed in French Oceania long enough for its crew to have acquired the local attitude toward life. Besides the Parisian cooking, there was always wine at the dinner table and brandy after dinner; to us it seemed more like a cruise ship than a warship, even if my assigned bunk was the floor of a lieutenant's tiny cabin.

Once underway the captain unexpectedly ordered all hands to battle stations for antiaircraft practice. The crew seemed astonished by this odd request, but with French shrugs began to take the canvas covers off strange-shaped objects on the deck. First they uncovered a long-forgotten table-soccer game; then they found the antiaircraft gun.

A large red target balloon was inflated while the gun was loaded and pointed skyward. On command the balloon was released, but as soon as it cleared the ship, it settled onto the water. After a moment of consideration, the order was given, "Fire, anyway," the gun was cranked down, and shells threw up spray in the general vicinity of the balloon. But it was quite safe, and after a few minutes was left bobbing in our wake. The gun was covered again; the exercise was declared a success, and the crew began playing the soccer game. Such is the effect of the South Seas.

The opening in the reef at Rurutu was barely large enough for the whaleboat to enter. The captain, rigid in dress whites, stepped off onto the small concrete pier to be welcomed by the smiling and bowing mayor who wore his tricolored sash of office over a white suit. A mob of shining-faced white-clad school children sang "The Marsellaise," everyone saluted the Tricolor, and a little girl placed a lei around the captain's neck, who then made a speech about the importance of this island to France. This was standard procedure.

Rurutu's principal village consisted mainly of a main street paved with grass bounded by low white-washed walls behind which were neat, old-fashioned wooden bungalows. There were no electric lights, very little plumbing (with the exception of a single water faucet in each front yard), and no autos. The only radio on the island, owned by the Chinese storekeeper, was powered by a generator driven by a tiny water wheel in his kitchen sink. This was one of the legendary missionary islands where, on Sundays, the girls would put on dark blue skirts, white blouses, broad straw hats, and shoes to march sedately into a nineteenth-century wooden church. There they would sit primly on the opposite side of the aisle from similarly attired boys.

On weekday evenings, the populace met in religious clubs to discuss biblical matters such as the type of outrigger and fish spear that Jesus used on the Sea of Galilee. On those occasions the principal speaker was required to wear shoes and a jacket as a symbol of dignity while the audience sat or slept on coco frond mats on the floor.

The four of us outsiders rented a house for a couple of weeks and hired a young girl named Rosalie to cook and keep house. She was inevitably accompanied by a couple of male relatives who acted as chaperons. These fellows perched on a pink-slatted park bench in our parlor watching our every move, pointing, laughing, and discussing our strange mannerisms in Tahitian and describing them with gestures to other villagers who stood in a ring outside the open window. We were celebrities.

Installing the tide gauge was not a major job. It consisted of a large vertical pipe with a small hole in the bottom so the sea could easily leak in and out, secured to the side of the pier. A float inside the pipe was connected by a wire to the recording mechanism mounted in a weatherproof box on its top. The local gendarme agreed to tend it and send the records back to Tahiti on the monthly schooner. The purpose of this instrument was not so much to measure tides for their own sake as to obtain more information about the seasonal shift in average sea level between northern and southern hemispheres. Eventually it confirmed that average sea level is about 15 centimeters higher in summer. Once it was in and working, we took time to look around more carefully and discovered that a school of lion fish, famed for the painful venom in their dorsal spines, lived around the base of our pipe where we had worked barelegged.

A week later the *Dumont d' Urville* returned to pick us up. It had been on a rough and stormy trip to Rapa Island, the southernmost island of French Oceania, where heavy seas had bent her rails and carried away some deck hardware. Once we were aboard, the sea flattened into a long swell on our beam, and the slender ship rolled hard. One soon grows to be part of a small ship; it becomes the reference point around which the universe moves. In the long evenings on deck, with the choice of watching old French movies or the night sky, we would wonder how the stars could swing so rapidly back and forth across the night sky relative to our fixed mast.

The next project was to install a tsunami recorder on Takoroa, the pearling island some 400 miles east of Tahiti that the *Baird* had visited on the Capricorn expedition 5 years before. Van Dorn made reservations for four of us on a trading schooner, the *Viatare,* and although our expectations for a comfortable trip were not high, we were dismayed when we inspected the accommodations. *Viatare* was a stout wooden ship about 120 feet long with the main cabin aft; ahead of it, between the masts, the ship's waist was buried under freight and green bananas.

The mainsail was furled and a tentlike tarpaulin stretched tautly over the boom to form a low tent above the cabin roof. There was a glint in the mate's eye when he assigned me the number one position; mistaking this for a friendly sign, I thanked him profusely. This meant the first sleeping place atop the cabin on the port or downwind side, but I did not recognize the implications of that until we were well underway. The remaining nineteen positions on the cabin roof were inhabited by my associates and an assortment of Polynesian and Chinese passengers.

The main salon had six curtained bunks along one side, which were occupied by a gendarme and a missionary with their families. There were also two tiny private cabins for the captain and the ship's engineer. The remaining open space was used for a galley and a dining table with eight places around it. Our stomachs quickly calculated it would take five shifts to feed everyone and groaned with the expectation of long waits for every meal.

Amid cheers and tears *Viatare* pulled away from the wharf at Papeete, cleared the harbor, and turned east in a reasonably flat sea. The myth that Polynesians are natural seafarers collapsed at once; some of our shipmates were instantly seasick. Others held off until the schooner rounded Point Venus and headed directly into the steep trade wind seas. Then it began to buck, shoving its bow into almost every wave and flipping the water on the foredeck upward and into a wind that blew the spray back over the rest of the ship. The sail on the foremast served mainly to steady *Viatare* against rolling; most of the motive power came from a diesel engine, and our net forward speed was only about 6 hours. In an hour or so the captain got seasick and disappeared into his cabin, leaving only the mate, the engineer, and the four Americans still mobile. The problem of waiting for food was solved.

The cook, also sick, steadied himself against a stanchion next to the stove and produced excellent meals that he was unable to eat. Our noisy table chatter would cause the gendarme and the missionary to peek out through the curtains of their bunks occasionally, smell the food, turn green, and withdraw again.

On deck the mate, an oversized but happy Tahitian in faded blue shorts and a tilted straw hat, lounged on a hatch cover and steered by turning the wheel with huge bare feet, all the while telling ribald stories in French. When it was time to retire, he suggested that we, who were assigned space on the steep sloping side of the varnished cabin roof, lash ourselves to the boom so we would not slide overboard in the night. Sensibly we took his advice. As the lucky holder of the number one position at the forward edge of the cabin, it was my privilege to screen those behind me from the flying spray and the concentrated diesel fumes.

The hard, sloping surface, the tether under my arms, the smell of diesel, the ship pounding, and the splash of wind-driven saltwater that struck me every few seconds made the cabin roof a seemingly impossible place to sleep. But rarely did it take more than two minutes to reach dreamland.

Two days later *Vaitare* made landfall on a fuzzy green strip of palms; in another hour a thin red line resolved into the roofs of Takaroa village. In the lee of the atoll the sea flattened; guitars and holiday clothes came out, and the captain appeared on deck again, shouting orders in an extra loud voice to cover his embarrassment. The ship slipped into the channel through the reef, and by the time it tied up, there was a line of local kids balanced on the bowsprit, eager to dive from its tip.

It was the first of March, the end of summer south of the equator, and the pearl divers were arriving. Pearling season brings excitement and entertainment, traders and camp followers; the village population triples to 600. Within a couple of days a whole new crop of buildings sprang up between the permanent ones, buildings that would shelter a pool hall, merry-go-round, and movie theater as well as become homes for divers and their families. All are made from a pile of used two-by-fours and a stack of rusty corrugated sheet iron lent by the Chinese store to those who would be its customers. And of course, there was an unlimited supply of palm fronds furnished by nature to be woven into walls and roofs.

During these first days of the new boom town, when living facilities were overloaded and even drinking water was scarce, we three Americans and our French interpreter were graciously provided with a wooden bungalow.

The diver's day begins at dawn when the world is still and conversation is by whisper. As the sun tops the palms on the eastern rim of the atoll, the outrigger canoes are lowered from their low stilts into the still water and pushed toward the channel. There four or five at a time will pick up a line from an outboard-propelled boat for a tow out into the broad lagoon.

We began at once to install Bill Van Dorn's special long-period-wave recorder on the reef along the edge of the village. This device measured the height of the water at any instant relative to the average height of water over about 15 minutes. Since the pressure sensor had to be underwater in a protected area, it was necessary to dig a pit in the coral reef near where the recorder would be located. I was appointed principal excavator, assisted by a sullen, much-scarred fellow named Tautu, lent by the gendarme who was holding him for unspecified crimes. He and I toiled skin-to-skin with pick and shovel for several days to make a hole some 8 feet across and 4 feet deep in the hard coral.

On the day our excavation was finished, Tautu was rearrested for stealing a wallet from a passing yacht. He was immediately put on trial, and it came out that he was a homosexual from a distant island who had killed a man in a knife fight. In the midst of his denunciation, the fellow dramatically dived through a courtroom window and escaped. From then on his uncertain presence on the small island terrorized those who had testified against him, including M. Labourie, our aged interpreter.

About the time Van Dorn's instrument was ready to record passing tsunamis, the *Marco Polo,* a small sailboat carrying two New Zealand adventurers around the world, tacked up the channel and tied up at the quay. Tig and Tony had come to see the famous pearl divers and the wreck of a great ship. When they had satisfied their curiosoity about the pearling, we sailed with them across the lagoon and anchored off an idyllic palm-covered islet on the northern rim. On its seaward side a sandy beach sloped down to a wide, hard, flat reef that was mostly dry at low tide. This beautiful desolation of blinding white sand and coral seemed designed to focus one's attention on the spectacular rust-red skeleton that loomed above us, the remains of the *County of Roxburgh.* In the great hurricane in 1906, it had gone up on this, the northernmost atoll of the Dangerous Archipelago.

The *County,* a 286-foot-long four-masted ship, had been over 50 miles north of Takaroa, heading west, when the gale struck. The sails were soon ripped away and, rolling heavily, the unmanageable hulk was driven south in spite of the crew's best effort to rig new sails and bring it round to the port tack. Just before dark they could see breakers on the reef ahead, and an attempt was made to lower the boats; one of these was upended, dropping a couple of men into the raging sea; the other was smashed by the waves.

With a tremendous crash the *County of Roxburgh* struck the outer edge of the reef, followed by periodic screeches of steel against coral as the ship was driven landward by the impacts of successive waves. A brave fellow named Wagner volunteered to swim toward the palm trees with a rope around his waist, intending to rig a breeches buoy through "breakers over 30 feet high." But "the backwash of a big comber sucked him back under the ship's bottom," and when his line went slack, those on deck hauled in a frayed end. The captain ordered the others below, because of the danger on deck that masts or yards would fall. In another hour the wreck had worked closer to shore, and the mate offered to make another try at swimming a line in. This attempt was successful, and as he reached shore, some natives appeared and helped him rig an aerial escape route with which the remaining crew were dragged ashore through the breakers over the rough coral.

Next morning the storm was over; the ship with all its masts and spars still in place lay on the beach where it remains today. The banged-up survivors concluded that, if everyone had simply waited aboard overnight, all could have stepped ashore without the loss of ten men. Half a century later, we fantasized on that storm as we stood inside the rusty red rib cage looking at the unobstructed length of the hull at bars of sunlight filtering through. The *County*'s builders would have been proud to know the main structure of their ship was still intact and unwarped half a century later.

We did not know it when our party left Takaroa, but our work there was all for naught. Either the long-period-wave recorder never recorded a tsunami or, if it did, Bill Van Dorn never got the record. Several months later Bill learned that the old Frenchman who had been left to tend the instrument had such poor eyesight that he could not tell if the device was working or even if the batteries were charged. So he was replaced by a beach bum who had good eyesight but apparently never used it to look at the recorder. Neither of the two ever sent in records anyway, and finally a hurricane in January 1969 leveled much of Takaroa village and wiped out the installation.

TAHITI

Rhoda and our daughter Anitra, aged eight, eventually arrived in Tahiti on the freighter *Waitemata;* we rented a beautiful new house made of woven bamboo and began learning the Tahitian life-style. Our house looked a little like a sheep dog with window openings that peered out from under a thick shaggy roof of coconut fronds. It was sheltered by coconut palms, surrounded by an acre of grass, and ended against a stand of huge ironwood trees along a private black sand beach. A stream along one side of the property flowed into a lagoon about half a mile wide between the dark volcanic shore and the white coral reef. The place was surrounded by a high, dense hibiscus hedge that constantly leaked Tahitians, as well as their kids, dogs, and chickens. These outsiders became part of our extended family and would occasionally give a cheerful "Iorana" as they entered our doorless house to chat or stare curiously at our strange ways. If we were away, they might borrow some utensil that would reappear a week or so later.

The purpose of taking a year's holiday was to get more time to read and write so, to make the most of it, I was up with the first light each day, sitting reading under the ironwoods by the water's edge with a cup of dark coffee and a chunk of long slim bread rescued from the mynah birds. Later there would be time to bicycle a couple of kilometers down the road to James Norman Hall's house on Matavai Bay where the *Bounty* had anchored, and exchange books with his library.

Sometimes when it was still dark, the old lady next door would scratch on the bamboo wall to wake me, and we would launch a canoe and paddle slowly out on the calm lagoon that rings the island. There she would harvest sea urchins on the shallow reef whose vana (eggs) could be sold for a few franks to hardy islanders, who seemed to enjoy the intense flavor of iodine.

When the sun was a little higher, the girl in the house beyond the hedge would pile her Tahitian records on the phonograph so that all day long rhythmic songs and throbbing drumbeats would waft through the neighborhood. Often we would bicycle to Point Venus past the wonderful smell of a vanilla plantation, to the place where Captain Cook and his assistant, Lieutenant Bligh of *Bounty* fame, had measured the transit of Venus across the face of the sun for the Royal Society of London. It was a calm and lovely life, every day ending with Rhoda and me sitting under the tamaracks drinking a rum punch and watching the sun set over Moorea.

The subjects of the stories I wrote there were the people of the five island groups of French Oceania (Society, Tuamotus, Isles Sous le Vent, Australs, and Marquesas). Many of these were about events long past, and this required a good deal of time in the local museum digging out historical material. The local newspaper, the *Messager de Tahiti* which has been published continuously since about 1840, was a good source of information. It was filled with news of whaling ships in the days of sail, naval battles in the outside world many years before, and long voyages in small boats. One day while I was gingerly turning the fragile, dog-eared pages of the copies of the archives, Aurora Natua, the museum director, said, "Why don't you buy new copies of the ones you're interested in."

"But this issue is April 1858, and the article is about the strength of the Union and Confederate armies in the U.S."

"It doesn't matter," she said, "you can still buy the old issues at the newspaper office. Five francs for any date."

Not really believing that was possible, I checked it out the same day by asking for several dozen copies of various nineteenth century issues, containing maritime stories of special interest. The clerk, surprised that anyone would prefer an old paper to the one that had come out the previous day, climbed a ladder by a wall of cubbyholes, each filled with a packet identified by the year, and found nearly every issue requested, many of them a century old.

There was much to do. A couple of young Chinese Tahitians who operated a store persuaded Joe Gores, a writer, and me to join them in the walk to the top of Mount Aorai, the high peak in the center of the island. "Only seventeen kilometers," they said, "and there is a road part of the way. If we start early, we can be back by dark." The road lasted for 2 kilometers; after that we climbed along a narrow path on the knife-edged crest of a divide between two deep valleys. It was easier than walking a tight rope, but one step off the trail on either side would have meant a long fall through very steep jungle. A thousand meters below on either side a glittering thread marked a stream.

As the sun began to set, we came to a small rock shelter. Here our Chinese friends seemed triumphant. "This is seventeen kilometers. We are more than halfway."

"We thought you said the top was only seventeen."

"It's really twenty-seven, but if we'd told you that, you might not have come with us. You'll be home by dark, tomorrow."

We all laughed at this Polynesian-style joke and then huddled shivering and hungry through the night. Before noon next day we reached the top, still following a succession of sharp-edged crests with a precipice on either side, occasionally climbing steep rocky outcrops. Although the summit is only 2,650 meters (8,694 feet) not many Americans or Chinese had been there before us. From it one can see down through the clouds to the blue harbor of Papeete, ringed by red tin roofs and white surf, all very tiny and far away. We scratched our names in an old French logbook stored in a biscuit tin under a rocky cairn and began a painful descent, limping home with huge blisters.

Then there was a trip to Makatea, an actively mined phosphate island, on the old trading schooner, *Oiseau des Îles.* It was like a trip in time to the beginning of the century. Makatea is a raised atoll, the opposite of Alexa Bank. Some crustal disturbance perhaps a million years ago raised it 40 meters vertically, so that the broad surface that was once a lagoon bottom is now a mid-Pacific mesa, surrounded by vertical limestone walls. On the downwind side, there is a beach wide enough for a small village of permanent residents. What makes this village different is that halfway up the cliff, immediately behind the wooden bungalows, there are caves in the raised coral where generations of dead are buried. Coffins, tilting at strange angles, are visible from the main street.

A cog railway runs to the top of the mesa, where a small town of tin shacks and offices built by the mining company sits amid a huge area of white coral pinnacles that look like headstones in a surrealist graveyard. The phosphate that once filled the openings between these spires to a depth of 5 to 10 meters has been removed by workers, who shoveled it into small buckets that were hoisted by hand and dumped into tram cars for transfer to a stockpile. Beyond the stockpile a conveyor belt moved the phosphate from the top of the mesa down to the edge of the reef, where a steel tower had been erected. Cables from the tower supported a long swinging arm that extended the conveyor belt out over water deep enough for ore carriers to moor under the belt's discharge.

Sailing back to Tahiti, the skipper of the *Oiseau des Îsles* talked to me about sailing in these complex island-dotted waters. He had studied modern navigation at the French merchant marine academy at Marseilles, but he said, "I steer by mana," and in the bright moonlight a mystical look came over his face, "as my father did. That is the way the ancient ones sailed. It comes from a strong belief in the gods of the sea. They give you the mana, the instinct, to find your way. I can locate any atoll in the Tuamotus at night and make a landfall opposite its village, because I believe." The owners of the ship might not have slept so well if they had suspected that he tested the power of his mana with their ship.

Out of the dark one night and into our brightly lighted house stepped two expatriate American neighbors, Joe Gores and Earl Nesbit. "You've screwed up again." was the opener.

"Who me?"

"All American scientists. Didn't you hear the news?"

"I guess not. What happened?"

"The Russians have launched a satellite. A sputnik they call it. It flies around the earth every couple of hours, beeping."

"And what did I have to do with that?"

"Well, you let 'em do it first. It's very embarrassing to have them beat us into space. What'll we tell our neighbors?"

Their neighbors were unsophisticated Tahitians like ours, but these barefoot people understood the impact of this Russian achievement on the outside world. It was important to them, and it played on our minds. I felt guilty, as though I should have been home helping someone do something about it.

Bora-Bora has at its center a high spine of volcanic rock with flanks that slope downward to a lagoon protected by an offshore coral reef. It was Charles Darwin's example of the third stage in the formation of an atoll, and a visit there was something of a pilgrimage. Rhoda, Anitra, and I traveled there on the trading schooner *Orohena,* and disembarked on a wooden pier piled with sacks of copra and swarming with easygoing Bora-Borans in bright cloth and plaited hats. We were the only outsiders on the island; there was neither a gendarme nor a hotel. However, a kindly fellow named Moetu offered us his new sleeping house for a couple of weeks. With some embarrassment, he said we would have to pay him for its use and suggested a sum, in francs, that amounted to about 75 cents a day. His family would sleep in the kitchen house, and we could use their fine beds with mosquito net curtains. All of us would share the outside shower, a barrel atop an enclosure of corrugated iron. The luxurious accommodations included woven bamboo outhouses balanced on a fallen tree over the lagoon.

Freely we roamed the beautiful island and lagoon, our only responsibility being that we, the only customers on the island, had to inform the proprietor of the only restaurant what we would have for supper that evening. The choice was simple, because all he had was steak and lobster, and the daily cost of food for all three of us was $1.50.

We paddled out to Moto Tabu, the tropical islet of everyone's dreams, bicycled around the island on an overgrown road pocked with crab holes, and I was the official photographer at a formal wedding (long lace gown, tuxedo, bare feet). Moetu led us up a steep jungle trail, through ferns with fronds that met far above our heads, to the top of the highest peak. Then later he demonstrated sailing a Bora-Bora-style outrigger canoe. These are quite different from those used by the Marshallese, being steered by means of a broad, short-handled paddle that is held against the stern by forces that seemed to defy the laws of hydrodynamics. When the breeze was lower than about 10 knots, the boat would slog through the water, but above that critical speed this sculpted log would rise up and skim the surface.

The islands of Oceania were a special heaven for anyone interested in the lore of the sea, and we might have lived out our lives there except that we were broke. While we were pondering what to do next, a cable arrived from John Coleman at the National Academy of Sciences: "Come home. Have new job for you."

Reluctantly we gave up our bamboo house under the palms on the black sand beach and began adjusting our minds to the less civilized society of Washington. We would miss the scent of vanilla, the drumbeats in the dark summoning dancers to practice for the fête, the sunsets over Moorea, the melodic himene, the mynah birds on the kitchen table, and all the things that made Tahiti more a dream than a place. The evening before leaving, there was time for one last visit to the grave of James Norman Hall on the hill to read the bronze tablet:

> *Look to the northward, stranger,*
> *Just over the hillside there.*
> *Did you ever in all your travels*
> *See a land more passing fair?*

As the small boat ferried me out to the flying boat through waters covered with the leis one throws away on leaving, I cried like a good Tahitian and swore to return.

That night the plane stopped at Apia, Samoa, and its passengers were put up at a hotel not far from the large blue-roofed governor's bungalow where Robert Louis Stevenson had once lived. Behind it a steep hill a few hundred feet high without even a trail to the top is surmounted by Stevenson's grave and his final verse, also recorded on bronze:

> *Under the wide and starry sky*
> *Dig the grave and let me lie.*
> *. . . . .*
> *Home is the sailor, home from the sea,*
> *And the hunter home from the hill.*

A light rain fell on the plain concrete vault, and twilight came as we pondered the values these two great writers had found in their Polynesian islands. They liked the calm, simple, unhurried life. So did I, for a while, but there was no chance to compete, to swim in the fast-moving mainstream of ocean science and technology. It was time to get on with my education.

HOME AGAIN

On my return, one of the jobs I took was scientific representative of the National Academy of Sciences and the American Association for the Advancement of Science on the production of a new CBS-TV science series called *Conquest.* This was an attempt, sponsored by Monsanto Chemical Company and hosted by Eric Sevareid, to make science interesting and understandable to the general public; it was a small piece of the new effort in science education that was part of the American reaction to the launch of Sputnik I.

The forty or so segments *Conquest* filmed covered nearly every aspect of science from open heart surgery to volcanoes and from animal imprintation to weather. The wide scope of the program required me to study new subjects and talk with their key scientists every week; it was a wonderful education.

As a responsible representative of the two leading scientific organizations on a unique television program about science, which was intended to influence the thinking of millions of people, I spent some time thinking introspectively about the nature of science and of the probable audience.

Science is the systematic investigation of the natural laws that explain the operation of our universe. Its objective is understanding and its rules require that any new theories put forward be accompanied by documented evidence so these can be reconfirmed by anyone who wishes to do so. Scientists do not accept anything on faith, which means the belief in something for which there is no proof.

The audience we especially wanted to reach was bright young high school students who were still undecided about how to spend their lives. *Conquest*'s goal was to interest them in some branch of science, preferably math or physics, so they would think about a career in those fields. Plainly our society needed more science-minded people to compete with the Russians and Germans and Japanese.

I was dismayed to learn that over one third of the U.S. population believes in such things as supernatural beings, ghosts, angels, devils, space aliens, numerology, ESP, miracles, and spells. It would be hard to explain natural laws and the need for scientific proof to that crowd.

Conquest's stories were about scientists with interesting personalities who had made recent advances in their specialty. The program showed how they collected the evidence and let them describe their work and findings in an understandable way. There was no way of measuring the effectiveness of the program in recruiting students, but we certainly tried.

My second job was as executive secretary of the Advisory Committee to the Maritime Administration, of which Vice Admiral Edward Cochrane was chairman. Ned Cochrane had been head of the Navy's Bureau of Ships during World War II and had been personally responsible for the rapid development of amphibious ships and small craft that were vitally important in a war where large numbers of men fighting had to be moved ashore across the beaches of remote islands. After retiring from the Navy, he headed the Maritime Administration and then had gone on to become dean of engineering at MIT. He had the ability to cut through masses of confusion to get to the guts of a problem, reduce it to basic parts, and assign people to solving each piece in short order. Soon a very illustrious committee of experts on shipping, harbors, and trade was assembled to advise Mar-Ad on how to set up a research and development program.

On a stormy day in 1958, the committee visited Hoboken, New Jersey, to see a new cargo-handling system. Sheets of rain blasted across the water and the wide, paved area beside the quay. Virtually every other operation in New York Harbor was shut down and the hatches covered, but the McLain Company's operation continued as usual. Semitrailers, each carrying one large sealed container, would stop alongside an unusual-looking freighter with open holds. A rolling cranelike mechanism on board would lift the container from the truck, drop it in a slot in the hold; then it would remove another container with goods headed for the New York area and set it on the truck for delivery directly to its destination. Three "gangs" of longshoremen, on what would otherwise have been a five-gang ship, swapped all the incoming cargo for the outgoing cargo and "turned the ship around" so it could sail again in about 8 hours.

As our group huddled under a corrugated iron awning, there was no doubt in our minds that we were witnessing the onset of a freight-handling revolution. Shipping time would be greatly reduced; the cargo had been made relatively immune to breakage, theft, and weather; the cost of stevedoring would be halved. The dozen small container ships then operating have since developed into fleets of container ships, many of which are larger than 40,000 tons.

Not long after that, there was a tugboat strike in New York Harbor; without tugs it was difficult and risky to bring large ships alongside a pier under their own power. This was especially true of the big passenger liners, which docked at the 51st Street pier in Manhattan. The committee asked me to look into alternative means of bringing ships alongside a pier, and I began a study of auxiliary maneuvering devices.

Having considered the possibilities, I reported back that there were several ways in which a ship could apply power to push itself around other than using its main propellers and rudders. These include active rudders (a small auxiliary propeller mounted on the rudder), bow thrusters (a crossways tunnel containing a propeller located near the bow), auxiliary trainable propellers (like large outboard motors), and vertical-axis or cycloidal propellers (four blades pointing downward under a ship whose attack angle can be changed as they rotate to exert thrust in any direction).

The tugboat strike was settled in a few weeks, and no ship owners installed any of those devices to avoid using tugboat services, but the information derived from that study was to be of great value in another project that was just about to begin.

My office at the National Academy of Sciences was adjacent to that of Bill Thurston, a lanky geologist who was executive secretary of the Earth Sciences division. He would occasionally drop by to bring me up to date on events that had transpired while I had been in Tahiti. Bill thought I might be interested to know that informal consideration was being given to drilling a deep core hole through the crust of the earth beneath the ocean—if that were possible.

I was very much interested.

VIII

The First Deep Ocean Drilling

THE IDEA OF DRILLING

The National Science Foundation, a government agency, supports our nation's basic research by contributing funds to projects proposed by accredited scientists. In the spring of 1957 a panel of experts it had appointed to review proposals in earth sciences adjourned with an air of mild disappointment after having reviewed some sixty-five ideas. The difficulty was that, even if these proposed pieces of research were all funded and carried to successful conclusions, no major advance in earth science would result.

Two of the panel members, geology professor Harry Hess of Princeton University and geophysics professor Walter Munk of the Scripps Institution of Oceanography, were bothered by the fact that none of the ideas for research attempted to make a breakthrough in any really important problem. They adjourned to the Cosmos Club to discuss what the earth sciences could do to make a great stride forward. Walter Munk suggested that they consider what project, regardless of cost, would do the most to open up new avenues of thought and research. He opined that taking a sample of the earth's mantle, the rock below the Moho, would be most significant.

The Moho is the boundary between the rocks of the earth's crust and those of its mantle. This abrupt change in shock wave velocity was discovered in 1909 by Professor Andrija Mohorovicic and named in his honor. Obviously it is of great importance to earth scientists to know what the mantle is made of since it comprises 85 percent of the volume of the earth.

Munk and Hess spoke of a deep drill hole, oil-well style, perhaps on a

194

midocean island, to sample the rock beneath the Moho. Neither one had any clear notion of how drilling is done or what its limitations were so Hess suggested that the idea be referred to AMSOC, the American Miscellaneous Society, for action.

AMSOC is a spoof of serious scientific societies; it has no officers, by-laws, or dues. Its divisions were Etceterology, Phenomenology, Calamitology, Generology, and Triviology. It claimed to maintain relations with the Committee for Cooperation with Visitors from Outer Space as well as with the Society for Informing Animals of Their Taxonomic Position. Anyone doing business with the Office of Naval Research's oceanography group could claim membership, but there were no membership rolls; any two members constituted a quorum, and any papers produced were forwarded to the central files, which looked much like incinerators. The best thing about AMSOC was that, without restrictive rules, the members could speedily develop new ideas. After a few meetings on deep drilling possibilities, the inner circle had enlarged to include Drs. William Rubey, Josh Tracey, and Harry Ladd of the U.S. Geophysical Observatory and Gordon Lill and Arthur Maxwell of the Office of Naval Research.

In September that year, encouraged by Roger Revelle, the International Union of Geodesy and Geophysics meeting in Toronto, Canada, adopted a resolution that said, in part, "the composition of the earth's mantle beneath the Mohorovicic discontinuity is one of the most important unsolved problems of geophysics . . . we urge the nations of the world to study the feasibility and cost of drilling at a place where it approaches the surface."

This was by no means the first suggestion for drilling to sample the interior of the earth or the deep, unknown rocks. Charles Darwin had proposed a hole at Funafuti in the Ellice (Tuvalu) Islands to confirm his theory of the origin of coral atolls, and in 1897 the Royal Society of London drilled 1,140 feet there. (The results were inconclusive, which was the reason why Harry Ladd drilled the holes on Enewetak for the U.S. Geological Survey mentioned earlier.)

In 1943, Dr. T. A. Jagger, director of the Volcano Observatory on Hawaii, proposed a grand program of core drilling under the ocean and sent out hundreds of copies to enlist support for it. It said in part:

Geologists should . . . drill 1000 core-producing holes in the deep ocean bottom each 1000 feet deep . . . with world-wide drill hole distribution. This project must not be permitted to fail for that would leave geology a speculative science as before, surviving by continental anatomy when its real function is global exploration of the two thirds of the surface that is under the sea.

Other well-known geologists, especially Professor Maurice Ewing, director of the Lamont Geophysical Observatory, were keen to "sample ocean sediments from top to bottom" and "dig or drill a deep hole through the crust of the earth." The general idea of deep drilling beneath the deep ocean was steadily gaining momentum, but none of the enthusiasts was sure that it was feasible.

In April 1958, some 200 geophysicists met in the great hall of the National Academy of Sciences to hear the AMSOC proposal. Dr. Harry Hess presided, and Gordon Lill described the scientific advantages of drilling to sample the mantle. Scientists are inclined to be conservative when threatened by new ideas so some opposition developed with the arguments going as follows:

"What good will it do to get a single sample of the mantle? The material beneath the Moho is probably not homogeneous and one sample cannot be expected to be representative. Ten or even a hundred holes may be needed before we will know what the mantle is made of."

To which Harry Hess answered, "Perhaps it is true that we won't find out much about the earth's interior from one hole. But if there is not a first hole, there cannot be a second or a tenth or a hundredth hole. We must make a beginning."

The second objection dealt with money: "This project will cost many millions of dollars—you cannot even estimate how much. If it is paid for out of geophysical research money, it will strip all other projects of funds for years. If that amount of money were divided up among the existing institutions, we would all be able to do more and better geophysics.

Roger Revelle answered that one: "I imagine that an argument like that was used against Columbus when he asked Queen Isabella for funds for his adventurous project. One of the Queen's advisers probably stepped forward and said, 'Your Majesty, it won't be important even if this crazy Italian does reach India by sailing west. Why not put the same amount of money into new sails and better rigging on all our other ships? Then the whole fleet will be able to sail half a knot faster!" This devastating analogy silenced that part of the opposition.

The third objection was "It's impossible to drill a hole in the bottom of the ocean in the foreseeable future. Nobody has any idea how it can be done. Why doesn't AMSOC forget about oceanic drilling until it has done some research on deep-drilling techniques on land?"

The answer to this was given by A. J. Field, an engineer from the Union Oil Company, who showed movies of the ship *CUSS I* (a name compounded from the initials of the oil companies who owned it—Continental, Union,

Shell, Superior) drilling an oil well at sea off the California coast in 200 feet of water with a full-sized oil-drilling rig. Admittedly the ship was a long way from being capable of drilling in really deep water, but it demonstrated new possibilities to everyone present. Almost until that moment the capabilities of floating drilling platforms had been closely guarded commercial secrets, and virtually no one present had heard of such equipment. But now they could see possibilities in this vessel, which looked as though it could be developed into the first deep-sea drilling rig. A wave of enthusiasm went through the audience, and they perceived the project in a new light. If American technology could go this far, it could, somehow, drill through the crust to the Moho. Why not? In fact, *CUSS I* itself looked as if it might be converted to drill shallow holes and sample the upper part of the sedimentary layers of the sea floor.

By this time most of those who had been on the fence were persuaded that the deep-drilling project was a better idea than it had at first seemed. I then proposed that the group assembled express its support by means of a resolution, similar to that which had brought bathyscaphe to the United States: "The project outlined by Hess and Lill is approved." It carried unanimously.

Armed with the resolution, Harry Hess appeared before the governing board of the academy and, noting that five of the nine members of the AMSOC study group were academy members, asked that it be taken over as a committee, able to accept funds for a feasibility study. They agreed, and Nobel Laureate I. I. Rabi, a board member, dryly commented, "Thank God we're finally talking about something besides space."

IS DEEP DRILLING POSSIBLE?

With the scientific support of the academy and geophysicists generally, NSF granted the AMSOC Committee (with me as its unpaid secretary) $15,000 for a feasibility study. I began at once to look into floating drilling rigs (there were five experimental ones in existence) and into possible test drilling sites.

One day while discussing drilling to the Mohorovicic discontinuity with Dr. William Thurston, executive officer of the academy's Earth Sciences Division, a new idea for a project name struck me. I said, "It's the Mohole."

He grinned, "Yep. That's it." From that moment on no one called the project anything else.

During the early Mohole work, I came to know and appreciate Maurice Ewing, commonly known as Doc. He was gruff but friendly, a big guy, bespectacled under a shock of bushy gray hair, a dedicated seagoing oceanographer who was a god to his students. I wished I had been exposed to such a professor of geophysics. He had been through some close encounters with the sea on the *Vema,* an ex-three-masted schooner flying the flag of Panama, sailed by a crew of Nova Scotians that had become a leading U.S. oceanographic ship. A few years before, it had encountered a winter gale off Cape Hatteras. When the ship rolled hard, four oil drums on deck came adrift; Maurice, his brother John, plus Charles Wilkie and Michael Brown, the mates, were trying to resecure them when an unusually large wave (every thousandth wave is over twice as large as average) washed all four of them and the drums overboard.

Coughing and spluttering, Maurice took off his shoes, all the while his scientific mind calculating how long it would take them to fall to the bottom in water a couple of miles deep. He could hear, somewhere nearby, first mate Wilkie choking harshly before he slipped forever into the abyss. The *Vema* was a mile away, and the prospects of rescue were not good. But the skipper climbed to the crosstrees to act as lookout (a very risky thing in a storm), and the sailing master took the wheel. The ship then moved upwind of John Ewing and drifted down to pick him up.

While this was going on, Maurice, too exhausted to swim and about to black out, heard a voice: "Doc, I could hold onto this barrel easier if you'd take hold of the other end." It was Mike Brown, the second mate. They held on until the *Vema* stopped upwind; as it drifted down, a line was heaved to the men but both were too weak by now to do more than hold on. The *Vema,* now lying in the trough, was rolling hard enough to put the rail under. As the men came alongside and the ship rolled back, helping hands along the rail scooped them out of the water. At this moment the ship's steering gear broke down. If that had happened a minute earlier, both men would have been lost.

Maurice Ewing was keen to have the first test hole drilled in "his" ocean, the Atlantic, perhaps in the Puerto Rico trench. The water depth there is about 35,000 feet deep, far beyond drilling capabilities even now, but that location was the subject of such ardent debate he decided to get some additional supporting data. He took the *Vema* on an expedition to obtain a continuous line of magnetic data across sea and land; I was invited along.

A deep sea magnetometer like the one we had used on the Capricorn expedition was still a high-tech item, and I talked to Ewing's expert on magnetometers. "I've just run the land portion of this traverse across the center ridge of Puerto Rico," he sighed. "It was slow work. The hardest problem was to find a nonmagnetic vehicle to carry the instrument and its batteries."

He let me think for a while about what kind of transport he could have used that would have no iron in it. When I gave up, he beamed triumphantly. "An ox-cart. Compared to using that, making measurements at sea is a pleasure."

After a few days of coring by day and towing the magnetometer at night, *Vema* returned to San Juan and tied up at a stone quay on the old waterfront section that Spanish galleons had used 300 years before. Art Maxwell and I went ashore together and stopped at a waterfront bar that was largely filled by American sailors from a visiting destroyer. An old gal sat in the corner selling tickets, and when we noticed that the ticket buyers would disappear up a back stairway, we suspected there was more to be had in this joint than beer.

Presently the swinging doors opened to admit a surge of sailors from a newly arrived Canadian warship, and almost at once the emotional temperature in the place started to rise. Presently one of the bar girls flung a glass of beer in some sailor's face, and a free-for-all began between the two ships' crews. Art and I watched in amusement at first; then as things got rougher, we dropped to our hands and knees and maneuvered under the tables toward the swinging doors.

Somehow we separated; I went on to a hotel, and Maxwell went to the airport where he caught a plane back to Washington. There he chanced to meet my wife at a party the next evening.

"Have you seen Willard?" Rhoda asked.

"Not since the fight started in the whorehouse," he answered and left it at that, knowing that I would get an extra warm reception when I arrived home. That's what friends are for.

In September of 1959 a grand international convocation of oceanographers met at the United Nations in New York when the regular body was not in session. There we presented the first official statement by the academy, *Drilling Through the Earth's Crust.* It was a thin booklet, but it laid out the scientific policies that are in effect today. This NAS publication, which I largely authored, said in part:

The AMSOC Committee believes that it is desirable to drill a series of holes into the strata beneath the ocean culminating in one that pierces the Moho and samples the mantle. The information produced will be unique since direct sampling and measurement is the only way to obtain much needed knowledge about the composition of the earth, its age and history, and the development of life. In addition to the immense scientific value, the oil industry will benefit from the experimentation with new ideas and equipment for deep drilling. The value of the information and experience which will be gained will far outweigh the cost of obtaining it.

Elsewhere in that publication the committee emphasized that no one could be sure about the composition of the "second" layer (with seismic velocities of 4.5–5.5 kilometers per second). Was it basalt, limestone, compacted sediment? Or the "deep crust" (third layer—6.5 to 7.0 kilometers per second)? Or the mantle (8.3 kilometers per second)?

The committee hoped to find fossils that would fill in the breaks in the land record and perhaps find pre-Cambrian sediments. It wanted to confirm indirect geophysical measurements by measuring rock conductivity, radioactivity, and temperature. It noted the possibility of a great, unexpected discovery, and it said, with considerable prescience:

> It is possible that the sudden revelation of the geologic history of the uppermost layers (by the cores taken in the test drilling) will generate so much scientific excitement that there will be a general clamor for a program of drilling at many places beneath the ocean. [Shades of Dr. Jagger!]

Phase I of the proposed work was to test drilling methods and drill as deeply as possible into the sediments in several places.

About a month later, my lead story in the *Scientific American* entitled "The Mohole" gave a more complete account of our hopes. The catchy title was picked up by the press and broadcast; suddenly the project became world famous.

Crackpot letters started to come in. Some writers said there was no need to drill, they could tell us what was in the center of the earth because they'd seen it in dreams; some worried we would puncture the earth, and it would deflate like a balloon when all the "ether" escaped. One man repeatedly telegraphed the Secretary of the Navy, "Stop the Mohole!" warning that if we punched a hole in the sea floor the ocean would drain out, and there would be no need for a navy.

ENGINEERING FOR UNKNOWNS

It is one thing for scientists to speculate on the nature of Venus or Mars and what they would do with samples of those rocks if they had them. It is quite another to go to these distant planets, take the samples, and bring them back. So it was with the first deep drilling. Some original engineering would be required to solve the bramble of problems that came from increasing the depth of water to be drilled by more than an order of magnitude.

When NSF granted funds to perform a detailed study of how to proceed, I was named technical director of the Mohole Project. My first task was to find a few special engineers to help out. Ed Horton, a lanky good-natured Yale man, was lifted from Standard Oil of California; Jack McLelland came with a new doctorate in mining engineering from a German University; Peter Johnson, a red-headed naval architect, came from Webb Institute; François Lampietti, a tall French, mathematically inclined mining engineer, was just being released by the Army; Bob Snyder was borrowed from the physical oceanography department at Woods Hole, and Chad Ohanian, an electronics engineer, was carried on the payroll of Union Carbide Corporation as a contribution to the project. All were young, smart, energetic, and bursting to set world records in a new field. They made up the AMSOC staff.

As the work progressed, we engaged some more specialized consultants to help with especially tricky technical problems. For example, Arthur Lubinsky, the leading expert on stresses in drill pipe, was contributed by Pan American Petroleum. Minneapolis-Honeywell lent us Phillip La Hue to work on electronic instrumentation problems, and Captain Harold Saunders (U.S.N., ret.), who was just completing a book that became the bible of naval architecture, agreed to help with stability problems and brought along Robert Taggart, who calculated anchoring stresses.

After an examination of the five existing drilling ships, we decided to redesign *CUSS I* for the test drilling even though it had been intended for drilling in water only a few hundred feet deep. It would require considerable modification, but it gave us a big leg up on ocean drilling.

The essential problem was to hold the drill ship almost directly above the hole in water over two miles deep for days or weeks on end. In water depths of a few hundred feet this is easily accomplished by a cat's cradle of anchor lines, but in the depth we wanted to start drilling (12,000 feet) anchors could not hold the ship close enough to a point above the hole. The reason is that the weight of the anchor lines causes them to sag in steep catenary curves. Even under considerable tension they would hang almost vertically and thus permit the ship to be moved about considerably by winds or currents before its drift was arrested.

The drill ship must stay close above the hole if the pipe is to hang straight; if it moves away from that epicenter, the pipe must bend both at the ship and at the rim of the hole in the bottom. We computed how far from the hole the ship could stray without overstressing the pipe; the distance was only about two ship-lengths or 5 percent of the depth of the water. No anchor system could hold the ship that close in heavy weather.

The objective of much of the engineering was to reduce bending stresses in the pipe caused either by the ship moving away from its proper position, by its rolling and pitching, or by vibrational waves moving up and down the pipe. The tension in the outer fiber (or skin) of the pipe on the convex side caused by a bend adds to the tension caused by the weight of the pipe below; if the sum of these is too high, the pipe will break. Unlike oil drilling pipe, which is supported by the sides of a hole, in the ocean the pipe would hang free. If it broke and fell to the bottom, it was gone for good.

Our engineering staff attacked several major problems that would have to be solved if we were to insure the future of deep ocean drilling by inventing new hardware and new techniques.

The first one was: How does one keep a ship, with pipe dangling from it, almost exactly above a hole a couple of miles below in spite of the tendency of winds and currents to move it away. For this I devised a scheme that used two ideas developed in previous projects. The first was the idea of equipping a ship with maneuvering propellers so it could jockey itself in the open water between two piers in New York harbor and dock itself without using tugboats. In a harbor it was easy for the pilot to see where he wanted to go, and he could apply thrust in any direction to ease a ship gently alongside a pier.

But in deep water the sea surface looks the same in all directions; there would be no visual clues to indicate where the hole is located or which way the ship was moving. For that I proposed using several taut-moored buoys something like those I'd built for measuring waves from the Mike nuclear shot.

If we put a circle of such buoys around the drill site, with a radius of, say, 500 feet, and we equipped a ship with four outboard motors (one in each corner), and we gave the pilot a suitable set of controls, he could adjust the amount and direction of propeller thrust and constantly maneuver the ship to keep it amid the buoys above the drill hole. In a report written in 1959, I first called this scheme "dynamic positioning" and the phrase is now used worldwide. "DP ships" have become so successful that others have tried to take credit for its invention.

Reducing dynamic positioning to practice on the Mohole Project required the help of other staff members. Peter Johnson was given primary responsibility for designing the buoys and installing a circle of them at each drill site. Chad Ohanian was assigned the problem of sensing the buoys with sonar and radar, and presenting their position to the pilot so he could hold the ship at night in stormy weather.

Captain Southerland of Murray and Tregurtha had the job of overseeing the production of four large outboard motors (200 HP diesel engines with 16-foot-long shafts and 4-foot-diameter propellers). And Robert Taggart was given a contract to build a console that would simultaneously control the thrust and direction of all the motors. I instructed him to build a "joy stick" type control that would move the ship in the direction the stick was pushed, while maintaining its orientation into the wind and waves.

In case something went wrong with the position-sensing system described above, Bob Snyder worked on an alternate scheme—a directional tiltmeter that could be lowered to the bottom on a cable and used to sense the drift of the ship away from the epicenter. (This only worked in water less than 1,000 feet deep, but it also was copied by the oil drillers.)

Our design team was also much concerned that if the ship were to roll heavily and/or get too far from the epicenter, the pipe could bend sharply at the rotary table. Because it was also under tension from the weight of the dangling drill string, it could kink and break in an instant. So Ed Horton devised a trumpet-shaped "guide shoe" made of heavy steel that opened downward from the rotary table and prevented the pipe from bending too far. When the ship rolled, the pipe merely lay against the smooth inside of Horton's trumpet. For extra insurance against kinking, he devised a round fluted kelly, instead of the ordinary square type, to transmit rotation to the pipe.

It was also necessary to reduce the bending stress at the rim of the hole in the bottom, where there could be more bending with the pipe under less tension than at the surface. For this a special tapered casing some 30 feet long was machined to give it variable flexibility throughout its length. This would be attached to a landing base, which rested on the bottom and was held in place by several hundred feet of casing in the hole below. McLelland and Horton tested its capacity for reducing bending by clamping it between two huge eucalyptus trees while pulling sidewise on an internal pipe with a tractor-crane.

After many tests, models, and mathematical contortions Lampietti, Horton, and Lubinsky designed a drill string that was tapered in steps with extra-heavy, high-strength steel at the top, intermediate strength steel in the midsection, and ordinary (80,000 pounds per square inch tensile strength) steel at the bottom. The bottom part of the string was something like that used by the shallow-water oil drillers. Below that there were several heavy-walled sections of pipe called drill collars that would act like sinkers on a fishline to hold the pipe straight. And at the tip there was one more collar that rested on the diamond drill bit and held it tight against the rock. Between the two sets of collars, we used a sliding spline called a bumper-sub, which would transmit rotation but prevent the ship's vertical motion from being transmitted to the drill bit.

There would be no way to change bits, and we did not know what kind of rock would be encountered, so only diamond bits would be used. These would last longer than ordinary rock bits if some really hard rock like basalt was encountered. All of the pipe, the collars, and the subs were specially made so that coring tools could be dropped down the center of the pipe into place at, or ahead of, the bit and retrieved by a wire.

The AMSOC Committee appointed various technical panels to assist the staff, but most of the people selected for these were either scientists with no understanding of engineering problems or representatives of ultraconservative oil companies, who were horrified by any variation in the way they did drilling on land. So one of the staff's biggest problems was persuading these subsidiary groups, who understood little about ocean engineering, to go along with our plans.

Even Global Marine, Inc., the contractor who owned *CUSS I,* had great doubts that deep-water drilling would be successful, and for several months after I had made a firm decision to use dynamic positioning to keep the ship above the hole, Global was still trying to persuade the Office of Naval Research to contribute several million dollars for a deep-water anchoring system. Even so, the AMSOC engineers set up an office in Global's headquarters and worked side by side with their engineers on specific changes that would be needed to convert their barge to a ship for deep-water work.

During the period while we were preparing to go to sea, I gave dozens of talks on the objectives and intentions of the project, two of them in Houston at the Rice Hotel and the Petroleum Club. After the latter, a couple of old timers from the oil patch came up to me and began:

"Son, did you ever see a real drillin' rig?

"Yessir."

"Did you ever see an outboard motor?"

"Yessir."

"You gonna try to hold a drill rig in the open ocean with four outboard motors?"

"Yessir."

They shook their heads as though they were talking to the village idiot and shuffled away.

That about summed up the attitude of the oil industry to our plans. Their idea of deep water was a few hundred feet; twenty times deeper than that was utter foolishness. Mostly they did not believe we would succeed, and neither did some of our own technical advisers. Those who did relied more on computations than on their instincts. The AMSOC staff was on its own.

The question of exactly where to drill first had been studied by Harry

Hess's Site Selection panel for a year and smothered with indecision. They did not seem to understand that our test of an untried system would have to be within 250 miles of San Diego so that we could exchange crews, send replacement parts, allow the press to visit, and return for repairs if necessary. Selecting the spot for the ultimate hole would come later. When the time to decide became critical, I suggested a site for the first trials in the San Diego trough only twenty miles from the shipyard. The water was only 3,000 feet deep but was a good place to debug the new hardware.

Most of the bottom area within a radius of 200 miles was sliced by faults, creased by crustal wrinkles, and pimpled with volcanic cones. Although our main objective was to prove that deep-water drilling could be done, we also wanted an unconfused bottom that would give maximum geologic information and ensure that we could reach the mysterious second layer. For the second set of holes in deep water, I proposed a site between Guadalupe Island and the Mexican mainland that met operating criteria and was only about 200 miles south of San Diego. Russ Raitt had already measured seismic velocities there, and Harry Hess thought it would be OK. "Why don't you check it out?" he said.

So in August 1960, a miniexpedition left on an old beat-up Scripps ship named the *Orca* to get some additional data. Guadalupe is a high rocky island made of folded metamorphic rocks. Its uplands are overrun by huge herds of goats, and its wide black beaches are home for thousands of sea elephants (which implies a lot of fish in the nearby waters). The small, rarely visited Mexican town on its southern end welcomed us with lobsters enough for a feast.

About 40 miles from the island, in water 12,000 feet deep, we took cores of the top 2 meters of mud, photographed a reddish-brown bottom covered with brittle stars and the tracks of small creatures, measured currents, and made seismic reflection studies with a gas gun. The place measured up to our specifications, the most exciting moment in the expedition being when François got into a bunk that had been primed for him with a large irritable lobster.

Our staff pushed ahead, solving one problem after another and published a detailed report on exactly what we would do and how. Based on it, the National Science Foundation agreed to fund the first deep-sea drilling. I was named Project Director by the National Academy of Sciences and technical representative of the National Science Foundation. We began ordering pipe, outboard motors and the other equipment needed and assembling subcontractors; the pace increased every day. Then the project was unexpectedly blindsided when, in November 1960, I went to arrange drilling contract details with the president of Global Marine.

Our staffs had been in intimate contact on this project for months preparing for this moment, but when we came to set the date of the tests, he said, "We can't let you have the *CUSS I;* the Shell Oil Company has it under contract for the next year."

A drilling contract without a date was signed two days before Christmas, but not until February 11 did the Shell Oil Company release the ship to us. Then began a mad scramble to convert *CUSS I* for deep-water work. We had two months, total, to buy equipment and ship it to San Diego, modify and mobilize the ship, try drilling in two locations, and demobilize again. It was a nearly impossible schedule.

We had made such a point about the possibility of the pipe breaking that often the question was asked: "What *will* you do if the pipe breaks?" In order to have a ready answer, we rented the best string of ordinary drill pipe we could find, magnafluxed it (tested it for metal flaws), and stored it in San Diego. It was not nearly as good as the new pipe of our own design, but it served its purpose. Now when the question arose we could say: "If we lose the first string, we'll come back for the second one." That seemed to satisfy everyone.

As soon as it was assured that sea drilling would start soon, I wrote to John Steinbeck, my friend of Monterey days, who was then in Barbados, inviting him to come along as project historian. He answered at once: "I remember the Monterey episode with pleasure, and I have a profound interest in your project. What little I know of it has been gleaned from the *New York Times.* Please send me some official material on the project's plans and intentions. Would you have any objection to my underwriting expenses by offering an account of the expedition to some magazine?" The long letter ended: "Finally may I say that I await your reply with considerable impatience. We may be coming into a period where thinking men are at least recognized as existing."

SHAKEDOWN IN SHALLOW WATER

On Sunday February 26, the outboard motors and their controls were ready to test; Bob Taggart and I felt a combination of apprehension and high hopes as we stood in the pilot house of *CUSS I,* tied up at a pier in upper San Diego bay, each trying to persuade the other to take the controls first. The lines were cast off, and the ship moved sidewise away from the pier; then slowly it backed out of the berth into open water. Once the ship was clear of obstacles, we turned it 360 degrees in one spot, rotating it around the derrick. Taggart tested the sidewise motions; I moved it up to a large navy mooring buoy and backed away again. Dynamic positioning worked as planned.

By March 3 the ship was deemed ready to go to sea and was taken in tow by a Navy tug. The San Diego entrance channel runs north and south so as the ship moved down it, and the Pacific swell struck it broadside. The previous night I had lain awake worrying about what would happen when we got in this position, and now my fears were confirmed. The period of the swell was 9 seconds, about the same as that of the ship, and because of this resonant condition, the ship rolled violently, several times to as much as 24 degrees. On one of these rolls, a guy wire restraining the 6-ton traveling block parted; there was a great crash as one of its guide rails high in the derrick cracked. Everyone on deck dived for cover.

There was no choice but to tow the ship back to the shipyard to repair the damage. This small accident early in the test was a blessing in that it probably saved us from a worse one later, but it shook the confidence of some of our scientists, who were inexperienced in dealing with heavy equipment at sea.

A few days later we were at sea again off Coronado, giving the pilots some practice at the controls that would hold the ship in position at night. Then we moved on to a position that was out of sight of land off La Jolla, where the only markers were two buoys with small flags, lights, and radar reflectors. We began to lower pipe, and the next day I was able to send a cable to President Bronk, who read it aloud at the annual meeting of the National Research Council. It said, in part:

At 0206 today, March 9, 1961, a drilling bit touched bottom in 3111 feet of water. . . . First hole was drilled to 115 feet. . . . Rotation of 30 rpm causes no pipe vibration. . . . Bumper-subs opened and closed properly, maintaining constant pressure of 6000 pounds on the drill bit. . . . We will continue drilling and coring until depth of 500 feet is reached. . . . All hands confident and happy. . . .

We believe that today's experiment at a water depth which is nearly an order of magnitude greater than the previous record clearly establishes the feasibility of deep sea drilling.

Signed: Bascom

The audience cheered.

Sometime in the night while the next hole was being drilled, someone brushed against a switch and changed the range scale on the radar so the pilot drove the ship half a mile off station. The driller then noticed that the hook load was 6,000 pounds light, so he tripped the pipe out to find that the bumper-subs were bent and the lowest drill collar, tipped with a diamond bit, had broken off. This was not much of a loss; we had replacements, and the DeBeers organization had donated the diamonds, but the press made it seem like a big setback, and some of our critics became nervous.

The fifth hole off La Jolla was the first attempt to set casing remotely (casing is large pipe used to line a drill hole). In this test the casing would be topped by a 6-sided steel landing base about 12 feet across, which rested on the sea floor; above it rose the special tapered casing intended to reduce pipe bending and simulate a reentry cone.

From the beginning everyone knew that for really deep coring in hard rocks, still a few years in the future, it would be necessary to retract the pipe, change the worn bit, and reenter the drill hole. That meant that a device to facilitate hole reentry by a drill bit would have to be implanted in the bottom. The structure we lowered on the drill pipe was over 200 feet long; it had to be drilled into the bottom and then released so the bit could drill on deeper. Several pages of blueprints were required to describe the order of assembly and the J-slot release mechanism.

It had been designed mainly by Ed Horton and Jack McLelland, and it worked on the first try. After unlatching the casing from the drill pipe, the bit continued on down to a depth of 1,035 feet.

Each hole had established a new record. Then *CUSS I* returned to San Diego for final refitting before going on to the deep-water site.

John Steinbeck arrived in San Diego about halfway through these preliminary tests and was transported out to the *CUSS I* on a crewboat named *Roughneck,* arriving about midnight. From the first moment he intently watched every operation, sweating out the difficult moments, cheering as we made progress, inspecting the cores with the magnifying glass that always hung around his neck. He was delighted to be at sea and endlessly prowled the ship, probing the diverse collection of personalities aboard. On looking the ship over he wrote:

> *Cuss I* has the sleek lines of an outhouse standing on a garbage scow.
>
> The deck is heaving and pitching. The men step like cats. There is nothing clumsy about them and as the steel sections of pipe rise and are screwed together and lowered, the drillers move with the timing and precision of a corps de ballet. They would throw me overboard if they knew I said this or even thought it.

Bunk space on the ship was at a premium, so he and I shared the ship's hospital. It was a good cabin, available with the understanding that if there were any injuries that required its use we would be ousted. Fortunately there were none.

When *CUSS I* returned to the shipyard for final touch-ups before leaving for Guadalupe Island, John stayed with Rhoda and me at our house on the beach at La Jolla, along with Dr. Frederico Mooser who would represent the government of Mexico when we returned to sea to drill off their coast. Steinbeck was a most remarkable observer; he looked at life through a variable-focal-length lens, starting with a wide-angle view and then zooming in until he could use his magnifying glass.

We gave a party, in which he met a couple of dozen oceanographers for the first time. Rhoda and I were still dressing when the first guest arrived, so John introduced himself to a lady who did not recognize his name.

"What do you do?" she asked.

"I'm sort of a marine biologist," answered the co-author of *Sea of Cortez,* and chatted for a while about tide-pool animals.

John could see directly into the souls of people at a single brief exposure, and after the party was over, he correctly assessed their personalities and ventured the opinion that there was "some bedroom trouble" in the community.

While in San Diego we learned that the minor mechanical problems described previously had unnerved some of the AMSOC Committee members. They thought we should "quit while we're ahead," especially because the project was operating in a glare of publicity. The engineering staff, of course, wanted to continue; after all, our design was made for operation in a water depth of 12,000 feet, and we wanted to demonstrate it would work there, in true oceanic depths.

An emergency meeting of AMSOC was called, and those present were evenly divided between stopping and continuing. I put in an emergency phone call to Walter Munk, who dashed down from La Jolla and cast the deciding vote in favor of going on to the Guadalupe Island site. Before adjourning, Chairman Lill made it quite clear that the object of the meeting had been to absolve the committee of any blame if the deep drilling failed. That was reserved for me, the Project Director. But I had confidence in our engineering.

GOING FOR BROKE

As *CUSS I* was towed out of San Diego harbor, en route to the Guadalupe Island site, two senior officers of Global Marine stood with me on the ship's flying bridge. One said, in a patronizing voice, "We won't be too unhappy if this scheme doesn't work, will we?" Plainly they thought the chance of success was small, as had the oil men in Houston. I growled back, "It damn well will work."

During the long tow down, I gave two lectures to the crew and scientists aboard on the scientific reasons for deep drilling and explained the careful engineering studies. Everyone seemed to understand that this was a one-shot affair; *CUSS I* would be operating close to the maximum capabilities of the positioning system and pipe strength; if someone failed to do his job properly, the project could end in failure. They all came through.

As the ship entered the circle of buoys Pete Johnson and Jeff Savage had planted, the mate hoisted the red, white, and green flag of Mexico on the foremast as a courtesy to that government, even though we were 120 miles off their mainland.

Almost immediately the wind increased to 25 knots; with the swell at 8 to 10 feet and a substantial current setting west, the pilots had to fight to maintain position throughout the first night. But early in the morning, the wind and sea calmed somewhat, and the drillers started lowering pipe. About noon on March 27, 1961, a drill bit touched down and drilled into the deep sea floor for the first time.

From then on the drilling superintendents (Don Woodward, borrowed from Texaco to represent AMSOC's interests on the drilling floor, Buzz Freeze and John Atwood of Global Marine) intently watched the wavering needles on the drilling console day and night. These needles indicated hook load, pump pressure, and torque on the pipe—the only means of determining what was happening below the sea floor. I would describe to Don Woodward what we wanted to do, and he would translate it into words that the drillers understood. Don was an old hand at solving tough problems, and he was very good at visualizing what was happening below. By midnight the first punch core, 110 feet into the soft sediments, was obtained; at the time the wind was gusting to 30 knots and occasional 14-foot waves were observed, so everyone's confidence increased considerably.

Bill Riedel, AMSOC's chief scientist, pronounced the material of the core to be a gray-green Miocene clay containing many microfossils. So far so good; we would go for broke on the next hole by adding a Christiansen diamond drilling bit and a "universal wire-line core barrel," which was capable of either punch-coring in soft sediments or rotary-coring in the hard rock of the "second layer." Steinbeck commented: "Our expedition should destroy the old and well-loved error that doers and thinkers are different breeds."

The pipe started down again, but by the time the bit got near the bottom, all four engines running at full speed were barely able to hold the ship into the wind. It was night now, and we were trying to decide whether to touch down and drill in when the pilots reported that the steering wheel would not turn. This made it impossible to change the ship's heading with the automatic controls. If *CUSS I* were to become broadside to the wind and sea, it would

surely be blown off station and roll heavily, with possibly serious conse-
quences. This was a real emergency, so in the dark pilot house Bob Taggart
and I crawled under the steering console between the pilot's legs, and while
I held a flashlight, he opened the panels that covered its mechanism and soon
found the trouble; a small idler gear had frozen. Fortunately the pilot was able
to hold the ship into the sea with the emergency rotation controls while
Taggart repaired the gear.

Then, in the face of heavy weather and against the advice of the contractor, we decided to drill into the sea floor. So began the second and most
important hole of the series. The disadvantage of starting a hole in adverse
weather could be more than offset by the fact that once the drill collars are
buried in the sea bottom they tend to act as an anchor and resist half the
sidewise drag of currents on the drill pipe. We drilled about 400 feet before
attempting to core; then, after experimenting with rotary speed, pump pressure, and drill bit advance, very satisfactory soft cores were obtained.

On April Fool's Day, at a depth of 560 feet below the sea bottom, the
advance of the drill bit slowed abruptly to 2 feet per hour. That meant
another, quite different, layer had been reached, and we were on the verge of
settling the long-debated question of the composition of the second layer.
Everyone, roughnecks as well as scientists, waited with anticipation for the
corer to be swung onto the walkway and laid out.

The answer was . . . basalt. And practically every scientist said, "Just as
I thought." The other theories evaporated.

Within an hour of retrieving the basalt, the *Roughneck* came alongside
with a new crew and the mail. The latter contained the first copy of *A Hole
in the Bottom of the Sea,* my book about the project. Although the manuscript
had been sent to the publisher nearly a year before, it turned out I had
accurately forecast what would happen. The joke aboard ship became "Let's
read the next chapter and see what will happen tomorrow." Nearly everyone
aboard signed it, crew and visiting scientists alike, including Roger Revelle,
Harry Ladd, Josh Tracey, Russ Raitt, Sir Edward Bullard, and John Steinbeck.

As for the new find, Steinbeck wrote:
The scientists guard the core like tigers. Everyone wants a fragment as
a memento . . . of the second layer which no one has ever seen before.
I asked for a piece and got a scowling refusal so I stole a small piece.
And then the dammed chief scientist gave me a piece secretly. Made me
feel terrible. I had to sneak in and replace the piece I had stolen.

Drilling in the basalt was slow, and it seemed likely that there was 3,500
feet more of it beneath the bit. So after drilling another 44 feet to make the

total depth exceed 12,000 feet, the hole was abandoned.

The fourth hole was drilled with a "logging" bit, whose center was a little larger than that needed for a core to permit the passage of various geophysical instruments. Twice, as the bit penetrated (at 103 feet and 502 feet), it was stopped and a long steel instrumented needle, the "temperature probe," was thrust out ahead of the bit into the green clay ooze. The data from it indicated to Art Maxwell and Dick Von Herzen, our heat-flow experts, that at this spot heat moved outward from the interior of the earth about twice as fast as the average for ocean bottoms. The actual temperature at the sea floor was 1.62 C, and that at 500 feet below the bottom was 22.7 C (70 F)—that is, the increase with depth was from nearly freezing to room temperature.

That hole bottomed 20 feet into the basalt, and the pipe was retracted until the bit was only 100 feet below the sea floor; that left us 500 feet of open hole beneath the bit into which sondes (geophysical instruments) could be lowered on logging wire. Various logs were run (resistivity, gamma, magnetic field, etc.), but there was little variation throughout the hole except for a single abrupt change when the basalt was encountered. This was done to find out if any anomaly existed in the soft sediment that did not show in the cores.

Finally Russ Raitt and George Shor of Scripps lowered geophones into the hole and were able to compare the actual (drilled) thickness of the sediments with those estimated from seismic reflection data at the site. They found that sound traveled more slowly in the soft sediments than previously believed, an important finding that would require revision of sediment depth estimates at many locations.

One evening, just at sunset, during the weekly change of drilling crews, a dozen or more men were waiting at the rail for the *Roughneck* to come alongside. It would ferry them to the flying boat that was used to transport men between Guadalupe Island and Los Angeles. To the west the sky was a brilliant blue, and the upper rim of the sun was about to drop below a clear horizon.

I hollered, "Hold up a minute, there's going to be a green flash."

These fellows, who had never before heard of a green flash, stared dutifully as the red sun dipped and saw a most spectacular flash of green rise a quarter of the sun's width above the horizon. "So what," they mumbled. "Big deal!" And piled aboard the crew boat.

Aboard the *Baird,* which we used as a standby scientific ship, Harry Ladd and Bill Riedel reconfirmed that the sediments were mainly Miocene (30 million years old) oozes with occasional layers of volcanic ash and an abundance of microscopic shells of plants and animals, half calcareous and half siliceous. The depth of the strata divided by the time indicated that the average sedimentation rate since middle Miocene time was about 2 centimeters per thousand years.

The fifth and final hole was drilled to test a "wire-line coring turbodrill," made by Neyrpic, a French company. This might be the most effective tool for long deep drilling through hard rock before the mantle of the earth was reached, and we wanted to check it out. Basically it is a long slender turbine just behind the bit, driven by the fast-flowing drilling fluid—in this case seawater—that was pumped down from the surface. For diamond bits, which are most effective at speeds on the order of 1,000 revolutions per minute, much higher than the whole pipe string could be rotated, it seemed ideal. On this test it raced downward through the soft sediments and efficiently took a short core of the upper few feet of the basalt.

We were already walking on air when we boarded the seaplane for home. All of the objectives, both engineering and scientific, of Phase I of the Mohole Project had been accomplished in two weeks on station.

The nervous nellies were silenced, and those of us on the front lines could get some sleep after a lot of twenty-hour days. John Steinbeck's final thoughts were these:

> *Cuss* I was a makeshift but even so it proved the contention of scientists that the work can be done. . . . And what a hell of a bunch of men are that motley crew of *Cuss* I, Project Mohole, National Research Council, National Academy of Sciences. We'll be all right, Jack, with men like these.

The engineering staff returned to Washington amid a blizzard of good publicity. There was a press conference and party in the great hall of the Academy, and a few days later President Bronk passed along a copy of a letter he had received from President Kennedy that said, in part:

> I have been following with deep interest the experimental drilling in connection with the first phase of Project Mohole. The success of the drilling . . . constitutes a remarkable achievement and a historic landmark in our scientific and engineering progress.

To which Bronk added:

> It was an undertaking that involved major extrapolations of existing engineering experience and practice. Our warmest good wishes for the future as the project moves ahead into its succeeding phases.

There was a party at the great hall of the Academy to celebrate our victory over the ocean depths and our detractors. I walked up to Dr. Bronk just as Rhoda was pointing out to him that the people who had complained the most about the methods and risks I had taken to get the job done quickly and inexpensively were those most eager to take the credit for the project's success. He turned to me with a look of mock astonishment and, feigning surprise, said, "Why, Bill, you didn't expect appreciation, did you?"

When the President's Science Advisory Committee invited me to describe what had been accomplished, they had a hard time believing that the total cost of everything we had done, including buying the pipe string, engines, and positioning system, chartering *CUSS I* for two months, and our staff's salaries for a year was only $1,536,500. This was a million dollars under the original guesstimate and probably as good a buy in science as has ever been made.

The staff wrote up the result in Academy Publication 914, *Experimental Drilling in Deep Water,* and took off for a few weeks' rest. *A Hole in the Bottom of the Sea* got good reviews, and lecture invitations flowed in, one from Minneapolis where I costarred with its symphony orchestra and was introduced by Dr. Athelstan Spilhaus, one of the great men of modern oceanography.

I also returned to the Petroleum Club in Houston to speak to the same group that had previously called our deep drilling scheme "science fiction." Now they said, "Hell, anybody in offshore drilling coulda done that."

Those were heady times. Rhoda, Anitra, and I put the top down on our Chevy convertible and drove across the country singing, "We'll be drilling to the Moho with a diamond-studded bit" to the tune of "John Brown's Body."

The State Department asked me to give a lecture tour in Europe, starting in Zagreb, Yugoslavia, the place where Professor Mohorovicic first defined the crust of the earth. In preparation I began to correspond with the great man's grandson, also a scientist. En route, in Paris, the U.S. Embassy informed me that my visa to enter Yugoslavia had not come through, and the day before the plane left, an official took me around to visit that country's embassy.

When we reached the ambassador's office, my official friend made an impassioned plea for an official visa; the ambassador thought a moment and then asked him to step outside. Then he said in a very kindly voice, "Son, you don't need the U.S. State Department; when you get to our border, just walk through the gate."

So I did and was met by an official in a long black limousine, who conducted me to see Dr. Mohorovicic. As soon as we talked, it was evident that he was not the fellow with whom I had been corresponding. "Ah," said my escort, "this is the *party* member. You must have written to the *other* one."

So off we went to meet Dr. Andrija Mohorovicic, another grandson now retired as a professor of geophysics, who was not in as good standing with the Communist Party. He too was pleasant to talk with, and he presented me with an original copy of his grandfather's historic paper, which described how the shock waves from earthquakes could be used to define the boundary between the earth's crust and its mantle.

My lecture to the Croatian Academy of Sciences in Zagreb was in a wide but not very deep room with five rows of high-backed seats that extended far to each side of the dais. The capacity audience consisted of lines of men in dark suits sitting rigidly upright. My talk was in English, translated by an interpreter, but enough of those present corrected the translation frequently to give the impression almost everyone understood English. They were most appreciative that an American national project was named after their local hero, and, after showing me Professor Moho's original lab and instruments, toasted everything in slivovitz, a plum brandy that one must beware of.

The National Science Foundation then decided that the National Academy of Sciences was not the proper home for a large project, in which the money would be spent more on engineering and operational matters than on science. The management of the project would be put out for bid. Our technical staff agreed that was logical and gave a bidder's briefing about the project to representatives from a hundred companies, assuming that whoever got the contract would be likely to hire us for our know-how.

Possible contractors formed groups and spent considerable money and effort preparing proposals. But in order to keep any of them from having an unfair advantage and to avoid any conflict of interest, our staff conversed with none of them. Instead we pushed ahead with the design of what we called an "intermediate" ship, capable of taking deep cores almost anywhere in the ocean and working out the next level of engineering problems, but *not* capable of reaching the Moho.

In October of 1961 I wrote a memorandum to the AMSOC Committee that described the views of the technical staff on how the project should proceed:

The ultimate objective of this project is to drill through the Moho and sample the mantle. We propose to reach that goal as soon as possible, proceeding by a course that will give maximum probability of success on the first try.

In the opinion of the staff and its principal consultants, it would *not* be prudent to *begin* by trying to design and build the ultimate deep drilling ship in final form with all the complications that are entailed. We are sure that such a ship can be designed as soon as sufficient data and experiences are obtained. Now we do not know enough about the stresses that will exist in pipes of various materials and configurations; about methods that can be relied upon for extended stays at sea; about hole reentry; about methods of handling pipe and casing; about the necessity for, and means of, using mud or cement in various circumstances; about logistics and communications; and about many relatively minor problems.

To attempt to work out all these problems simultaneously for Moho depths is to invite gross inefficiency and long, expensive delays. The difficulties of building a machine capable of drilling to as much as 35,000 feet from a ship in the ocean are monumental. These begin with the problem of making a drill pipe that will hold together while conducting power and drilling fluid downward for seven miles, while being flexed and jerked at the top. No such pipe now exists. Minor modifications of existing oil field equipment cannot do the job required.

Our opinion as engineers is that the proper way to proceed is to build an experimental drilling ship of modest proportions and use it to develop ideas and equipment and to work out logistics problems at a relatively small cost. We propose that this ship be equipped to reach downward 20,000 feet with a drill bit. (Note that this is half a mile deeper than the Soviet land record; equal to the U.S. land record of only ten years ago; deep enough to reach the 3rd layer at sea in many places.)

This is the intermediate ship concept. The hull we propose is a C1-M-AV1 (339' by 50' by 21', 7400 displacement tons). This is nine times larger than the Scripps' tugs, four times larger than the newest U.S. oceanographic ship (not yet built), and two and one-half times as large as *CUSS I.* The ship we propose will be self-propelled and self-contained, capable of operating far from land without consort ships.

Thus we are proposing a ship which, although it may not be capable of sampling beneath the Moho, will open virtually all of the oceanic layers above the deep crust to a new kind of exploration. We think it is likely that, viewed with the perspective of time, this work will prove to be more valuable than the reaching of the Moho.

In this memorandum we were trying to nudge the AMSOC Committee away from the "one deep hole" concept that many members favored toward our view that the objective should be to explore the crust of the earth beneath the ocean. Half of the ears this fell on were deaf, and the AMSOC Committee split neatly between the two views.

A long silence then settled on the staff, as we waited in a sort of limbo for NSF to name a successful bidder. To reduce the general nervousness about our fate I wrote a famous quote from Charlie Chan, the master detective, high on my office blackboard: "Softly, softly, catchee monkey." In other words, keep quiet, and we will be rewarded for betting our reputations and our careers against the rough Pacific and winning. Why worry about the whims of bureaucracy?

My leadership was described by friends as "daring and devoted, although sometimes willful"; my critics were mainly annoyed because practically everything turned out as I had forecast. Others noted that "I did not suffer fools gladly," and there was a plentiful supply of those.

When ten responsive proposals came in, a special NSF panel carefully evaluated them, and on October 20, declared that one made by Socony Mobil was "in a class by itself." They recommended that NSF begin negotiations at once with that company. But then politics reared its ugly head. In fifth place, well back of the leaders, was a Texas contractor named Brown and Root, of Houston, with virtually no experience in offshore drilling (and the only one not to include the AMSOC technical staff in its plan).

Brown and Root did, however, have two great advantages; it was a long-time supporter of Lyndon Johnson, then vice president, and of Congressman Albert Thomas of Houston, who was chairman of the House subcommittee that reviewed NSF's budget. So NSF appointed an internal group to review the proposals again, and this time, wonder of wonders, Brown and Root triumphed.

In a masterful political stroke Dr. Alan Waterman, head of NSF, publicly announced that NSF would negotiate solely with this "management company" on February 28, 1962. By an odd coincidence, the following day Dr. Waterman was to appear before Congressman Thomas to defend an NSF budget that included $5 million for the Mohole. In the words of *Fortune* magazine, "Congressman Thomas, often cool to NSF's proposals, said warmly that this particular budget was a work of art."

We had the satisfaction of having been the first to drill for science in deep water, but we had lost to politics and bureaucracy and were now out in the cold. So the five of us engineers, Jack McLelland, Edward Horton, Peter Johnson, François Lampietti, and I, set up a new company, Ocean Science and Engineering Inc. Using money raised partly by mortgaging my house and full of naive optimism, we went into business for ourselves.

A year later Brown and Root decided *not* to build an intermediate ship along the lines we had proposed; instead they hired "acres of engineers" to design an ultimate drilling platform. Their budget climbed every year, and in 1966 after some $55 million had been spent and no further holes had been drilled, it was obvious that what had been an inexpensive and sensible project under my direction had become a disaster. The Congress then canceled it, ostensibly because of the cost of the war in Vietnam. So, in spite of a very promising beginning, the United States got "No hole" for its money instead of Mohole.

In 1968, long after our group had left the scene, the Deep Sea Drilling Project was established by a consortium of universities and, with the "intermediate" ship *Glomar Challenger,* began the general exploration of the sea bottom we had recommended in 1959. It eventually drilled 1,092 holes at 624 sites in all oceans except the Arctic, recovering 96,000 meters of core. It is sad that deep drilling could not have been started six years earlier with the money that was thrown away on the Brown and Root contract.

At the twenty-fifth anniversary of the first deep ocean drilling, in 1986, it was noted that no drill bit had yet come close to reaching the Mohorovicic discontinuity and sampling the earth's mantle. In 1987, *Nature,* a leading scientific magazine, announced that a new group was being formed to drill to the Moho. Stay tuned.

IX

Diamonds from the Ocean

THE SKELETON COAST

"No one returns from Johannesburg empty handed." So went a bit of diamond folklore passed along to me with an inscrutable smile by Frank Christiansen, the diamond bit maker, when I asked how he had persuaded the DeBeers group to donate the diamonds that became Mohole drill bits. The diamonds had arrived by ordinary mail, uninsured, in a plain package with no return address. Not knowing what the little bundle contained, I tore open the blue inner wrapper and a couple of handfuls of diamond crystals some 5 millimeters high spilled out. These greasy-looking stones shaped like two tiny pyramids attached base to base (octahedrons) trickled and bounced across the desk. Astounded, I called the U.S. Customs Bureau, who sent an inspector around, and he informed me that was the usual, legal way to ship uncut diamonds.

Diamond mining had always seemed to me like a romantic business, and now rumors reached us that gem-quality diamonds were being mined in the ocean off South West Africa. Since I was an ex-mining engineer and president of Ocean Science and Engineering, Inc. (OSE), a new company that needed work, both personal interest and company hopes attracted me to Jo'burg. There I would be able to find out if there was a place for our company in the waters off southern Africa.

An exchange of introductory cables made it possible to start at the top. The DeBeers-Anglo American Corporation headquarters occupied two large buildings on opposite sides of Main Street. One of these contained the offices of "the chairman," Harry Oppenheimer, and the financial group that held the reins of about 140 companies, many of which were in mining. Harry had served with the South African Tank Corps in North Africa during World War II, and after several terms in parliament, where he promoted political reforms and made liberal speeches in Afrikaans, he withdrew from politics to give his full attention to business. Charming, modest, full of common sense, Harry Oppenheimer was one of the world's richest men.

The building across the street had a fortified top floor, occupied mainly by diamond sorting and storage facilities. The floor below contained the executive suite and dining room presided over by Ted Brown, a bluff stocky fellow with a light brown mustache and a twinkling eye who seemed to be responsible for new developments in many directions. He had a way of stating his views that ended with him leaning forward and earnestly asking, "Don't you agree?" It was hard not to.

From our first meeting Ted Brown was very cordial and invited me to become a regular visitor to his exclusive luncheon group. It was not uncommon to meet ambassadors, distinguished industrialists, and the chairman himself while enjoying superior French cooking. Ted told me about their problems with the undersea diamonds. It seemed that a fellow named Collins was mining diamonds in the ocean just offshore of their big coastal diamond mine and probably it would be necessary for the DeBeers group to buy him out. Some of the questions were, What was the value of the undersea territory? How many diamonds were there? How deep? How could they be mined at a profit? It was evident that some new technology would be needed.

"I think you'd better fly up the coast and get a feel for the territory. There's nothing like it on earth. Have a look at our mine; then we can talk about how to proceed. Don't you agree?"

I agreed. DeBeers would furnish a company airplane to fly me nearly 1,000 kilometers north from Cape Town along a shoreline that could not legally be overflown by most aircraft because it included the coastal diamond concessions. The operating company was CDM, as Consolidated Diamond Mines of South West Africa is generally known; its small single-engine machine picked me up at Cape Town, and the pilot described the flight plan: "We'll stay in Oranjemund tonight at the mouth of the Orange River, and then push on north to Wallfish and Swakop [Walvis Bay and Swakopmund] where we'll stay tomorrow night. Then back to CDM and Jo'burg." Sounded great to me, and we took off to see the "diamond coast."

Two hundred kilometers north of Cape Town, civilization was behind us. This was the coastal veldt, and the pilot flew at about 500 meters altitude a little offshore so I could take pictures of the shoreline. This was southern

Namaqualand, a flat rangeland covered with sparse sagebrush reminiscent of Wyoming, gently sloping from bluish hills in the distance to low bluffs at the water's edge. The shoreline was generally straight, interrupted occasionally by a few rocks offshore that created sandy beaches facing into the southwest swell. About a mile inland, a dirt road ran parallel to our flight line, and once in a while a puff of dust indicated a vehicle passing. Otherwise these open spaces were deserted.

After 300 kilometers more, in which we flew over an occasional cluster of sheds used for fishing camps, we came to the "city" of Port Nolloth. This pathetic little town was a splatter of a hundred houses and sheds, where lobsters were packed and frozen for shipment. A hundred meters out of town the flat veldt began. It reached far in all directions, broken only by a string-straight road headed for the distant hills.

The port's only defense against violent seas that originated in Antarctica was a line of rocks barely awash 200 meters offshore. These culminated at the south in two large rocks that had caused a wide sandy tombolo to form, which created a small, not-very-well-sheltered harbor. Boatmen would have avoided it if there had been any choice, but this was amid the best lobstering ground off South Africa, and it was a day's run up or down the coast to the next harbor. Port Nolloth served as a port only for fishing boats who were willing to risk running in behind the rocks and anchoring in a half-open roadstead.

Another 40 kilometers, and we could see a glittering lake separated from the surf by a thin line of sand. This, the pilot explained, was the mouth of the Orange River. "The outflow is dammed by sand thrown up by big storms. When it lets go, you can hear the roar for miles." Abruptly he turned landward and presently Oranjemund, the city of diamond miners, was under our wing. The dusty desert was kept at bay by a fence and a high brushy hedge; inside a rigid rectangle of fencing, houses and buildings were neatly laid out and fronted by trees and green grass.

That night the manager regaled me with stories about diamond mining and stealing. "There's a saying in this business: If you're mining diamonds, someone is stealing them! We have to be on guard every moment. And it's hard on anyone who gets caught. The law puts 'em to work building the breakwater at Cape Town."

When we parted, he closed with, "When you come back I'll get someone to show you around the mine." Then as an afterthought, "There's no need for you to leave early tomorrow morning; if you want to see the coast, you'll have to wait for the fog to lift. You could get in a game of golf while you're waiting. We have a golf course down by the river, but you have to watch out for the red cobras". Luckily I was not a golfer.

Immediately after taking off, the plane flew low over the mine workings, where large power shovels, trucks, and conveyors were moving sand overburden from the diamond gravels on a grand scale.

A little north of the active workings, the coastal dunes were striped with deep cuts at regular intervals, perpendicular to the shore. The pilot explained that these were prospecting trenches. Inside the main cut is an inner channel in the gravels, exactly 1 meter wide, which goes to bed rock. The trenches are half a kilometer apart, and with the data on diamond concentration they provide, mine production can be forecast precisely.

Another hundred kilometers up the coast we came to Chameis Bay, a north-facing bay created by a rocky point that sheltered a wide, curving sandy beach. Beyond the protection of the point the shoreline became a rocky ledge that sloped upward from the high tide mark. "I'll show you something," said the pilot. "See. That's the wreck of the 77." He pointed to a large rusty barge high on the ledge and swooped in low so I could take a picture. "Six months ago Sammy Collins was using it to bring up a thousand carats of diamonds a day from the bottom of this bay."

Engine noise made it too difficult to get detailed answers to the obvious questions, so I saved them for later, and we flew on to a new kind of terrain, naked rocky hills and valleys. Until this point the landscape visible from our vantage point above the edge of the sea had been like a prairie, with sagebrush and ostriches the only sign of life. Now it became surrealistic and weird, a suitable background for a horror movie starring giant lizards. Rocky peaks hundreds of meters high sloped upward in steps, sometimes glistening with the special sheen of rocks blasted smooth by wind-blown sand. The valley bottoms were largely barren except for an occasional wandering crescent-shaped sand dune. The fierce windstorms for which this coast is known had blasted most of the sand and pebbles away except for those which had been trapped by cracks in the bedrock.

This was the place where the earliest coastal diamond mining had been done. The method was for a line of twenty or so native laborers, elbow to elbow, to cross these barren valleys on their hands and knees, with a match box slung around their necks, picking diamonds out of the cracks with their fingers and putting them in the box.

Next we flew over Pomona on the rim of Elizabeth Bay, a diamond ghost town for over forty years, where two lines of brick houses and shops stood gauntly against the wind. In some, the bricks had been softer than the mortar so the blasting sand had eroded the bricks away leaving the mortar intact as a ghostly openwork wall. Great piles of blue mussel shells and a rusted still for converting seawater to drinking water hinted at how the miners had survived in this forbidding region.

Then on to Lüderitz, a quiet harbor surrounded on all sides by low smooth rocky hills. By far the best harbor on this rough coast, it was first used by Bartholomew Diaz in 1486. He had left two small supply ships here with orders to trade with the natives for food and replenish the water supply while he sailed on to discover the Cape of Good Hope. When he returned, those crews were dead of thirst; there were no natives and no drinking water for over a hundred kilometers in all directions. The highest rock in town was known as Diamond Hill, not because diamonds had been found there but because residents of a small hotel that once stood on its crest tossed out empty water bottles that shattered on the rocks below and sparkled in the sun. Both the bottles and the water they contained had been imported from Germany. There was no other source.

From Lüderitz on north for over 400 kilometers, the coast was made of sand that extended inland farther than the eye could see. This is the Skeleton Coast, where shipwrecked sailors left their bleached bones because it was impossible to walk far enough to reach water. Below us there were large breakers, a narrow beach, and then the "great wall" of sand rising steeply for several hundred feet. Once camel and oxen supply trains had worried about being trapped against this steep sand slope by high waves and rising tides. Beyond there was nothing but yellow sand, dry and trackless.

The Namib desert is a place of terrible beauty and savage desolation. Its stark landscape of wandering dunes is inhabited only by a few exotic animals such as the gemsbok, a horse-sized antelope with long straight horns and beautiful markings on its head. How herds of an animal this size manage to make a living in this virtually waterless and plantless wasteland of drifting sand is a mystery. Plants are few and rare; some of them have hollow stems that seem to be made of rock. Even Bushmen stay out of the western dunes of the Namib.

Flying along that coast, we passed over a strange flat structure not far offshore known as Hollam's Bird Island. Sea birds like to nest on rocky offshore islets where they are safe from predators. Their droppings form valuable guano deposits, which are rich in nitrates but inconvenient to mine or transfer to boats, so Hollam decided to cover a group of rocks with a flat surface. A wooden floor on steel stilts makes it easy to shovel the guano. It took a few years for the birds to become accustomed to the island; then they made it into a profitable operation.

Occasionally the coastal dunes gave way to wide flat sand-filled bays, called pans, that extend several miles inland. Deep in one of these, a strange dark shape projected from a dune. My pilot said, "That's the stern of the *Edward Bohlen,*" and he turned hard right to circle the rusting hulk of a large ship. "She went up about fifty years ago. Since then the shoreline has moved seaward. For a few years its cabins were used to house diamond miners, and they say it was very odd to see a line of portholes gleaming in the midst of the desert at night."

The sun was now low and red; we had seen no sign of human life for several hours, and I began to feel an uneasy empathy with shipwrecked sailors. Then, just at dusk, a tiny black train with its firebox glowing and headlight blazing came chugging down a narrow-gauge track through the dunes, much like a scene in the movie *Lawrence of Arabia.* A few minutes later the lights of Walvis Bay twinkled ahead; there were only a few dozen of them amid blackness that reached to the horizon, but on that coast these seemed the equivalent of New York's.

The Diamond Coast and its consort, the Skeleton Coast, were as Ted Brown had said, "like nothing else on earth." On this first two-day lesson in coastal diamonds, I had flown along 600 kilometers of coast where the sea bottom was strewn with possibly diamondiferous gravels. These waters were studded with rocks over which great waves broke and bounded by the world's biggest sand pile. Except for Lüderitz, there was no really good shelter for ships from violent South Atlantic storms, and it was most of a day's run from the main areas to be prospected. All OSE had to do was think of some practicable means of estimating the value of undersea diamonds there.

SAMMY COLLINS

There is an extensive folklore about the diamonds of South West Africa, and in the wry words of an earlier commentator, "None of these odd tales have lost anything in the retelling." The following versions are as close to the truth as can be discerned from the various versions circulating.

Diamonds were first discovered far inland in 1867, along the banks of the Vaal and Orange Rivers, and the rush was on. The story of how the "digs" shifted from alluvial workings along those rivers to open pit excavations at diamond pipes and thence to underground mining is a fascinating one that has often been told. But relatively little has been written about the coastal diamond deposits. Quite logically the earliest river miners followed the Orange River toward the sea, but as soon as it entered its gorge through the mountains, they found no more diamonds. Nor did an extensive search of the gravels where the Orange discharged into the sea turn up any stones, and for the next few decades not much effort was spent looking for coastal diamond deposits.

In that period Bismarck, chancellor of Germany, declared all the territory between the Orange and Kunene rivers (southern boundary of Angola) to be under the Kaiser's protection. However, in order to protect the natives properly, it was first necessary to subdue them. This required that a railroad be constructed from Lüderitz Bay to the interior. One of the track maintenance foremen, August Stauch, instructed his native workers to show him any pretty or unusual stones they found along the roadbed through the rocky desert. In August 1908, Zacharias Lewala, one of the track gang who had previously worked in the diamond mines at Kimberly, brought him a stone saying, "Master, this is a diamond." Stauch had studied enough minerology to agree with that identification, and soon the track gang discovered a few more.

Stauch reported the find to the railroad authorities, but they did not take him seriously. "Diamonds," they said, "are dug from the ground, inland, a thousand kilometers from here."

"Do you mind if I obtain a permit and mine them myself?"

"Not at all," replied the railroad boss. "Go right ahead."

So Stauch did just that. In a matter of months, he found the barren rock valleys where diamonds waited in the cracks to be picked up by the lines of natives on their hands and knees. In a few years he was a wealthy man.

By 1911 the country's income from diamonds was 24 million marks, and the German government set up its own selling organization to compete with the DeBeers group at Kimberly. A huge area that stretched far beyond the known diamond deposits was declared forbidden territory. The *Sperregebiet* ("prohibited area") extended 80 kilometers inland along 600 kilometers of coast. It was patrolled by soldiers on camels, who were permitted to shoot diamond poachers on sight. Adventurers who survived both the desert and the soldiers had a very small chance of finding diamonds.

The area is still under rigorous security control. No aircraft may fly over it or boat land or car drive through it; all mine employees leaving are subject to X-ray examination and search. It is still a criminal offense for an unlicensed person to possess an uncut stone in South Africa or Namibia.

In 1915 a methodist missionary named William Everleigh published the first book in English that described the resources of South West Africa. In a chapter entitled "The Diamond Fields," he says, "It is not at all unlikely that new deposits will be discovered and it is believed that diamonds were found off Pomona during undersea dredging operations but these activities were terminated by Imperial Decree. As the gems are found along the coast and on the islands off the coast, it is not unreasonable to infer that they lie on the sand of the sea-bed. Here is an opportunity for an enterprising syndicate."

After the end of World War I Ernest Oppenheimer gained control of the diamond holdings of the defeated Germans, consolidating these into CDM. A few years later his men began digging the trenches that uncovered the diamond terraces near the Orange River.

On returning to CDM after that first flight up the coast, my first question was about Sammy Collins and the wreck of the *77*. Some of staff there laughed a bit derisively at this lone Texan from Port Lavaca who had challenged the great DeBeers corporation. They told me what they knew about his operations, and here is that story, complemented by things I learned later from others, including Sammy himself. What they did not quite say was that, because of his brash attitude, free-spending habits, and general good humor, Sammy had already become a folk hero, known as the Texas millionaire to readers of the *Cape Times.*

He had first come to Oranjemund in 1960 when CDM decided to build a pipeline out to sea so that fuel oil to run the mine and the city could be pumped ashore from tankers. The Collins Submarine Pipeline Company of London was one of several companies invited to examine the area and bid on the pipe laying. But soon after its president arrived, he clashed with the conservative mine manager. Sammy, flamboyant and outspoken, asked embarrassing questions. Apparently they had a conversation something like this:

S.C.: "Why don't you fellows mine the undersea diamonds?"

M.M.: "There aren't many diamonds under the sea, and it's too rough to mine 'em there anyway."

Then Sammy Collins, in a rerun of August Stauch's words to the railroad a half century earlier, said, "If you aren't going to mine them, do you mind if I do?"

The manager, exasperated by Sammy's attitude and eager to get rid of him, replied as had his predecessor: "Go right ahead."

So Sammy caught the next plane to Windhoek, the capital of South West Africa, formed the Marine Diamond Corporation, and obtained a concession to mine diamonds right in CDM's front yard. Then he obtained a surplus British oceangoing salvage tug, named it the *Emerson K,* and began prospecting for diamonds.

The system was to anchor the tug at random locations outside the surf zone, lower a 6-inch suction hose overside to the bottom, and pump up samples of the sand and gravels. These were sieved and sorted on deck, and occasionally a diamond was found—a small encouragement but hardly enough reason to start a mining operation. However, when the prospecting ship reached Chameis Bay, 100 kilometers north of the river, the prospecting ship found enough stones to justify building the *77*, the first diamond-mining barge.

Soon *77* was bringing up a thousand carats a day from the bottom of Chameis Bay. Like 90 percent of the coastal stones, these were gem quality, but they were small, the median size being about half a carat. There is no big demand for small diamonds, and ones that size were worth about $25 a carat. The Collins operation appeared to be a high-risk, low-return operation, but it attracted great attention to the possibilities in undersea diamonds, and before long five concessions were taken up by other companies, none of which seemed to be making any serious attempt to prospect or mine.

Sammy's operation was going well until a great storm arose that lasted three days and produced breakers said to be 40 feet high. The *77* was torn free of its four anchors and tossed high on the rocks at Chameis Bay, where I had seen the wreckage. It must have been a rough ride, and I could imagine the grinding crash when it struck, but there were no casualties. Later, after I got to know Sammy, I saw a photo on his office wall of that barge in the breakers with the sea at its worst. Thirty feet of the after end of the *77*'s bottom was exposed, projecting out into the air as a huge wave passed beneath.

All this had happened before I arrived in Johannesburg; now DeBeers was pondering how to proceed. Except for the data on diamonds actually brought up by the Collins ships, there was not much information about where the undersea stones were most plentiful. Somehow this huge undersea area would have to be sampled and evaluated geologically before anyone could make a sound decision.

Next morning at CDM one of the geologists took me on a tour of the mine, starting with an area where mining had laid the bedrock bare. Two distinct terraces had been carved into the metamorphic base rock of Africa long ago by waves using pebbles, cobbles, and diamonds as abrasives. When I looked along the surface of a terrace at a low angle, it seemed as flat as a table, but from above, the surface is seen to be sliced by a series of parallel surge channels 1 to 10 meters apart and about as deep. These channels were sculpted into graceful shapes by the abrading materials, as were the high and dry, undercut seacliffs against which the terrace ended. I clambered down in one and found a cluster of barnacle shells plastered on its wall, which were later carbon-14-dated at 28,000 years, indicating one occasion between ice ages when the sea covered these channels. Near the Orange River these two actively mined terraces are 10 and 30 meters above present sea level, but to the north they converge at the lower level and eventually disappear. Apparently this was caused by the gradual tectonic uplift along the gorge of the river.

Generally the diamonds were more likely to be concentrated in the bottoms of the surge channels or in potholes along with heavy mineral sands. The previous January, one pothole less than a meter across was found to contain several hundred stones averaging half a carat. Other nearby potholes had few stones. Production temporarily halted in the area while the chief geologist, Dr. Arnold Waters, tried to figure out why. If there was any explanation besides chance, the evidence had long since vanished.

Outside the workings are huge piles of sand—former dunes covering the gravels, which had been moved aside. Once the gravels were cleared of sand, they were scooped up and taken to a processing plant where their diamonds had been extracted. The final step in mining comes when the bedrock is scrupulously cleaned with brushes by Ovambo laborers to recover any diamonds left by the machines. These men are paid cash rewards for any stones they find, and it is not unusual for a man to pick a month's pay from one crack.

Generally the larger stones are found nearest to the mouth of the Orange River where the median size is about 1.7 carats; to the north the size decreases until at 100 kilometers the median size is down to 0.6 carats. This is pretty good evidence that the diamonds came down the Orange River and that over millions of years the smaller, lighter ones were gradually moved northward by littoral currents caused by waves from Antarctica striking the shoreline at a slight angle. An occasional storm from the north temporarily reverses the direction of the littoral drift and accounts for relatively few stones found south of the river.

Size variations are very important because the value of a diamond crystal is very much related to its size; that is, a 2-carat stone is worth four times as much as a 1-carat stone. Stones smaller than half a carat have comparatively little value because some 40 percent of the weight is lost in cutting.

Other bits of information came to light whose importance was not immediately apparent. One was that the coastal stones would not stick to the grease tables that are ordinarily used to separate diamonds from heavy pebbles at the inland mines. Second, the coastal stones are about 90 percent gems, whereas those found inland are only about 20 percent gems. The industrial stones that make up the remainder have the same physical properties as the gem stones but any theory of diamond origin must explain this distribution.

A diamond's most special property is hardness; they are ten times harder than corundum, the next hardest mineral. Diamonds almost always occur as well-formed crystals, and because of their hardness, they are rarely damaged by natural processes. If chance puts a tiny diamond and a large quartz boulder in a pothole together, in a few hundred years the pothole is larger and the boulder has become a pebble, but the diamond is as it was in the beginning, its crystal edges still sharp. Later, in the DeBeers sorting room, I was shown almost spherical stones that appeared at first to be worn or frosted but on inspection with a glass, these were seen to be covered with many tiny crystal faces.

Excited by the crash course in diamonds and my head buzzing with theories about diamond origin and distribution, I returned to Jo'burg, still wondering what I could propose. At least I had learned enough to carry on a sensible conversation.

THE *XHOSA COAST*

Ted Brown's greeting was warm and enthusiastic. "What d'you think of the coast of South West?"

"A fascinating place, rugged and deserted. I sure thank you for the chance to see it."

"It's a tough place to work. Especially at sea."

"So I see, but that's the business OSE is in. Exactly what would you like us to do?"

"You looked at the old sampling trenches along the beach? The ones started by Sir Ernest?"

"You mean the one-meter wide trenches through the gravels to bedrock where the sand was cleaned away."

"Just so. We'd like a set of trenches just like those made undersea. We think that would give us a good valuation on the property. Don't you agree?"

I didn't disagree, but it sounded like a really tough assignment in a region where the surf zone was too rough and too wide to permit access from the land. I thought for a moment and made a counterproposal: "Let me suggest a way of getting started. No one knows how thick the undersea gravels are, or how far out to sea they extend, or what kind of rocks underlie them and whether there are wave-cut terraces like those on shore. Why don't we start with a seismic survey and find out?"

Ted agreed, and that broad proposal was soon converted to a management contract that required Ocean Science and Engineering to assemble a staff, find a proper ship, locate seismic equipment, and set up a navigation system. First we had to find office space. Our ship would inevitably operate out of Cape Town, and in a conversation with Harry Oppenheimer, I mentioned that we needed an office there. He said, "Use mine for a while. Since I'm no longer in Parliament, I don't need it." It was, of course, a beautiful location with a second floor bay window overhanging St. George Street and the park. Before Harry, it had been used by Sir Ernest Oppenheimer and, I believe, Cecil Rhodes.

Before long OSE had set up a South African subsidiary and engaged a bright and energetic geologist, Dr. David Smith, as managing director. The most difficult problem was to find a suitable ship. There was a very meager choice of ships in southern Africa, and it would be both expensive and time-consuming to find one abroad and bring it in. Then someone mentioned a tiny freighter that we suspected had been used in the smuggling trade along the African coast, because it could enter small harbors or rivers not ordinarily watched by customs agents.

The *Xhosa Coast* was about 130 feet long, driven by a single screw, and it had crew quarters that would have given a submariner claustrophobia, but it had two great advantages. There were two fair-sized cargo holds, and it was available immediately.

Space was needed where geophysicists could work and live. That was obtained by sandblasting the main hold and filling it about one third full of carefully laid bricks to keep the empty ship from bouncing in big waves. Over the bricks a floor was laid, a huge table installed on which charts and seismic tapes could be laid out, and some bunks were added. A new 5,000-gallon water tank in the forward hold permitted the ship to remain at sea for several weeks and also served as ballast.

Our subcontractor for probing the underwater rocks was Alpine Geophysical. While its "sparker" system for sounding layers beneath the bottom and its precise navigation system were being installed on the ship's bridge, I spent a few days at the CDM mine, thinking about the problems of undersea diamond-hunting.

In the early morning, standing on the beach peering out through fog that hung over the water, one could feel the sand shake and hear the roar as each unseen breaker crashed. A few seconds later, an onrushing wall of broken white water 6 to 8 feet high would materialize through the mist. Except for the fog, conditions were not much different from those on the Oregon coast in winter. Nor did the surf appear any lower when the fog lifted.

I lay awake for two nights, wondering about the wisdom of risking a shipload of men in these rough, foggy, inshore waters. Was it madness to send the old *Xhosa Coast,* loaded with bricks, water, noisemakers, and geologists, on a long, crooked track close along the Skeleton Coast? In order to get some statistics I sent a man to copy weather data from the log of the *Emerson K.,*

Sammy's prospecting ship. It had already worked months along this dangerous coast, and after reading the laconic accounts of sampling amid heavy
swell, I decided the risks were acceptable.

The inshore waters were not well charted, and there was always the
possibility the ship would hit an unrecorded rock. Our skipper was required
to spend long hours on the bridge as the ship cruised slowly just outside the
breakers. Eventually when the *Xhosa Coast* did strike a rock and a small hole
opened in the bow plates, he threatened to quit, but Dave Smith convinced
him he was a hero of hydrography, and the rock was named after him:
Merten's Reef.

Alpine Geophysical's system for probing the bottom consisted of towing
two lines behind the ship. One ended in a spark gap through which a bank
of condensers discharged about once a second creating a small explosion; the
other line was a string of hydrophones that listened to the reflection of the
shock waves from each layer beneath the bottom. When the returning signals
were recorded, it was possible to distinguish between the top of the gravels
and the surface of the rock beneath, the amount of separation indicating the
thickness of the gravels.

The problem of knowing the ship's precise position was more difficult.
While working off the CDM mine, it was easy to set up radar transponder
beacons at each of the existing survey monuments. These beacons were tuned
to accept the ship's radar signal and return a pulse in a slightly different
frequency that would show as a bright dot on the ship's radar scope. A
technician would then adjust two range rings on the scope until they were
exactly tangent to the two bright dots. Then he would press a button, and the
ship's position would be pricked in a chart of the coast by a moving needle.
These methods sound archaic now, but they were state-of-the-art technology
in 1963.

North of Lüderitz in the great dunes, there were no survey markers, so
it was necessary to run our own survey lines across the sands of the Namib
desert to locate the ship's position at sea. Land Rovers or jeeps can cross the
desert slowly, but in order to keep ahead of the ship the survey parties used
a Skylark helicopter. It would leapfrog the surveyors from dune to dune, in
5- to 10-kilometer jumps. Theodolites would be used to measure precise
angles, and tellurometers, a kind of electronic tape measure, would get distances. The trick was to be able to see the next station through blowing sand
or fog or shimmering heat waves. Even though the foresights and backsights
were marked by mirrors that reflected the sun, it was not always easy to see
the points.

The helicopter had a short operating range and used a lot of fuel, which had to be brought in by a small plane called a Beaver. It would land on the wide flat pan near the old wooden buildings of a long-deserted diamond camp and roll the gasoline barrels out the door. Then the Skylark could go where it pleased over the Namib's sea of sand, skipping from the tops of huge dunes to the pans to the beaches and back to camp again. A week in that camp, sleeping on the ground and cooking over a fire, gave us a feeling for the life-style of the old diamond miners. Slowly the insistent noises of civilization faded from our memory.

Rhoda flew up with us to pick up a man at Swakop airport, an installation made of and surrounded by sand in all directions. It consisted entirely of a wind sock and a line of white rocks alongside an unpaved flat area. Rhoda and he exchanged places. He had arrived by Land Rover, and she had never before driven one. We showed her the gears and pointed west across the wasteland of sand, telling her to follow the tracks to town. Then we took off for the desert camp. When we returned a week later, Rhoda had not only found the town but made friends with some of the wealthy caracul breeders who vacationed there; thereafter she referred to the place as the Riviera of South West Africa.

The quiet and isolation of the Namib sets one thinking. Just to see what it was like I walked out on a vast pan that Kipling would have described as an "illimitable plain" until the camp was but a speck in the distance. The sun blasted down on a hard flat surface that had once been a lagoon and now was studded with oyster shells, inexplicably standing on edge. I tried to imagine myself a first-time explorer, a shipwrecked sailor, or a hunted diamond poacher as I trudged through the heat. But, knowing we had a helicopter, I was too secure to be scared by my imagination.

At night there was usually a ring of bright jackal eyes just outside our camp. Jackals looked like friendly dogs, and one day Jack Mardesich, our chief engineer in South Africa, and I tried to catch one with a jeep. We had speed and endurance, but he had a shorter turning radius and knew more about the desert. After playing with us for a while, he deliberately led us across a patch of soft sand. The jeep went in over its hubcaps and stopped abruptly, almost sending us through the windscreen. The jackal sat watching us dig for a while, then trotted off wearing a crooked smile.

Driving along one beach, Jack and I discovered that clams about 4 inches long, known locally as white mussels, inhabited the wet sand at the water's edge. In an attempt to estimate how many of these might be available for commercial harvesting, we drove about 10 miles along that shoreline, stopping at random points. At each stop I would dig my hands with fingers spread into the wet sand and make a single scoop, bringing up an average of seven mussels each time. We were impressed, but what we did not know was that these were the small ones; later, someone demonstrated that the clams found by digging a foot deeper with a shovel were twice as large and just as plentiful. There are surely enough clams in that one beach to start a new industry if it were not inside a diamond concession.

One night while the survey party was working at Meob Bay, a strong windstorm blasted away much of the surface sand on the beach. Next morning, our men picked up hundreds of green copper coins on the beach, each perched atop a little cone of red garnet sand. Each of these bore the legend VOC and a date, showing they were minted by the Dutch East India Company in 1746. Among these pennies I picked up a Mexican pillar dollar, a large silver coin dated 1743. Then the hunt intensified, and for a few hours our surveyors paced the beach like zombies concentrating on the small patch of sand just ahead of their feet. We knew we were close to the untouched wreck of an East Indian, but there is no time to spend on silver when you're after diamonds.

Occasionally the surveyors would come across desert adders, deadly snakes that lay buried with only their eyes above the sand so they are almost impossible to see. As one of our survey helpers, an old desert hand who had once collected snakes for zoos, told me about them, I noticed his left hand was missing its index finger. "Oh that," he said, "I got careless, and one of them hit me. So I whipped out my sheath knife and whacked off the finger. Damn near died anyway. Their venom works fast." When I asked if there was any protective action the company should take he snorted, "You might get shovels with longer handles."

A small helicopter with a plastic bubble cockpit was sent to take Mardesich and me down the coast to another desert camp, but that day the usual morning fog did not clear. From above, the low white clouds stretched across the horizon, and it was impossible to see anything on the ground, so in order to find the next camp, the pilot elected to fly under the clouds along the shoreline. He would follow the narrow beach, keeping the great wall of sand on the left and the surf on the right, both barely visible through the vapor. At an altitude of less than 30 meters and at a speed of 50 knots, we headed south along this perilous corridor through the fog. The pilot concentrated intently on following the ribbon of beach below; Jack and I peered intently over his shoulder into the onrushing mist looking for rocky obstacles that might project abruptly from the sand.

We were all tense, knowing that the flight was an unnecessary risk (we could have waited for the fog to lift) and that, if anything went wrong, we would be instantly dead. But after a while it became hard to concentrate as the sand and surf flicked past on either side. Our eyes began to glaze over.

Suddenly there was a millisecond of terror. Dead ahead of our chopper, on this remote beach in no-man's land, a drilling rig loomed. We all shouted at once. The pilot reacted instantly, and our flying bubble zoomed just high enough to clear the top of the mast by a meter or two. White-faced, the pilot eased us back to our previous altitude; but for a while our hearts raced and our breath was short. That was about as close as a fellow can come to "going on the great adventure" and live to tell about it.

We never learned why a drill rig had suddenly sprouted on a beach that had been deserted a few days earlier.

Back in Cape Town, Dave Smith and the geologists converted the geophysical data accumulated by the *Xhosa Coast* along 1,425 survey miles into useful charts. Our sonic work had discovered a gorge in the bedrock beneath the sediments off the mouth of the Orange River some 70 meters deep. This was not unlike the undersea gorges off the Nile and the Congo, indicating a much lower stand of sea level long ago. The sparker had also found evidence of two kinds of previously unknown sedimentary rocks lapping up on the flanks of the African basement rocks offshore. One of these was named the Osesa formation after our company's acronym. There were seacliffs, broad depressions in the base rock filled with sediments, and layers containing gas bubbles that blanked out the sound reflections from the deeper rocks. We also found lots of gravel deposits, but they were patchy, and each would have to be examined for diamonds. But exactly how?

THE *ROCKEATER*

After six months of exploration work, our company had developed enough working knowledge of sea conditions and alluvial diamond deposits to rethink the problem of how to dig undersea prospecting trenches. We came up with an alternative idea. Instead of a trench across the bottom, OSE proposed drilling lines of large diameter holes and bringing up the material for processing on the drilling ship. This would give similar results to that of the trenches—that is, vertical-walled holes with known volumes of gravel that could be converted to diamonds per cubic meter of gravel.

Sammy Collins had by this time built two more diamond mining vessels, the *Dimantkus,* an LST conversion, and the *Colpontoon,* a barge. The De-Beers group, eager to move rapidly, accepted our proposal. We were given five months to create a new kind of prospecting ship. Back in the United States we summoned our best talent: Ed Horton, Pete Johnson, and John Marriner, all ex-Mohole men, and started to work.

What's big and red and eats rocks? A big red rockeater, of course—just what we needed for the Diamond Coast. We began by buying a hull 186 feet long, built as a U.S. Army freight ship for World War II use but still in mint condition. With a South African government representative aboard, it was towed out beyond the three-mile-limit, where the United States flag was lowered and the South African flag raised. The conversion of this now-foreign ship began the next day in Long Beach, California, at a shipyard OSE later bought.

Two months later to the day, *Rockeater* sailed for Cape Town, its official home port, looking like a new ship. It was under the command of Captain Cook and crewed by South Africans. In that short time, a center well had been installed and a derrick built above it, complete with drilling equipment. Four large anchor winches were operating, a big hydraulic system had been added to furnish power for the drilling and deck machinery, a large outboard propeller was mounted on the stern and a bow-thruster added forward (in addition to the two main screws), and additional living quarters had been constructed. On the bridge a new console directly controlled the engines and automatically steered the ship.

In Cape Town a local contractor set to work installing a diamond-processing plant below decks: screens, bins, cyclones, jigs, and sorting table. That would take another month. Only then did we learn the significance of the completion date we had been given months before. The DeBeers board of directors would be in Cape Town on that day, and the members wanted to see a demonstration by the new ship that would drill for diamonds. If there is anything that experience with newly built machinery teaches, it's this: "Never demonstrate anything thats not been thoroughly tested." Although there had been no chance to debug the drilling system, our chaps dared not tell the directors we were unprepared.

An hour's run out into Table Bay, before the ship was properly anchored, it began to drill its first hole while Ted Brown, Chairman Oppenheimer, and the distinguished board members looked on. When the pump that was intended to lift diamond gravels aboard was started, one of the first items brought up was an octopus. One of the black crew grabbed it off the screen, intending to eat it raw, but the creature squirted from his hand and landed on the head of a distinguished lady observer on the deck below.

After that things got worse. The ship moved from a point above the hole, and the pipe broke, pulling the swivel to one side. That sheared a bolt on the cross head that fell 40 feet to the deck, narrowly missing Baron Rothschild, who skipped quickly from the drilling deck to the safety of the bow. Now the loose crosshead wedged, jamming the hoisting system. Humiliated, our fellows had no choice but to leave the bit stuck in the sea bottom.

As a demonstration the occasion was a disaster, but the board reacted with reasonably good humor and indicated they had been entertained adequately. The next day, without visitors aboard, the bit was retrieved, and the ship began serious drilling trials. A week later it was busy drilling the undersea equivalent of prospecting trenches.

Rockeater's system was to lay a forward anchor on each side of the proposed line of holes and back shoreward, paying out anchor wire until the incoming waves were beginning to turn light green under its stern just before breaking. There, at the outer edge of the surf, using a special bit that Ed Horton had designed, the first hole would be drilled through semiconsolidated sands and gravels until solid bedrock was reached. The bit was a steel tube about 12 feet long and 26 inches in diameter; its perimeter was rimmed with hardened steel teeth. Just inside the teeth a row of nozzles directed jets of high-pressure water against the gravel below, blasting it free of the bottom. The upper part of the tube thus became a suspension chamber containing a slurry of gravel supported by the turbulence created by the jets below.

Above this chamber there was a telescoping section that compensated for any vertical motion of the ship's. Above that the tube narrowed to a 6-inch pipe jacketed by an air chamber; hundreds of perforations in the pipe wall allowed compressed air to enter the pipe and then rush surfaceward carrying water, sand, and rocks with it. This combination of jets and air lift was sufficient to lift cobbles from 100 feet below the surface up into the derrick and then push them horizontally through another pipe to the sorting screens.

The bit did not rotate continually in the same direction but reciprocated, turning first one way and then the other like a giant cookie cutter. This prevented hoses from fouling and reduced the chance of overstressing the drill pipe if the bit hung up.

At the screens, sand grains smaller than 2 millimeters and rocks larger than 16 millimeters (about ¾ inch) were immediately shunted back overside along with the excess water. Visitors aboard found some amusement in watching the oversize rocks bounce across the top screen on their way to the discharge chute. These stones were clean and easy to identify; they ranged from broken chunks of metamorphic rock to stream-worn granite cobbles. Occasionally a new observer would watch their mesmerizing dance on the screen for a while and say, "Aren't you fellows concerned that a big diamond will be thrown back?"

To which Ed Horton would reply with a straight face, "What really worries me is the ones larger than six inches that can't come up the pipe."

Three drill holes exactly covered 1 square meter of surface and usually produced about 1 cubic meter of proper-sized gravel, enough for a mill run. Having collected its gravel sample at one station, the ship would take in on the forward anchor wires, move itself seaward 100 meters, and drill three more holes close together. Finally, when the gravel bed ended or the water depth was over 60 meters, the line of stations was complete.

After each set of three holes the plant below decks would start processing the gravel, putting it through a cyclone first and then jigs. The cyclone was a conical container in which the gravels, mixed with a heavy solution of ferro-silicate, were spun by jets. The effect of rotation in this medium was to exaggerate the effect of gravity so that the lighter stones could be drawn out at the bottom and the heavier ones, including the diamonds, would flow off at the top. From here the heavier gravel went to the jigs; in these, shallow containers about 18 inches in diameter with a mesh bottom are "jigged" (oscillated) up and down, causing the heavier stones, including the diamonds, to migrate toward the center.

Periodically the trays are removed from the jigs, carried to a sorting table and flipped over for inspection. There are thousands of pebbles in this pancake-shaped mass of stones, illuminated by a single bright light from above (like those in jewelry stores). Any diamonds present are most likely to be in the center, atop a core of darker stones, and they shine like little searchlights.

On one occasion OSE vice president Larry Brundred tried his hand as a diamond sorter; he was methodically examining each pebble and had carefully set aside five bright, clear stones about which he was uncertain. Then he came on a real diamond; its sparkle was so much greater that after one quick look he swept away his previous selections. Diamonds are unmistakable, and the nice thing about this kind of prospecting was that it produced an answer in diamonds on the site.

Although there was always a security guard watching the sorter and all stones found immediately went into sealed packets inside steel cases, we learned that some of our diamonds had turned up in the possession of one of the friendly night ladies of Lüderitz. We arranged that when *Rockeater* next stopped at that port, the IDB (Illicit Diamond Buying) police would immediately board the ship for a surprise search of the crew, so we could catch the culprits. But for some reason the police arrived an hour after the ship docked, strolled aboard, and asked a few casual questions. Predictably, they caught no one.

The loss of a few stones was negligible, financially speaking, because *Rockeater*'s real product was information. The serious loss was that we never learned how many stones were taken or from which holes they had come. Since the results of such prospecting are multiplied by a factor of as much as a million to determine the value of a mining property, this meant that we may have greatly undervalued some underwater areas. On the next trip to Cape Town, the entire processing crew was changed.

The officers and scientists aboard *Rockeater* were white, the drilling crew was black or African, and the ship's crew was "colored," meaning Asian or mixed breed, in accordance with South African classification. They were all good fellows but inclined to play practical jokes, and managing them was not easy. For example, when the drilling supervisor brought his pet monkey along on one trip, it was too tempting a morsel, and soon it was captured, cooked, and eaten by the black crew. This, of course, started a ruckus that lasted several days.

On another occasion Jack Mardesich and the ship's radio operator rigged a cherry bomb in the chief engineer's carefully built radio-controlled ship model, so it would blow up while he was playing with it. The chief retaliated the next night by slipping a ten-man rubber boat, rigged to inflate itself automatically, under the radio operator's upper berth mattress. When our radioman got in bed that night, the raft quickly filled with air, shoving him hard against the ceiling beams and holding him there until he was rescued the next morning. Luckily for all of them, I did not learn of these events for several years.

While *Rockeater* was drilling for diamonds, the oceanographic ship *Vema* with Maurice Ewing aboard put in to Cape Town for repairs to its electric generator system. I stopped around for a talk that evening and found the ship dark, quiet, and almost deserted. A deck hand led me by flashlight below to the great man's cabin, where he sat going over charts by candlelight.

We talked the evening away about such matters as the world's changing attitude toward the theory of continental drift, new and puzzling magnetic data that showed huge offsets in crustal rocks, the possibility of finding really ancient sedimentary rocks in deep water and, of course, diamonds.

Maurice was a crusty character, probably the best seagoing geophysicist alive, overflowing with original theories about the earth. He had found himself in a wide-open field and charged ahead, not afraid to be wrong, occasionally. As we sat there chatting by candlelight in his old ship with the background sound of small waves lapping against the hull, a strong feeling came over me that in our very different ways we were both pushing back the frontier of oceanography. I remembered the time he had been washed off the *Vema* in a gale and how he had supported me during the Mohole project infighting. Ocean scientists were lucky he was still among us.

Maurice's parting words were, "Go for the big diamonds. Don't take any smaller than walnuts." I couldn't bear to tell him those were the ones that bounced across our tailings screen.

Rockeater was usually accompanied by the small geophysical ship *Klipbok*, which towed an acoustic profiling system. This device, a 3.5 kilohertz high-resolution sonar capable of penetrating 50 feet of sediment, was developed by Dr. Clyde Lister and became known as Lister's Lizard. It was a very convenient way of finding patches of gravel to be drilled amid the areas of barren bedrock. The sediment thickness data it produced was used to estimate the volumes of gravels available for mining.

By the end of 1964 *Rockeater* had drilled 3,074 holes in the Marine Diamond Concession, totaling 22,226 feet deep, and had found 129 diamonds. This must have been a world record of some kind.

Dave Smith and his crew of geologists, including Drs. John Hoyt, George Shumway, Ben Oostdam, Robin Harvey, and Willie Van der Merwe, struggled to reduce the findings to geological charts that showed where every hole was drilled and how it related to undersea topography and geology. That was followed by an intense statistical analysis of the number of stones per square meter, per cubic meter of gravel, and per group of holes. The scientists used moving averages, triangles of influence, and several other correction factors to come up with a Final Adjusted Estimate of the Mineable Caratage: 10.5 million carats in that concession.

Sampling and calculating the quantity of mineable diamonds on the Tidal, Southern, and Atlantese concessions consumed another year. While *Rockeater* was drilling the Tidal Concession off the Skeleton Coast, a great storm arose, by far the largest its crew had ever seen. There was not enough time to run for either Lüderitz or Walvis Bay, so the ship moved in behind a small guano island to take advantage of whatever shelter it afforded. Once in its lee, the ship set its four main anchors and hung on in heavy swell for nearly three days. On the second day they heard distress calls from Sammy Collins' diamond mining barge *Colpontoon* far to the south in Chameis Bay. A couple of its anchor lines had parted, and each wave was driving it a little closer to the beach.

Presently Sammy came up on the radio directing his tug *Collinstar* to go into the rocky bay and save the drifting barge. Thus began a real-life four-hour drama of salvage, in which the shouts were frequently punctuated by bursts of "galley music" (cussing). Eventually the tug and its intended tow collided, flung together by a great wave; the tug sank with all hands, and the *Colpontoon* went on the beach. The loss of the men and the tug was very sad, but when the sea became calm again, the *Colpontoon* hooked up its many air compressors to its punctured pontoons and was towed back to Cape Town balanced on a mass of air bubbles. There it was moved into a graving dock. When the dock gates were closed and the dock pumped out, a coelacanth (a so-called fossil fish, very rare) was found in the water around the barge. Believe it or not!

When the storm subsided, the men aboard the *Rockeater* could see that the outlines of their protective island had been changed considerably and much of the guano had been washed away by overtopping waves. Some of the men went ashore in a small boat and found two mummified bodies dressed in armor, which had been exhumed by the storm. They buried these again (unfortunately without taking pictures that could be used to get their age and nationality) and then returned to diamond prospecting.

OSE had hoped to become involved in the mining of the undersea area, but the Anglo-American-DeBeers group, which was now one of our shareholders, offered us another deal. They would finance a new jointly owned company to be called Ocean Mining, A.G., of Zug, Switzerland, to prospect for other minerals under the sea around the world. Under the direction of François Lampietti, our group moved on to prospect for tin off Thailand, Malaysia, and Tasmania, gold off Alaska, titanium and zirconium off Australia. Those projects were fun too, but without the glamour of diamonds.

X

Ocean Science and Engineering

BEGINNING ANEW

When we first formed Ocean Science and Engineering in 1962, it was not at all clear what our principal business would be. The first problem was to get enough income to eat while we decided where to focus our attention. To accomplish that the other four fellows, Horton, McLelland, Lampietti, and Johnson, had taken a contract to advise Brown and Root on the continuation of the Mohole Project. Their ideas and suggestions were not accepted but the income held the wolf at bay while I found some other projects for OSE.

One of the first things I tried, even before the diamond prospecting, was a proposal to the Central Intelligence Agency and the U.S. Air Force based on an invention of mine for finding and recovering objects lost on the ocean floor. The idea had been conceived on Christmas Eve, 1961, when I was contemplating sea trade in ancient times and pondering the problem of how to locate and raise ships lost a couple of thousand years ago in the depths of the Mediterranean. In January 1962 I learned that the Soviet Union was lobbing intercontinental ballistic missiles into the central Pacific. The nose cones had been observed splashing down in the ocean near Palmyra Island, where the water depth is 18,000 feet, so I modified my ideas to fit that problem. Neither the Soviet Union nor the U.S. Government believed that these could be retrieved from such deep water, so my unsolicited proposal for recovering the nose cones created quite a stir in the intelligence community.

The idea was to use a center-well drilling ship equipped with dynamic positioning to lower a long drill pipe at whose tip a side-looking sonar would be mounted. A new version of this kind of sonar made by Westinghouse for the Navy's mine hunters did a marvelous job of using pings of sound to create shadowy pictures of objects such as crab holes and old tires on shallow harbor bottoms. Presuming that a similar one could be made for deep water work, I proposed that such an instrument be towed at a height of about 100 feet above the deep bottom. If so, it would readily detect a nose cone within a distance of many hundreds of feet on each side of the ship's path. Nowadays, the deep-water use of such equipment is routine, but then it was a new idea and, as such, was viewed with some suspicion.

When a likely target was observed protruding from the bottom, the ship would stop, lower the pipe tip until it was within television camera range of the bottom, perhaps 10 to 15 feet, and make a detailed inspection of the object. Then, if the object seen was indeed a nose cone, tongs on the end of the pipe would grasp and retrieve it.

OSE's original proposal showed profiles of the sea floor in the drop area and gave some estimates of the sizes and weights of nose cones, along with calculations of how fast one would fall through the water and how deep into the mud bottom one would penetrate. (A 9-foot-long nose cone weighing 1,500 pounds would sink about half its length into the bottom.)

The men I talked to inside the in CIA and Air Force were plainly delighted with this entirely new idea for intelligence gathering, and they gave the project the code name of "Sand Dollar." They arranged SECRET level meetings at Andrews Air Force Base in Washington, D.C., and Edwards AFB in California, where I explained my ideas for deep-water salvage to groups of officers and suggested using a geological drilling program as a cover story. At each meeting we discussed in increasing detail how the objective could be reached. The participants acknowledged the "proprietary" rights of OSE that were claimed on the cover of the proposal, and it looked as though we would get a contract of some kind, at least to develop more detailed plans. But, after a few weeks, the contacts abruptly ceased, and the intelligence people refused to return my calls. A letter from General Philip Strong, the deputy assistant director, said the CIA could not afford the project without the help of the Department of Defense. Something had gone wrong, but it was seven years before we learned what had happened.

Both the Mohole and the undersea diamond projects had been interesting to the press, and they generated considerable good publicity for our new company. We had begun operating at a time when the public was being deluged by stories about the promise of "inner space." The ocean, some writers claimed, was the great new frontier; fortunes would be made from its vast resources, and the Federal government would soon spend hundreds of millions of dollars on contract research as it was doing on the space program.

One result of this publicity was that, in board rooms of many major corporations, the directors asked how their company could participate in this new and vague—but prospectively large—industry. Should they set up a new division or invest in an existing group with an established reputation? Some of those who chose the latter route contacted us, offering to invest in OSE. The first would-be participant was DuPont, a very attractive corporate partner. However, because we were so involved with the DeBeers–Anglo American group in diamond exploration, we thought it best to offer the latter a "symmetrical" deal. The result was that each of these companies bought the same amount of OSE stock for the same price at the same time.

Once we had financial backing by two of the world's major corporations, the word spread fast. Over the next few years the value of our stock steadily rose, and other major corporations invested and joined our board of directors; these eventually included Alcoa, Amerada-Hess, Southern Natural Gas, and Fluor.

OSE had not been eager to sell its shares to major corporations; in all cases the new investors had approached us wanting to buy in. We responded that contracts and projects were harder to find than financial backing, and we added a condition to stock sales: Each new corporate investor would sponsor an appropriate project with OSE that might be expanded into a business. In the case of Alcoa that turned out to be the construction of the world's largest aluminum ship, for Fluor a new company named Deep Oil Technology, for DuPont a submarine pipeline down Delaware Bay, for Anglo-American an undersea mining company.

OSE also searched for business opportunities in old industries where we thought a shot of new high-tech ideas might be helpful. We built an underwater dredge to replenish eroded beaches with offshore sand; we worked on an automatic shucker that would speed up harvesting the big beds of calico scallops off Cape Kennedy, Florida, and we took over California Shipbuilding and Drydock Company in Long Beach.

OSE probed for opportunities in many fields. It designed and modified ships to perform special missions, and studied waves in various parts of the world. It worked on specialized problems of the oil industry connected with deep-water or rough-water drilling, made geophysical studies of remote areas of the world, and operated ships. All of these projects appeared more profitable in our advance estimates than they did in practice, and although we made valiant attempts to find good undersea mining property, to sell heavy marine and undersea oil machinery, and to operate research ships, these were only mildly successful.

These efforts in many directions made the company look forward-thinking, wide-ranging, and busy. But none of us knew much about business; we did not have the necessary instinct for converting these trial projects into really profitable operations.

In 1969, OSE guided by its treasurer, Willard Bernardin, made a public offering at $12.75 a share (relative to the 5 cents a share that we had paid before it split). Abruptly all of our original group were rich, at least on paper. We were riding the crest of the wave. *Life* magazine did a long story, describing me as the "Trailbreaker of the deeps, part Columbus and part Barnum" and revealing that I had become, by the luck of Wall Street, the first to make a million bucks out of the new oceanography. These paper riches had no affect on my life-style; it did not occur to me to spend it.

BUOYS, SHIPS, MONSTERS, AND WAVES

The fact that some of our projects were not profitable does not mean they were not technical successes or that we did not learn a lot about the ocean and have a lot of fun working on them. Our backers wanted a good shot at the new technology, and they got it. But the attempt to keep the corporation at least breaking even by finding contracts reminded me of Chairman Mao's expression, "Like a blind man catching sparrows." I seemed to snatch futilely in all directions, only rarely snagging one of the few contracts erratically flitting about. Someone once asked Jack Wrather, the oil man-entrepreneur who successfully promoted the *Queen Mary* and the *Spruce Goose* into a major tourist attraction what Bascom did. He thought a moment, grinned, and said, "Everything."

It was a nerve-wracking business, contracting to develop complicated hardware on virtually impossible schedules. Sometimes equipment had to be guaranteed to work and built for a price that had been fixed before we were aware of all the problems. Some of our competitors were large companies who could afford large losses to establish themselves in the same business. Sometimes we lost our shirts. Ocean equipment or resource development is not a business for the unsure or faint-hearted. For example, let me describe WEBS.

WEBS (for Weapons Effects Buoys System) was probably as complicated a set of buoys as has ever been built. It was intended to measure certain characteristics of a nuclear explosion at sea. Those were the days before solid-state microcircuits, when electronics racks were bulky with tubes and required more power, when telemetering systems were relatively primitive, and when no one was sure about the ability of the electromagnetic effects of a nuclear explosion to obliterate magnetic memory. The explosion was to be in deep water far from land, and its effects were to be detected by sensors mounted on the tip of a spar buoy and recorded by a dozen oscillographs housed in an underwater buoy. The electronics were to be made by a team at Sandia, New Mexico; our job was to design and build the buoys and heavy hardware.

Under Jack McLelland's direction, the OSE team set about designing a buoy that would hold an electronic rack 2 feet square and 16 feet long containing the dozen or more recording oscillographs. It was to be moored 100 feet below the water surface where it could protect the delicate instruments against a 1,000 pounds per-square-inch shock wave from the explosion. This buoy was a rather formidable cylinder, 20 feet long and 4 feet inside diameter, circled by a dozen rings made of H beams to withstand the pressure. This buoy was larger than most research submarines and it had many tons of excess buoyancy; it was intended to be taut-moored in a vertical position to a heavy clump anchor made of 20 railroad wheels. Getting this cumbrous piece of hardware installed at the proper depth in deep water was quite a problem.

It required a large spherical buoy from which an electric winch holding the mooring cable was suspended. On command from the surface, the winch would turn, winding in the anchor wire and pulling the sphere downward, towing the cylindrical buoy behind it. The cylinder, in turn, towed the spar buoy down. When the flotation chamber of the latter was barely submerged, with only its tall slender mast projecting above the water holding the sensors, the three buoys were in a vertical line, ready to measure the explosion. Electric power for all these buoys was supplied by an electric power cable to a catamaran boat whose hulls carried racks of storage batteries.

The commands to the buoy were sonar signals from the ship that did the installation. Controlling buoy depth with sonar is easy for one buoy in the open ocean, but we were to have six such systems within a few miles of each other, where dozens of ships using sonars of many frequencies might be operating. That required us to use signals specially coded for each buoy (like individual combination locks), so no other sonar would accidentally order these buoys up or down.

This sketchy outline hints at the complexity of the four-buoy WEBS system. Six of these systems had to be installed at various distances from ground zero by Navy personnel aboard a Landing Ship Dock a few days before the explosion. This was a substantial undertaking, but somehow OSE delivered a workable system, and the Navy installed it properly. As far as I know, everything worked well.

Other projects had to be done on an emergency basis. One of these was the reconversion of a geophysical ship for Gulf Oil Company. Dr. Hollis Hedberg, Gulf's chief of exploration and a supporter of our group at the academy during Mohole days, had persuaded his company that it needed a ship to hunt for new oil fields on a worldwide basis.

At that time it was common practice for an oil company to obtain survey rights on a block of land and then engage a geophysical contractor to make a seismic survey of it to look for possible oil structures. Subsequently other contractors might be engaged to make gravity or magnetic surveys of the same areas, each using different ships and navigation systems. Sometimes it was difficult to reconcile the results of several separate surveys with each other.

Hollis Hedberg's idea was for Gulf to use one survey ship that could make all the geophysical surveys simultaneously. This would avoid having to match up several sets of ship tracks and make it much easier to work out the relationships between, for example, seismic and magnetic anomalies. The ship would be permanently employed, moving from country to country around the world, sometimes in a clandestine manner. The cost of data would be lower, and there would be less chance of information leaks.

Gulf was in a hurry. It wanted the new ship to be operating within three months, and they had come to OSE because we could respond rapidly. There was no time to build a new ship, so they offered to finance the purchase and remodeling of a ship that OSE would own and operate in their behalf. It would be called the *Gulfrex,* for Gulf Research in Exploration, and it would receive its sailing orders by short wave radio directly from Gulf's headquarters in Pittsburgh. The scientific crew, except for Western Geophysical's seismic party, would work for Gulf; the officers and crew would work for OSE.

We signed the contract and then set about finding a hull to rebuild, beginning by calling dozens of ship brokers all over the country. After ten frantic days we heard that a large new fishing vessel had been tied up because its draft was too deep to pursue fish schools into shallow water. It was available for inspection at Reedville, Virginia, the "menhaden capital of the world."

The next day, a rainy Sunday, Skip Johns, OSE's marketing manager, and I drove down for a look. The ship was slim and gray, 220 feet long, with twin screws, looking more like a navy vessel than a fishing boat. The construction was rugged, and the engine room contained two almost new Fairbanks-

Morse diesels. Its huge holds, intended to hold fish in icy brine, could be converted to whatever we wanted. After a short test run, OSE made the owners an offer they did not refuse.

Peter Johnson and Ed Burgess, our naval architects, drew a set of plans, and we moved the ship to the West India Shipping Company yard in West Palm Beach for two months of around-the-clock conversion effort. While the fish holds were being filled in with cabins for scientists and a gas-gun (to make seismic impulses) was being installed, we assembled an international crew and began adding the electronic data recording systems.

Gulfrex did not quite make the deadline, but at least its outside looked finished during the christening ceremony. As champagne splashed against the bow, I paraphrased Kipling:

> *"The gull shall whistle in her wake.*
> *The blind wave break in fire.*
> *She shall fulfill Gulf's utmost will*
> *Unknowing its desire."*

Everyone cheered, but they were not permitted to see the unfinished electronic recording laboratories with the hundreds of dangling wires waiting to be connected. It was another two weeks before the ship left for its first long cruise around the world. *Gulfrex* surveyed coastal shelves as it went ranging from West Africa and South America to the China Seas and Australia. Once in a while, it would make port to trade crews and ship off a large crate of magnetic tapes containing the data about prospective oil fields to Gulf's headquarters.

I suspect that *Gulfrex* surveyed the undersea lands of a number of countries without their knowledge and greatly improved Gulf's bidding position during lease sales. At any rate, the ship worked very well for about five years, until it was replaced by a new ship, the *Hollis Hedberg*.

Another of OSE's ships was the *Oceaneer*. It was one we had built as an example of how much capability could be incorporated in a vessel only 100 feet long. When it was not under charter to some other company or busy testing new equipment, sometimes we would use it to take friends out on scientific holidays. One of those occasions was a sea monster hunt that I wrote about at the time.

First, you establish that monsters exist. That's easy, not only because tales of sea monsters go back 2,000 years, but because recently they had been photographed by a Ulysseslike character with a flowing white beard, named John Isaacs. For years, Isaacs and an associate named Dick Shutts experimented with various ways of building a baited camera that would attract deep-water fish to a camera. The cameras steadily grew in size until the last version was large enough to be called the monster camera. What does one photograph with a monster camera? You've got it!

The first lowerings of the camera to the sea floor were made 80 miles southwest of Ensenada, Mexico, where the depth is about 6,000 feet. The camera with its lights and batteries is held in position a few feet above the sea floor between an underwater buoy and a baited ballast can that rests on the bottom. Every 15 minutes it photographs the ballast can below and whatever fish are around it. After a couple of days, a clockwork mechanism releases the ballast, and the camera floats to the surface.

After an hour or two in clear deep water, furious activity develops, with hagfish, sable fish, and grenadiers clustered around the bait. Then, in 6 hours or so, the party ends abruptly when these are replaced by a brute that scares the others away. This hulk is part of a huge animal that often extends well outside the camera's view. Gill slits make it clear the great fish is a shark, old, scarred, and probably wise to have survived so long.

Probably it has never seen the blue haze of surface water and the underside of the waves; in turn, probably no man has ever seen such a creature on the surface in these latitudes. Dr. Carl Hubbs, the encyclopedia of ichthyology, tentatively identified it from the photos as a variety of Greenland shark, like those occasionally caught on the surface near Thule. Perhaps, he thinks, such animals are widely distributed around the world, but live unknown in very deep, cold waters and come to the surface only in the Arctic.

On a subsequent lowering some 50 miles away, another such beast was photographed. Again the head extended beyond one side of the picture and the tail beyond the other, but the scarred flank was clearly visible. This time the pattern of scars was different, proving this was another animal. Two monsters such a distance apart meant there must be more, and, of course, that was an irresistible attraction to a fisherman like Isaacs.

We rigged the *Oceaneer* for a monster hunt. John Isaacs was chief scientist, Dick Rosenblatt went as principal ichthyologist, Don Wilkie represented the Scripps Museum, I was raconteur, Alec Winton was captain, and Dan Brown went along to do the hard work.

The hooks had been made specially by an astonished blacksmith who ties trout flies as a hobby. They were over a foot long, with an 8-inch throat and secured to the main lines by quarter-inch wire leaders about 30 feet long.

Bait was a bit of a problem. Rumor has it that Greenland sharks have been caught with nearly whole reindeer (except head and antlers) in their stomach. Reindeer being in short supply in San Diego, Isaacs decided he would fool the monsters with horsemeat. Horsemeat is tougher to find than one would think, but by swearing it would not be fed to humans, Isaacs got several hundred pounds from the San Diego Zoo. Then, reasoning that sea monsters may prefer seafood, other hooks were baited with mackerel and bags of anchovies. That seemed to cover all tastes.

The plan was this: The ship would return to one area where photos had been taken and fish there with 7,000 feet of half-inch nylon line. A couple of hundred feet of its lower end would lie on the mud bottom trailing eight hooks, suitably baited. The upper end of the line would be secured to a buoy. The *Oceaneer* would move off a dozen miles, and fish in the same manner with the ⅜-inch steel wire on the main winch. Monsters beware!

The winch line was on bottom about 10 hours on the first try. Nothing. So we went back to the buoy that held the nylon line and wound it in. The bonito was gone, and the horsemeat was frayed by the nibbling of small fish, but there were no monsters. We tried a total of five lowerings, and although our techniques steadily improved, the results were the same.

After the first cast we wound wire, hooks and all, on the main winch drum; once, while pulling it in, one of the large hooks hung up on the winch frame and was partly straightened out. When we returned to San Diego and the curious came aboard to ask if we'd caught any monsters, Isaacs would suggestively massage this unbent hook, shake his head, and murmur darkly, "Not quite." The lubbers would depart, much impressed with the strength of sea monsters.

If you work around the ocean, wave problems are always with you. My brother Bob, an architect and structural engineer who was head of our coastal engineering group, studied them from the point of view of their effects on coastal structures. On one occasion he analyzed wave data and used the results to make a remarkably accurate forecast of some most unusual waves.

In 1964 Raymond International was preparing a bid for the construction of a long pier off the coast of northern Australia that would be used to load iron ore onto bulk cargo ships. It was important that the pier deck be above the highest wave crest, and OSE was engaged to choose the proper design wave.

For a year we maintained two observers who camped at remote Cape Preston measuring waves daily and recording the high water marks left by storm waves. A Scripps scientist was engaged to draw refraction diagrams showing how wave height would be increased by the effect of bottom topography if waves from certain directions arrived. Another consultant assembled statistics on the probability that typhoons would pass through the region during the highest tides, which are called Indian Spring Highs.

Bob's job was to assemble these data and justify the height of the wave that our client, Raymond International, should use as a basis for designing the pier. In his final report, after giving considerable thought to the statistics of waves, the frequency and duration of typhoons, on-shore wind setup, and the probability that the largest waves would arrive at the highest water level during the life of the pier, Bob suggested a design wave height and a maximum crest elevation. These sounded high for what ordinarily are placid waters but that's the way the figures came out.

Raymond used that seemingly improbable wave for their design and submitted a price based on it. Understandably, they were annoyed with OSE when they lost the contract to a competitor, whose consultant had recommended a much lower design wave.

Years afterward Bob ran into Raymond's project manager, who told him what finally happened: The successful bidder began construction of a pier that was designed to withstand the lower waves, and the structure was well along when the conditions that Bob had forecast actually arrived! The unfinished pier was destroyed by high waves and, since this was a fixed-price, performance contract, the wreckage of the first pier had to be cleared away and a new pier designed and built with an increased design wave, all at the contractor's own expense. In hindsight, Raymond was delighted that OSE had saved them from making a costly mistake. But, of course, OSE and Raymond had lost work we should have had.

SURVEYING IN VIETNAM

When David Smith returned home from managing OSE's South Africa operations in 1966, I asked him to stop in Vietnam on the way and see if he could find some work for the company there. The war had been going for several years, and there were rumors of problems relating to ports and harbors. Dave obliged and was able to obtain a contract from RMK-BRJ, the consortium that managed all construction, to make hydrographic surveys of the amounts of material moved by contract dredges.

Many Vietnamese ports and their approaches were too shallow for large cargo ships; substantial areas of bottom had to be deepened. Dredges from several countries had been hired and put to work on an emergency basis, and when there were no before-and-after surveys, the dredgers were able to charge on the basis of their own estimates, which usually were high.

These odd-shaped excavations in the mud were located in strange-sounding places like Nha Trang, Qui Nhon, Da Nang, Cam Ranh Bay, and the Mekong River, some of which were occasionally contested with gunfire as the fortunes of war shifted. OSE's job was a form of civil engineering; the technical problems were simple, the weather was pleasant, most local people were friendly, and our men did not get shot at too often. But it was not easy to run an efficient operation there.

All we had to do was make a set of profiles across an underwater area before and after dredging, correct the depths for river level or tide, and calculate the volume of mud or sand that had been moved. But things that were easy in the United States could become a struggle lasting weeks in Vietnam. Revisions of plans would require us to move our men out of an area before a survey was completed; often when they returned to finish it, the original stakes had been bulldozed out, new buildings built across lines-of-sight, pilings holding tide gauges had been knocked over, and monuments blown up by the Viet Cong.

OSE had pleasant, airy offices on the fourth floor of a slender new building in downtown Saigon. Although that city was no longer the Pearl of the Orient, the slim ladies with wide-brimmed sun hats and long white billowing gowns called *ao dai* still attracted much attention. The old sidewalk cafés had retreated inside to avoid the possibility of a grenade being lobbed from a passing car, and the sidewalks were crowded with black marketeers. Hotels fronted by heavy wire mesh and sandbagged posts were guarded by "white mice" as the local police were known. Awnings had signs: "No Bicycles Within 50 Feet," because bikes with large seat pads made of plastic explosives had been left in such places in the past. During our three-year stay, the original herds of bicycles gave way to noisy motor scooters that added a blue haze to the previously clean air.

Although the city was surrounded by the war and the fighting was only a few miles distant, life was not unpleasant there. Ho's revenge lurked in the drinking water, but good wines and excellent meals were available several places including the famous My Canh floating restaurant, a gangplank's length from shore at the river bank. Its informal entertainment consisted of watching the local people bathing in the river alongside. Even more stimulating was the possibility that it would be attacked by the Viet Cong at dinnertime as it had been twice before with considerable loss of life.

After dinner, it was customary to have a drink on the roof of the ritzy Caravelle Hotel, where one could watch the star flares hanging in the sky along the horizon and speculate on the progress of the war. When Rhoda and Anitra joined me there for a few days, they enjoyed the excitement of wartime living although they were treated as competition by the local ladies of the evening.

OSE operated two small ships in Vietnam, the *Tam Huu* and the *Indian*

Ocean. We needed these to serve as transportation, living quarters, and offices in the harbors our survey party visited. They gave our surveyors a clean safe place to stay in places where such facilities were often unavailable. Some of the ports where we worked, which became famous during the war, were little more than shanty towns on sandspits.

On one of my early visits to the Saigon office, I had been told that no suitable vessels were available at any price. However, our need was desperate, so I took one of our small outboard-driven boats out into the river at Saigon to look for some kind of a vessel we could use. Long lines of freighters were moored in midstream discharging their cargos into lighters, and we cruised slowly about, watching the procession of junks and river craft go by. Finally one caught my eye; we came alongside it in midstream, and I climbed aboard, pirate style, while it was underway.

The *Tam Hu* (Three Friends) was about 90 feet long, stoutly built of mahogany, powered by two Grey marine diesels, with a small deckhouse aft of the hatches and a large open hold used to transport rice. The laborers moving rice bags seemed amused, and the captain, one of the three owners, was astonished when my first words were, "What'll you take for it?"

There was a price on everything in Vietnam, and over the next few days we agreed on one and moved the *Tam Hu* into a shipyard for conversion of the cargo hold into living quarters and offices.

At Cam Ranh Bay, a large, almost landlocked harbor, one of our small survey craft was a small scruffy barge supported by pontoons and driven by an outboard motor hung from a planklike projection at the rear. The deck was often awash, and the roof flapped in the wind as did a very tattered United States flag, but it supported an echo sounder and that was all we needed. The other boat was a clean, dry, fast fiberglass skiff which we used in places where it might be necessary to make a quick getaway if we attracted bullets. To discourage anyone on shore from taking a casual shot at the boat, the fellows ostentatiously carried a machine gun to indicate retribution was available if required.

Except that the worn-out engines broke down occasionally, the *Tam Hu* was reasonably satisfactory as a movable hotel-office. Our system of operation brought it into Vung Tau, the closest coastal port to Saigon, about once a week for supplies. On those days we would send an old beat-up Chevrolet down to meet it, the back seat piled with groceries, mail, and supplies.

On one of those trips, the driver was fixing a flat tire at the edge of the road when an old Vietnamese in a conical straw hat and faded black pyjamas walked up and squatted alongside our man. When this fellow set down the bag he was carrying, it was allowed it to fall open, revealing an automatic weapon amid the rice. In reasonably good English he spoke to our driver: "We have been watching you drive past every week, and we want to inform you that there is a charge for using this road. You owe us. . . ." (and he named a sum that was about equivalent to $15 a trip). The driver quickly got the idea and paid with alacrity. He was thanked and politely invited to stop each trip and pay the toll. Which side of the war the toll collector was on, or if he was an independent operator, we never knew.

Our other small ship, the *Indian Ocean,* was chartered for use on the Mekong River near Can Tho, where we had a party surveying some river deepening done by a dredge named the *Jamaica Bay.* I went down to visit that operation for a few days and get a feel for the situation. The wide muddy river carried a great deal of traffic; large solid-looking junks, low in the water with painted eyes and shaggy roofs, plowed determinedly along in both directions. We assumed that most of them were on our side in the war.

The levee along the river had a good dirt road on its top, but the bridges over inflowing side streams had been blown up years before. Because the road now carried only foot traffic, it was an excellent place to lay out a survey base line for triangulation of points well out in the river.

The area was said to be "secure," but security in Vietnam was a statistical thing like a weather forecast. Only a week before, the dredge *Jamaica Bay* had been attacked by hooded men in black pyjamas, who boarded it at night and killed some of the crew before blowing holes in the hull with explosives. Salvage work was in progress, and an increased effort was being made to protect a second, smaller dredge that had moved in to take over its work.

On the far side of the main river channel, there was a jungle-covered island where our surveyors needed to set up a transit and measure some angles to close the triangulation net. The Vietnamese commanding officer said, "OK, you can go there at 0800 tomorrow morning for an hour." At 0700 his troops opened up with machine-gun fire on the corner of the island our party wanted to visit, and the bullets stripped away a considerable part of the foliage. Then his troops landed, firing, and took up a perimeter while our fellows quickly set up their transit and turned the needed angles. Then everyone retreated to our side of the river again. Whether or not there had been enemy soldiers or anyone else on the island, we never knew.

After we had been there a few days the American army arrived, set up a line of sandbagged machine-gun emplacements across our base line, and began to construct a camp on the sea of mud pumped ashore by the dredges. Within twenty-four hours Vietnamese entrepreneurs arrived and set up temporary buildings containing bars and brothels within spitting distance of the guns. The previously deserted road became busy. American soldiers cleaned their machine guns and chatted with party girls in tight silk dresses; local women and children plodded along with yokes over their shoulders supporting heavy baggage; salesmen hawked cokes and booze.

Our survey party had its own special guard, a single American soldier, who carried a rifle and watched while we worked. He was insistent that we be back inside the American line before dark: "After that, the GIs will shoot at anything that moves." He also reminded us that the local kids are good indicators. "When they cluster around, happy and jabbering, you're safe. When they start looking sullen and keeping their distance, watch out; there may be trouble."

One day, two of the surveyors, Pajack and Murdaugh, were sitting on the road alongside their transit making calculations when they heard an odd noise and looked up. No one was visible but an American M-76 hand grenade was bouncing along the road toward them. It actually struck one of them in the leg before they scrambled to their feet, took two steps, and dived for the ditches alongside the road. An instant before they reached safety, the grenade exploded and peppered their backsides with dozens of shrapnel wounds. Firing a .25-caliber Beretta pistol into the air to discourage further attack, they managed to crawl along the ditch until a passing jeep gave them a lift to the American lines. Having spent a week in the hospital, these two returned to work, considering themselves to be lucky to have survived and gotten a paid holiday.

After long days in the sun tracking our survey skiff as it crisscrossed the dredged area, the transit operators on the road would climb down the wreckage of a ruined steel bridge to the water's edge and wait for a dugout canoe to ferry them out to the *Indian Ocean,* our home away from home. It was a 60-foot-long wooden boat that had once been owned by the Royal Navy and looked like it. A double-ender with a small deckhouse and rigged with awnings about 8 feet above the deck, stem to stern, it seemed right out of "The Road to Mandalay," and was run by a Britisher whose personality matched. It anchored in the wide Mekong just north of Can Tho and east of the main ship channel, half a mile from the projecting hulk of the sunken dredge.

My assigned sleeping space on the *Indian Ocean* was the roof of the bridge, padded by a single blanket and open to the stars above. As a visiting dignitary, I had been awarded this coolest spot on the ship to sleep; it was surrounded by a waist-high canvas-sided pipe structure that enclosed emergency engine controls and steering. If it rained too hard, I could duck into the bridge below for shelter.

One night after watching the dark shapes and dim lights of the river traffic for a while under stars that "leered from their velvet skies" I fell into an exhausted sleep. Sometime later I was awakened by something going bump in the night and rolled onto my back. A bright full moon had arisen, and silhouetted against it, a dark head leaned over the rim of the railing above me. Instantly a shot of adrenalin brought me to a state of alertness. I imagined the head belonged to a black pyjama-clad fighter or perhaps a river pirate but, having no weapon to defend myself, I held perfectly still. When he made the first move, I would kick upward with both feet from my prone position. After a few seconds of tense waiting the hooded head resolved itself into a searchlight mounted on the rail covered with a protective cowl of black canvas; in the dark before moonrise I had not noticed it.

Annoyed with myself for being spooked by a dark shape, I stood up and looked about once more. Now the great river mirrored the moonlight, and the lines of moving junks came into sharp focus. They were going about their own business, and I had been the victim of my own romantic imagination. Yet with the war so close it could be worse not to be alert. That was the worst problem with Vietnam; it was impossible to tell who your enemies were or which situations were really dangerous.

OSE had its share of personnel problems, and on one occasion I stopped to explain to the Navy officer who oversaw our operations that a couple of our men had cracked up, and the work would be a little tardy until they were replaced.

"Don't tell me your problems," he answered testily. "One of my men committed suicide this week and another went home in a straitjacket".

I tried another tack, "Do you have a psychiatrist"?

"Not now. That was the guy in the straitjacket".

Vietnam was an educational experience, but we cussed it as we learned. Eventually the *Tam Hu* was mined and sunk as it lay at anchor in the Saigon River; OSE left when the dredging was completed, a year or so before the war wound down to its unhappy conclusion.

TENSION-LEG PLATFORMS

Edward Horton, after obtaining a degree from Yale and serving a hitch as a Navy officer on submarines, returned home to do graduate work in petroleum engineering at the University of Southern California. His special interest in 1956 (before Mohole) was in inventing a steady support for oil-drilling rigs in deep water. Ed's thesis subject was tension-leg (TL) platforms, a name he first used to describe a type of platform he had invented that was something like a large version of the taut-moored buoys I had built in 1951. He proposed that a buoyant square of large air-filled tubes be held beneath the sea surface by four "legs" made of heavy cables attached to large dead-weight anchors. This pontoon would provide an almost rigid platform on which a drilling rig could be mounted.

Like some of the other inventions by OSE engineers, this one came before its time, better anticipating the needs of 30 years in the future than the time when it was presented. Before the offshore oil industry would be ready for tension-leg platforms, which are most suitable for depths greater than 600 feet, it would have to drill and produce oil from shallower waters using the less expensive compression leg platforms (steel towers) and semisubmersibles. But eventually deep-water drilling will rely on the TL platform, which is just now coming into its own.

In 1956 Ed was primarily concerned with finding out how such a platform would move when subjected to wave action. The response of complex tubular structures to the orbiting motion of water particles in waves of various frequencies is not a simple calculation, and some experimental data was needed to confirm or deny whatever answers mathematics produced. No wave tanks of suitable size were conveniently available so Ed invented one.

In his experiment a foot-square model of the TL platform that was to be subjected to wave forces was hung from his garage ceiling. It was held in place laterally by springs that could be changed to vary the stiffness of its response. Around this model he constructed a wave tank about 2 feet square and 4 feet deep with a window in one side. The base of the tank could be rotated around a horizontal axis so that the tank and the water in it moved to simulate wave orbits. (This is the opposite of most wave tanks which are fixed and only the water moves.) When the water in Ed's tank moved relative to the model, strain gauges measured the changing wave loads on the platform.

The whole idea, including the experimental setup, seemed like a good one, but the equations that Ed, assisted by Walter Munk of Scripps, was using never properly predicted the wave loadings measured by the strain gauges. Later they realized that they had neglected a term for virtual mass—the mass of water adjacent to the columns of the platform that moves and responds as though it was part of the platform.

Ten years after those early experiments, when OSE was looking for projects, Ed Horton again brought up the TL Platform idea. It made sense to me, and we used it as the basis for organizing a new company along with some other ideas that Ed and Ray Walker had for completing and servicing deep-water wells. The time seemed right, because the offshore oil fields of Southern California were just beginning a boom, and the best production seemed likely to be in the depths of water we were setting up to handle.

Deep Oil Technology was the name of the new company that would be owned jointly by the Fluor Corporation and OSE. Its objective was to develop equipment for producing oil in deep water (we had in mind depths of over 600 feet), and the proposed market was at our door.

By this time OSE owned California Shipbuilding and Drydock Company and that gave us space on the waterfront in Long Beach. There, alongside the pier in water 50 feet deep, Deep Oil Technology built a full-size mock-up of a deep wellhead with huge hydraulic valves. It was complete with a submersible work chamber that descended on vertical tracks made of cables and moved on horizontal pipe tracks at the bottom to position itself precisely opposite each wellhead. Prospective buyers could descend in the SWC and personally connect jumper pipes, and adjust valves with our multiplexed control system.

Of all the schemes that were proposed at the time by various companies, I believe we were the only ones that built an undersea working model. Everything was going as planned until tragedy struck in the form of the Santa Barbara oil spill.

From early the first day of the spill in November 1969, I was at the site and in the air above it with a movie camera, recording what happened. In a few days OSE built a bubble barrier across the harbor mouth so that boats could go in and out, but oil could not enter. The spill caused a considerable mess, depositing a coating of black crude on the white boats, the rocks of the breakwater, and some beaches. Convicts were brought in to spread chopped hay (to absorb the oil) and then rake it up again. A few birds and intertidal animals died, (relative to the local population) but the environmental damage reported by scientific studies was surprisingly small and short-lived. Probably greater environmental problems were caused by the concurrent Ventura River flood that wiped out a whole marina. The real problem was psychological; there was a public outcry against any further oil development in the Santa Barbara Channel, and our hopes of a substantial nearby market for the newly invented equipment vanished.

Later Ed Horton was able to overcome the suspicions of the major oil companies and organize a group of them into a consortium that financed the construction of a "model" TL platform. It was a triangle of large tubes a hundred feet on a side that was installed off the western side of Catalina Island. The model performed as hoped, produced valuable data, and looked like it would be the beginning of big business. Further, Ed was able to demonstrate that multiple riser pipes could be used in deep water without mutual interference. But OSE was forced to drop out, because we could no longer pay our share of the development costs, and Fluor showed little enthusiasm for continuing. Eventually Continental Oil, one of the companies that had been involved in the testing of Ed's large model, developed the ideas further in a huge platform built for the North Sea and took much of the credit for the concept of tension-leg platforms.

When Ed Horton had first proposed the idea, others in OSE began thinking about other applications of it. One possible use of TLPs is as a support for a bridge where the water is too deep for ordinary bridge piers. We were considering doing a design study for a number of places, including Puget Sound and the Strait of Messina, when Andrew Stancioff, an imaginative staff member, proposed the Strait of Gibraltar, a truly grandiose idea.

The narrowest part of the waterway between Europe and Africa at Gibraltar is a little over 8 miles wide, and the water depth is as much as 2,800 feet. Current shear in the strait between surface water and deep water, which move in opposite directions, can be 5 knots (the upper water always flows into the Mediterranean and the lower water always flows out into the Atlantic), winds to 60 knots have been recorded, and there is considerable ship traffic. The idea of a bridge in this great gateway was so bold that it had never before been proposed as far as we knew; it would not be easy to make a credible design.

Using tension-leg platforms for bridge piers only helped with part of the problem; deciding on the superstructure was equally difficult. For that my brother Bob suggested a cable-stayed girder arrangement. Since he was the only qualified structural engineer among us, the job of making the preliminary bridge design fell to him.

In a cable-stayed girder bridge, each tower has parallel, symmetrical, slanting cables on each side to support the girders used for the roadway. This contrasts with a suspension bridge in which the roadway girders are hung from cables suspended between two towers. The idea of hanging a section from each tower and then linking the sections together is a relatively new one, but when our study began, at least four bridges with spans of more than 1,000 feet between towers had already been built.

Bob's span was to be 1,200 feet and the height of the roadway above the water would range from 150 to 250 feet. This would permit all ships to pass under it and between towers. The highway would be capable of carrying 1,000 vehicles per hour each way, and the four lanes would meet all United States standards.

From the first it was evident that the appropriate construction materials would be aluminum alloys, because these would be light, flexible, and require little maintenance. Once that was decided, Alcoa agreed to sponsor the design study. Bob encountered dozens of very sticky engineering problems, and about once a week he and I would go over the progress. Although we shared credit for the design my part in the work consisted mainly of listening to the problems, suggesting alternatives, and encouraging Bob to try other solutions. Eventually we came up with a workable plan.

The most novel aspect of this design was that winds and currents would cause the bridge's taut-moored foundations to move about slightly, whereas all previous bridges of this scale had been erected on rigid foundations. This meant it would be necessary to provide sufficient flexibility in the superstructure to allow for the small motions of the tension-leg platforms.

The underwater pontoons that provided the buoyancy consisted of an equilateral triangle of horizontal cylinders 45 feet in diameter and 200 feet long, from whose corners three 30-foot diameter columns rose through the surface of the water. Above these a tapered monolithic tower rose 450 feet to support the bridge cables. Only one fourth of the 26,000 tons of buoyancy in each pontoon was required to support the loaded bridge; the excess buoyancy provided sufficient tautness to restrain lateral movement caused by exceptional winds and current loadings. Provision was made inside the vertical cylinders for adjusting mooring cables and on their outside for controlling collision damage.

Each of the three vertical steel mooring cables holding the pontoon down was comprised of seven 6-inch diameter cables, which in turn were to be made of seven strands of 2-inch diameter bridge cable. The total breaking strength of the cables on each pontoon was 36,000 tons. The clump anchors to which each cable was attached would be steel tubs 50 feet in diameter filled with pelletized iron ore weighing 10,000 tons.

The bridge span that carried the roadway consisted of a pair of hollow aluminum box girders 12 feet deep and 50 feet wide; these were connected by vertical webs every 40 feet to resist deflection by wind loads and spread the stress between girders. Parallel steel cables supported the span for 500 feet out from each tower in each direction, raising the question of how to connect the ends of adjacent spans in such a way as to permit angular motion between the two and take care of temperature expansion. Bob's solution was to hang a 200-foot-long section between the ends.

The bridge would have been a heroic structure, bridging the gap between Tarifa, Spain, and Punta Cires in Morocco. We had no doubt about the technical feasibility of such a structure, but at the time there was no need for such a roadway that was commensurate with its cost.

In addition to demonstrating the value of aluminum in marine conditions, our widely published artist's conception helped point out that long-span bridges using tension-leg platforms to support cable-stayed spans are feasible.

We estimated the cost at $520 million in 1969 dollars and made a stab at the long-term economic feasibility of such a structure. It was apparent that any advantages would be long term, because there were no major cities or road networks near either end, and ferries took care of the existing traffic quite well. One Moroccan whom I queried about the use of such a bridge asked, "How much would it cost to take a camel across? That's what we drive."

Some years later while passing through Algeria, I came across a small item in a French newspaper there reporting that a committee representing several nearby countries was studying the matter of a Gibraltar Bridge. Some day great bridges of this type will be built, and we hope our pioneering efforts will be remembered. Already our design has been commemorated by *Esquire* as one of six heroic projects in an article entitled "When Men Think Big."

RAISING A 727

In the early morning darkness of January 13, 1969, a Scandinavian Airlines System DC-8 approached Los Angeles Airport from the west over Santa Monica Bay with its wheels down, ready to land. Somehow the big plane came down in the bay several miles short of the end of the runway. The largest part of the cabin including wings and nose compartment stayed afloat; the tail section broke off and sank almost at once. Most of those on the plane survived, but fifteen people died.

OSE, with ships and salvage equipment stationed in our shipyard in Long Beach, at once contacted SAS and the National Transportation Safety Board (NTSB) to get a salvage contract, but there were delays, and it was four days before our ship *Oceaneer* began the search for the lost parts of the aircraft. Using a Cubic Autotape system for precise navigation and a side-looking sonar, it took two days to find two chunks of wreckage in water about 350 feet deep and a mile apart. Large white buoys were left to mark these objects, which later turned out to be the wheels and the tail section. To make sure we maintained our salvage rights, a skiff with two men on it was left moored to one buoy while negotiations with the owners dragged on.

While SAS still vacillated, a small storm came up, and it was nearly eight o'clock at night before the *Oceaneer* was able to return to the buoy to pick up the wet and hungry fellows in the bouncing skiff. By then the wind was gusting to 20 knots, and the larger waves were 5 feet high; visibility was not much more than a couple of hundred feet.

As we ran back toward Marina del Rey, small craft began passing us, headed seaward and shouting that they were going to "look for survivors." "What the hell does that mean?" the skipper said, and switched from the marine band to a public radio station. According to news broadcasts, a second aircraft had gone into the bay somewhere near us. We had neither seen nor heard anything. Two crashes in a week, both in the ocean, at an airport that never before had a serious crash, defied probability. *Oceaneer* turned seaward again to help search.

The plane was a United Airlines Boeing 727 with thirty-seven people aboard; it had been traveling at around 200 miles per hour when it hit, and there were no survivors or substantial floating fragments. We and other would-be rescuers poked around futilely in the dark at what seemed to be the crash area but found nothing. It was an incredible coincidence that the *Oceaneer,* probably the best equipped deep-water salvage ship on the coast, was already at the crash scene.

By the next evening United and NTSB had set up a command post in a hotel near the airport and engaged us to find and salvage the wreck. They also contacted Jacobson Brothers of Seattle and asked them to bring their special salvage rig, the J-Star, that had been used to find a downed aircraft in Lake Michigan a year before.

Starting at the crash position estimated by the Coast Guard, *Oceaneer* towed a side-looking sonar through miserable weather 12 hours a day for ten days in water depths of 500 to 1,200 feet. Then one night Ed Kruger radioed in to the command post. "We've found your airplane." The voice on the loudspeaker was calm, but it brought everyone to his feet. "The plane disintegrated, but there's a pattern of objects on the bottom in 950 feet of water. The television cameras are just going overside."

Company representative Carl Christensen answered coolly into his microphone: "We want positive proof. Get us a TV tape showing the company name, serial numbers of parts, some evidence that will leave no doubt that you've found the right plane." Word of the find spread, and by midnight others joined us at the hotel room to speculate and fidget impatiently. Floodlights shone on tables littered with telephones, ashtrays, and used coffee cups. Carl, wearing his lucky red sweater, paced quietly, throwing a bold shadow on starkly lighted walls that were plastered with charts of the bay, showing the tracks the ship had searched and lists of telephone numbers in large red letters where key people could be reached quickly if some major find was made.

A couple of hours went by without a word, and Carl, who had used considerable willpower to resist calling the ship, suddenly stood up with a big grin. "I'll bet they're just swinging the evidence on deck."

A moment later Kruger's voice came over the loudspeaker: "We've just landed a piece of your airplane on deck." Everyone cheered.

Santa Monica Bay is a big place, and we had found the plane in water nearly 1,000 feet deep quite a distance from the location originally indicated. Now the ship would use the salvage technology the Jacobsons had developed to retrieve lost torpedoes for the Navy. Anyone who has tried to lower a TV camera on a wire to inspect the bottom knows it is very difficult to get it pointed at or focused on any specific target—or to get enough light in the bottom water to see more than a few feet. The J-Star solved those problems, at least for depths to 1,000 feet, in places where currents are low and bottom visibility is good.

The splatter of parts covered a circle nearly 1,000 feet in diameter. Anchoring the *Oceaneer* required us to lay out an equilateral triangle of 800 pound Danforth anchors outside the wreckage pattern, each with an anchor line to a large buoy. The *Oceaneer* then moored to all three buoys simultaneously, so that, by adjusting the length of the three mooring lines, the ship could be positioned over any spot within the triangle.

Once the ship was ready, the J-Star cage, a cube of steel tubing about 4 feet on a side containing lights and cameras, could be rigged. It was lowered on half-inch wire rope until it was a couple of feet above the bottom. Then its motion in any direction could be controlled by four slender steel wires that ran from four small gurdy winches on the ship's rail down through pulleys on the cage and thence outward for several hundred feet to small anchors placed north, south, east, and west.

In order to adjust the camera position, the operator would take in on the west gurdy winch while letting out on the east one, etc. In this way the cage could be moved a few inches or many feet in any direction; when the lines were all taut, the cage was held rigidly. The TV camera and lights inside the cage could then be pointed from the surface by a pan-and-tilt mechanism. Large objects projecting above the bottom beyond visual range could be spotted with a "searchlight" type sonar. Then the TV camera would be moved along the sonar beam until a visual inspection could be made. Beneath the J-Star cage a set of tongs was attached so that objects seen by the TV camera could be picked up.

The first job was to map the main pieces of wreckage on the bottom without disturbing anything. The company and NTSB had experts aboard the *Oceaneer* who knew 727s thoroughly and had catalogs with the serial numbers of each part on the plane, but the destruction caused by the crash was so complete that these fellows could not always be sure what part they were looking at. There were thousands of small pieces of twisted aluminum, lumps of machinery, snarls of colored wires, and occasionally bits of human wreckage. But the water was clear, and the television picture was sharp enough so that it was often possible to read the serial numbers of small parts as they lay on the bottom.

The mapping took several days, and when it was finished, the next instruction was "Get the engines." These engines were around 4 feet in diameter and 12 feet long; they weighed 4,800 pounds each, and they had landed some distance apart. But with charts of their exact locations on the bottom, it took less than two days to locate all three engines, get the ship in position over each one, grasp it with the hydraulic pincers under the J-Star, and bring it almost to the surface. Then divers would attach slings around the engine so *Oceaneer*'s overside crane and big winch could set it on deck. In a few more days the tail assembly, the wheels, and some other objects were brought aboard.

After the large items were salvaged we engaged the *Corsair,* a commercial fishing boat, to trawl up the remaining pieces. It was quite a sight to see its tattered black net come out of the water festooned with odd-shaped pieces of twisted aluminum.

All the pieces were brought to OSE's shipyard and "reassembled" to discover why the plane had crashed. The answer, NTSB ultimately decided, was that a tire had overheated during the takeoff run, and when the wheels were retracted, the heat triggered a fire-warning light in the wheel compartment. One of the pilots misread the light as an engine fire, and when he switched that engine off, the overloaded remaining generator quit, plunging the cockpit into darkness. The instrument lights also went out; at night amid the storm and without instruments the plane had plunged into the sea.

DEEP SEARCH IN DISTANT SEAS

As I related earlier, on Christmas eve in 1961, I was at home thinking about how to find sunken ships—especially ancient ones—in deep water. It occurred to me that the best way to search would be with a side-looking sonar if its transducer could be steadily moved along about a hundred feet above the bottom in water 1,000 to 15,000 feet deep. The great depths posed several technical problems but the one that concerned me most was how a searcher would examine any object found projecting from the bottom and determine whether it was a ship. If the sonar tranducer were towed on a cable it would be extremely difficult to stop for a detailed look, but if it were mounted, along with the television camera, at the tip of long pipe, the ship could stop and maneuver (using dynamic positioning) until the object was directly below and could be inspected.

This was a whole new line of thought, and my ideas came thick and fast, tumbling over each other. I noted that deep bottoms should be easier to search and inspect than shallow ones, because they were smoother, had lower sedimentation rates, held fewer extraneous objects that would be likely to interfere with a sonar search, and because the water was clearer. In ten minutes my rough notes contained the essential ideas, and before midnight I hollered down the stairs to Rhoda to come up and witness those notes about the new invention.

After the aborted CIA proposal and a subsequent one to NASA, I had obtained a patent on this deep-search method, but set its implementation aside temporarily. However my hobby of reading ancient history continued, and one day, while thinking about Jason's voyage around the Black Sea in 1200 B.C., the significance of the name "Black" struck me. The water of the upper 100 meters of the Black Sea is clear, full of fish and much like those of other great seas, but *its bottom is black because there is no oxygen there.* The blackness is caused by sulfide compounds; the deeper part of that sea is a *reducing* environment. There is no dissolved oxygen there and few animals can live without dissolved oxygen in the water, especially including the marine borers that devour a wooden ship in the Mediterranean in a few decades. Without teredos or *Limnoria* to eat the wood, microbes to consume soft material, infauna to disturb sediment layers, and any oxydation of materials, a wreck should be practically perfect.

It dawned on me that, if Jason's *Argo* or other ancient ships had sunk to the bottom of the Black Sea, they would survive forever. So might many other things that have never yet been found on other very old shipwrecks: decking, superstructure, rigging, cloth, even bodies. At least that was a possibility. The absence of oxygen in the water is critical to wreck survival! If an ancient shipwreck could be found in the Black Sea it would be worth a hundred times one that had sunk in the Mediterranean where the only wood surviving is that buried in the mud where there is no oxygen.

Here was a great opportunity for modern archaeologists; if an ancient ship could be found in an anoxic environment they would be able to examine wooden objects—not just those made of ceramics, stone, and metal—that had been in use before the Christian era. I became even more excited when I learned that a Woods Hole ship had found a "biological membrane" in a Black Sea core that was dated at 70,000 years by counting yearly sediment layers.

There were drawbacks; the Black Sea is deep, and it would be a huge place to search without any clue about where to begin. One could only guess where an early ship might have sunk, and even the most efficient search ship might have to hunt for years before finding anything. The area to be examined would have to be reduced.

Then I learned that the depths of the Sea of Marmara, in Turkey between the Bosporus and the Dardanelles, are also without oxygen. It is much shallower than the Black Sea (only 1,000 meters versus 4,000) and more accessible. A rough calculation indicated that there could be one ancient ship per square mile on its deep bottom. That meant a higher density of targets for a search in a relatively small area.

Why would anyone want to recover an ancient ship? For me, driven by an interest in classical history, I simply wanted to know more about how people lived than one can learn from Homer's writings or the archaeology of cities or from the ship remnants found in shallow water. With a completely furnished ship some of the questions about sailing and commerce in the ancient world could be answered. A ship is a microcosm of the civilization it serves and the times in which it sailed—a time capsule containing the things needed for living at sea. A special kind of treasure.

I had no wish to own any of the artifacts or to make any money out of a salvage operation; indeed, anything recovered would have to remain in Turkey. I just wanted the fun of finding and raising an old ship. Recently built ships such as the *Monitor,* the *Titanic,* or the *Vasa* held no interest for me. We have their original plans; we know what they were doing and how they were built; many copies of the artifacts they carried are still in existence. My objective was to reach back beyond well-known history by studying complete ships that represented a time for which all such records have been lost.

In order to find old ships, it would be necessary to explore the bottom in parts of the world where sea traffic had been heavy. The depths of the warm (13 degrees C), highly oxygenated Mediterranean are not much more likely to produce a complete wreck than its shallow waters. One could expect to find piles of amphorae, but no wood or soft parts not buried in mud—although there is always the remote chance of finding bronze armor and weapons, or a statue wreck like that at Antikythera, or a stone temple being moved by sea as at Mahdia. But the possibility of finding a complete wooden wreck in the Sea of Marmara or the Black Sea is pretty good, and I received a very satisfactory response from archaeologists who wanted to go on an expedition with me after my article on the subject appeared in *Science.*

Some may be curious why I did not favor the use of small submarines for this kind of work. The answer is that it is more difficult and expensive to do useful work with such submarines. Because subs are built to be neutrally buoyant, they cannot lift much of a load. They cannot stay below very long, or search a large area systematically, or even see the bottom very well except with a television camera (a person looking at the bottom through the view port of a submarine gets the feeling he's in a conical pit). It is impossible to illuminate a large enough area to see very much at one time, and there are inherent risks involved in launching and recovering small submarines in rough weather. The threat of bad weather means that operations are curtailed on days when a surface ship would continue to operate.

Finally, small submarines need to be supported by at least one surface ship, and the combination is expensive. When one divides the total cost of an operation by the actual hours of useful submarine work on the bottom, the answer is many thousands of dollars per hour, an expense that can be borne only by a government or very large organization.

Searching and inspection without people at the bottom is much cheaper, it can continue indefinitely and it can cover more territory. If something heavy is found that must be lifted, a surface ship is required anyway. In any case, submarines are not the convenient, fast, maneuverable, good-visibility platforms that science-fiction movies make them out to be.

All of my ideas for a deep-search-and-recovery ship came together in a proposal to Alcoa to build such a ship for use in Mediterranean-Black Sea archaeology. Alcoa agreed to finance its construction to demonstrate the possibilities in marine aluminum. My article in *Science* (October 1971) and a subsequent book *(Deep Water, Ancient Ships)* describes that ship, *Alcoa Seaprobe,* which could reach downward 12,000 feet to examine the bottom with sonar and television. The bottom, thousands of feet beneath the ship, could be seen on a television monitor on the bridge, and the pilot could use its cycloidal maneuvering propellers to move the ship in any direction to examine and hover above artifacts on the bottom. Objects weighing up to 100 tons (including small ships) could be grasped with tongs and lifted to the surface. Unfortunately I was not able to obtain sufficient financial support to take it on a deep-water archaeological expedition, and the ship has since been scrapped.

However, my dreams are intact and better focused. I still have hopes for an expedition, and I firmly believe that some of the great archaelogical finds of the future will be the recovery of complete ancient wrecks from an oxygen-free environment. They will contain things no archaeologist dreams of finding except in the reducing bogs of northern Europe.

SUING THE CIA

In 1973 glowing stories about the capabilities of a new deep-sea mining ship named the *Hughes Glomar Explorer,* which would be operated for Howard

Hughes by Global Marine, Inc., began to circulate. Newspapers reported that this mystery ship soon would go to work "scooping magnanese, copper, and nickel from the depths of the ocean to add to the fortune of the world's wealthiest recluse." My thoughts at the time were: "He really is crazy if he thinks he can make money at undersea mining".

Then the project was forgotten until abruptly in March 1975 a headline in the *Los Angeles Times* announced: "CIA Recovers Part of Russian Sub. Hughes 'Mining' Vessels Used in 6-year Project." The cover story had been blown, and what was supposed to have been kept secret had been blabbed to the press by some crew members. The account in *Time* magazine described the pipe-handling ship as 36,000 tons, 618 feet long with a center well 199 $\times$ 74 feet and a lifting capability of 4,000 tons. It claimed that when a G-class Soviet submarine had suffered an explosion and sunk in 1968, U.S. Navy listening devices determined its location within a 10-square mile area. Project Jennifer had been born when the Russians searched for the missing vessel far from the actual site, leaving an opening for the U.S.. The salvage had been successful in that the front third of the broken submarine had been retrieved. Maybe.

Within hours my telephone began to ring with calls from persons who knew about my patent on this kind of deep search and recovery and were familiar with the *Alcoa Seaprobe*. All said essentially the same thing, "They've stolen your idea." One call was from Ruth Ashton Taylor, a well-known CBS-TV news broadcaster who had previously interviewed me about the *Seaprobe*. She was around with cameras on the same day to ask about my connection with this new recovery. I had none, but I was able to show the original model of ship-grasping tongs designed so the *Seaprobe* could pick up ancient ships.

Next I went to see my lawyer, George Wise, taking along the 1962 proposal to the CIA (to salvage nose cones with a similar ship), the correspondence with senior officers of the CIA, and my patent. I approved of and applauded the submarine salvage project and did not begrudge the use of my patent. All I wanted was a word of thanks for the use of my ideas on an important national project. (At the time I was a member of the Navy's Research Advisory Committee.)

Attorney Wise wrote to the CIA saying that we first wanted to "ascertain whether just compensation can be arranged without litigation." Its general counsel replied that they had no record of receiving my proposals and that the persons with whom I had dealt were no longer associated with the agency. Neither of those statements turned out to be correct.

Within the next month stories came out in *Science* and *Newsweek* suggesting that the project was based on my ideas for the Mohole and comparing the *Alcoa Seaprobe* with the far larger *Glomar Explorer.* In August, after a futile exchange of letters, we brought suit for $100 million against the U.S. Government in the U.S. Court of Claims.

I began with the naive idea that the Court of Claims system was set up to provide even-handed justice, a view that was derided by some attorneys who knew the ropes. On the contrary, that court sees itself as a defender of the U.S. Treasury against all comers. The trial was held in camera, behind guarded doors that kept all except participants out, before a judge whose opening statement included the remark that "if the CIA owed you they would have paid you." He did not intend to be humorous.

This is not a rehash of the case; many of the interesting details are still secret, and although the press has guessed some of them, they have never been confirmed or denied by the government. But the attitudes of some of the participants are of interest.

The judge would not permit the prior existence of the *Alcoa Seaprobe* to be used as evidence, although he did view a movie of it operating off Florida. Details of the secret, proprietary briefings I had given to the CIA and Air Force in 1963 turned up in the private diary of a Global Marine vice president in California on the same day they were presented in Washington, D.C.. (This was an obvious breach of security, as well as a black mark against the integrity of those involved.) One CIA attorney admitted privately that, when my drawing of an ancient ship being raised from the deep with a huge set of grappling tongs appeared in *Science,* it gave agency persons quite a shock (they were in the midst of trying to design such tongs to raise the submarine and thought their cover had been blown).

During the trial the lawyers and witnesses on both sides spent the days in the courtroom and the nights in preparation. This necessitated a "secure" area to work in downtown Washington, and the opposing attorneys were assigned rooms side by side in the old OSS (Office of Strategic Services) building in the Watergate area. Each party was given one of a pair of four-drawer locked and guarded filing cabinets that stood side by side. Late one night, while using our files, I came across an unfamiliar envelope in our file and began going through the material it contained.

One item was a memorandum from the government's chief engineer on the project to other senior officials involved that said something to the effect that "it is obvious we are using Bascom's ideas and patent. We should pay him off now to prevent a problem later." It was a flat admission of guilt by someone who was in possession of technical expertise. Suddenly I felt rich. Somehow a groggy attorney for the government had mistakenly put one of their folders in our filing cabinet. The next morning my attorneys called for a private session with the judge to discuss how this admission of guilt could be introduced into the trial. But the judge ruled that because the chief engineer was not a lawyer he was not competent to judge whether the patent had been violated and would not admit the memo as evidence.

It was clear this fellow was going to defend the U.S. Treasury against me, regardless of the evidence. Before the end of the trial, the CIA did find my original proposal and locate some of the persons I had originally talked with; they confirmed what I had said. The person who leaked the information I had presented to the Air Force meeting was named, but no action was taken. Other papers were found that said the project leaders privately agreed that "no one connected with the Mohole Project would be consulted." If I were to inquire, my questions were to be answered by a person at Hughes they called the Great Dissembler.

There was little doubt that I had been conspired against, but this trial had become a purely patent case. The only matter the judge would consider was whether all eight steps of Claim One of the patent application had been done by the government—a requirement for patent infringement. Actually there was no doubt that the first seven had been violated, and most of the arguments dealt with the last step. In the end the judge ruled against my claim, and I did not even get a word of thanks from the government for inventing the method. Instead medals were presented secretly to those who lifted it.

XI

Finding a Treasure Galleon

SEAFINDERS, INC.

In 1970 Ocean Science and Engineering, Inc., was struggling with two new subsidaries on the east coast of Florida. One operated two steel scalloping boats, each with a processing plant aboard that would automatically shell the scallops at sea; its objective was to harvest profitably the extensive beds of calico scallops off Cape Kennedy. The other, Ocean Dredging, Inc., assisted by Caterpillar Tractor Company, was trying to develop a submarine dredge that would travel on the bottom, unaffected by wave action. We expected there to be a considerable market for bypassing sand around tidal entrances on the Florida coast and for replenishing the sand eroded from resort beaches if one could find a good all-weather solution to the sand pumping problem. For a number of reasons, neither company was ultimately successful.

At any rate these test operations required me to spend some time in Florida along the coast between Fort Pierce and Pompano Beach, a region where Spanish treasure was a common topic of conversation. This included the area where the Real Eight Company, founded by Kip Wagner, had found several of the galleons that had been blown ashore by the great hurricane of 1715. That group had successfully salvaged a lot of silver, mostly in the form of pieces of eight, plus some gold trinkets, and set up a public treasure museum at Melbourne, Florida.

With my mind on suitable targets in New World waters where the *Alcoa Seaprobe* could practice deep salvage, I asked if Real Eight had any data on deep-water wrecks. They did not but referred me to Robert Marx, one of their seasonal employees, who was said to know a lot about Spanish wrecks. When I first stopped to see Marx, he was just getting back from the Little Bahama Bank, where he was searching for a galleon named the *Maravillas*. He told me some of the story of that ship and admitted he was obsessed with finding it, even though several well-financed and equipped expeditions had failed to locate it.

My specific interest was in old sunken ships, not necessarily known to contain treasure, that were well beyond the depths reachable by divers but within the grasp of the *Alcoa Seaprobe*. Marx knew about Spanish ship losses, and he agreed to try to dig out some data on specific wrecks that sank in deep water and might be a suitable target for the new search ship. With that objective I arranged for him to get a grant for a study of deep salvage targets in the Caribbean-West Indies region.

Eventually he suggested several possibilities for deep wrecks; these were in the Mona Passage west of Puerto Rico, in the Florida canal near Key West, and off Cartagena, Colombia. The wreck of the *San José* off Cartagena was a particularly interesting possibility because it was said to be the richest galleon ever lost.

The *San José* was the flagship or *capitana* of a fleet of seventeen ships that sailed from Portobelo, Panama, in 1708. The king was broke as a result of the war of the Spanish Succession, and he had ordered the fleets transporting New World treasures to sail home at any cost. The result was that the *San José* carried an unusually large amount of registered treasure (11 million pesos in silver and gold) and was defended by sixty-four bronze cannons. As the Spanish fleet sailed north along the coast of Colombia, *San José* was in the lead, about to round the tip of a group of islands before turning into Cartagena, when it came upon three English ships waiting to give battle. Because the Spanish ships seemed to have greatly superior fire power, their admiral decided to sail on through, and a gun duel began between *San José* and the flagship of Commodore Wager, the British leader.

In a battle that raged well into the night, the *San José* caught fire. Eventually its stores of gunpowder exploded, immediately sinking her and taking nearly 600 people down with her. At the time of the explosion, the two ships were so close that the heat from the *San José* scorched the faces of those on the Englishman's deck. By reading Commodore Wager's account of the battle and taking wind and geography into account, one can approximate the site of the sinking. There the water depth was about 1,000 feet, just right for the *Alcoa Seaprobe*.

At the time I was a member of the Princeton University Geology Department's advisory committee. One of the other members was William Weaver, a Wall Street investment banker who had been involved in a number of successful mining ventures. We talked about the possibilities in hunting for treasure with a ship intended for archaeology, and he suggested that a company be organized that would give maximum tax advantages.

By July 1971 I had resigned as president of OSE (but remained chairman of the board) and was relatively free to engage in other affairs. Back in Florida I traveled down the coast with Marx, meeting many of the well-known treasure hunters, including Norman Scott, Tom Gurr, Bert Webber, and Mel Fisher, all of whom persisted and eventually found treasure. These fellows were chronically broke but full of hope and wonderful stories, something like the ones I used to hear from prospectors in the mining business. There was always a mother lode or lost galleon out there somewhere, inviting the unwary to adventure: "For $10,000 you can have 10 percent. . . ." or whatever.

At least once in each of those meetings, these experts grinned and repeated the wry advice often given to newcomers to their trade: "Never trust a treasure hunter." Beneath the surface humor and general amiability, greenhorns were warned that honesty and truth had limits in their world.

There were also tales circulating about an underwater "causeway" off the Bahamas; it was used to support the opinion of some that the lost city of Atlantis was in the neighborhood. I thought that was a nutty idea, but I was curious to see for myself whatever was there. The so-called causeways turned out to have been beaches at a lower stand of the ocean, whose sands had been consolidated by sea chemistry. Later the hard surfaces had cracked in an interesting pattern, making them look something like huge paving stones carefully fitted together. One can see the same thing on beaches still at sea level in the Tortugas islands.

Later that year Marx and I, accompanied by our wives, met in Seville, Spain, mainly to visit the Archives of the Indies, where the records of the old galleon fleets are kept. On request its librarians will bring out thick piles of old unbound papers that deal with certain subjects or dates. These are penned in early Spanish and kept together by folders 6 or more inches thick, tied by ribbon. The researchers sit at long tables in a dark high-ceilinged room, painfully interpreting the faded scratches. Our request to open the tall dark velvet curtains and let in light enough to take pictures was answered with an astonished glower, but finally this strange whim was indulged. Local translators and document finders can be engaged to help as needed. Eventually one usually can find some written material on specific ships or dates, mainly because the Spaniards were bureaucrats par excellence and made multiple copies of most important documents, such as ship manifests and letters to the king.

In October 1971 Seafinders, Inc., was founded. It was a Subchapter S corporation, meaning that there were fewer than ten owners of a high-risk enterprise and that all expenses could be written off the contributor's income tax if it failed to make money. Bill Weaver, who had assembled the group of wealthy investors looking for income-tax write-offs, was chairman; I was president and chief operating officer.

One of our early moves was to engage attorneys in Bogata, Colombia, to obtain concession rights from the Colombian government to search for and recover the *San José*. Eventually, after months of politicking and maneuvering, our lawyers did obtain a much-signed and stamped piece of paper that looked official and said Seafinders had been awarded the rights we asked for. Whether or not this document was really valid was uncertain, and there was no way to find out without spending a lot of money and actually salvaging something from the wreck. I had no doubts that a systematic search with the *Alcoa Seaprobe* eventually would find the *San José,* but that might take months at a rate of at least $10,000 a day. This was far more money than Seafinders had available, but that wasn't the worst of our problems.

When Marx and I visited Cartagena, we learned that it is still a pirate stronghold; it was the place where any treasure found would have to be brought ashore and placed in the same government warehouse from which dozens of U.S. Army trucks had been stolen the year before. If the Army couldn't protect its property, how could we? Furthermore, because of the country's "patrimony" laws relating to historic objects and the attitude of the local officials we talked to, it was plain to see that it would be a very long time, if ever, before any treasure left the country. Nor would anyone there have looked surprised if, after we had done the hard work of locating the wreck, some well-connected person had arranged to have our throats slit and begun salvaging the wreck himself. That was way the system seemed to work. But the *San José* is there somewhere, waiting, her cargo now estimated to be worth on the order of half a billion dollars.

We backed away from Colombia and increased our effort to obtain a concession from other countries, including Jamaica for rights to hunt on Pedro Bank, where some galleons were known to be lost, or for several sites in the Bahamas. Clearly there would be long delays in getting salvage rights in any of the highest priority places.

Time was passing, and Seafinders needed a specific place to work. In

March of 1972, while talking with John Jones, president of the Real Eight Company, I learned that his company had six "pinpoint" leases on a part of the Little Bahama Bank. These had been taken up in 1968 but were still valid, so Seafinders agreed to sublease these for 10 percent of any treasure found. Unlike an exploration lease, which may cover a large area to be searched, a pinpoint lease is a dot on the chart where something is alleged to have been found, surrounded by a circle with a radius of 1 mile to take care of uncertainties in navigation. Together these six pinpoints covered about 20 square miles, where there was deep sand that we hoped covered the wreck of the *Maravillas.*

It was a convenient place to work, because it was close to Florida ports, and it had the big advantage of being in a country where there were no taxes. As with most other countries, the rule was that the government got 25 percent of the material salvaged, but with the others taxes then raised the government's total take to 50 percent. The best deal was in the Bahamas.

The disadvantage of hunting for the *Maravillas* was that at least five previous expeditions, including a couple with Marx as a member, had searched the suspect area with modern equipment, including helicopters and magnetometers, and found nothing. All that one could see from the air in this region where the bottom was easily visible was white sand, and sometimes a patch of darker coral. If much of the wreck had survived, it would have been found long before; whatever remained would be only the hard and heavy materials buried under the sand.

Bob Marx had two pieces of information that focused his interest on this particular ship. One was that there were no iron cannons on the *Maravillas;* this meant that it would not have a strong magnetic signature and could not easily be detected. Second, he had once talked with an old Bahamian who showed him some silver bars hidden in the bottom of a barrel of flour and revealed that these had been taken from a wreck somewhere on the bank.

After a couple of months spent in assembling equipment, poking around in old ballast piles on the Bahama Bank, and learning from Bob about potsherds as wreck indicators, we were ready to go to work. By then I had done some homework on this specific wreck and on the Spanish flota system in general, learning from writings by Marx, Mendel Peterson, and others; from traveling in Spain; and from visiting museums.

THE WRECK OF *NUESTRA SEÑORA DE LA MARAVILLAS*

For most of the years between 1530 and 1800 the Spanish crown continually moved fleets of ships along a route that began and ended in Seville, home of the board of trade. Once the passengers and supplies were loaded, the ships descended the narrow Guadalquivir River for 60 miles to the lagoon at Sanlúcar, not far north of Cadiz, to wait for the fleet to assemble. When all the ships were ready, the *capitana* fired a signal gun, hoisted her mainsail, and led a procession of ships through the channel in the sand bar. When all ships were ready in the outer bay, the fleet set sail southward for the Canary Islands.

From the beginning these ships were beset with misadventures, including collisions, fires, groundings, and pirates waiting to pounce on stragglers. Often the ships were overloaded, filthy, and smelly, the crews and passengers underfed. The captains, who sometimes knew nothing of the sea and bought their positions in the hopes of recouping that cost by smuggling, were often incompetent.

It was important to demonstrate Catholic religious fervor, so special chants were used to bless the light of day, raise the anchor, hoist the sails, and encourage the wind to blow. There was a pretty good chance the ships would make it to the New World, but the risks for individuals were great, and often voyagers died from disease on the miserable two-month trip.

From the Canaries the ships would head southwest with the trade winds until they reached latitude 15 north, at which they sailed due west until the peaks of Guadeloupe in the Leeward Islands rose before them. After a few days there, replenishing water, fruits, and vegetables, the ships split into two groups. The Tierra Firme fleet sailed southwest to Cartagena and thence to Portobelo on the isthmus of Panama. There it unloaded outgoing cargo and picked up the treasures of the Indies, including Potosi silver and Peruvian gold.

The Nueva Espana fleet continued on northwest to Vera Cruz to exchange its supplies for the silver of Mexico and goods from the Orient that had been shipped across the Pacific by galleon and then carried across Mexico on muleback from Acapulco.

When the exchanges were completed, a process that took months, both fleets sailed independently to Havana, Cuba, where they formed into one large flota, often composed of fifty or more ships, which could protect each other against the English and Dutch pirates that lurked along their homeward route. After about 1570, all treasure was carried on galleons, a specific kind of ship that was much larger than the caravels and faster than the naos. The galleons were specifically armed with bronze cannon, because when these were fired they would not burst, as cast-iron cannons often did, destroying their own crews.

On leaving Havana the galleon fleet would sail out into the Gulf Stream and north along the Florida coast to Cape Canaveral (Kennedy), whence they would set a course northeast. They were instructed to sail until they sighted Bermuda and then to turn eastward, following a great circle route to the Azores and Sanlúcar, their place of beginning.

On return to Spain, the smaller ships were towed up the river to Seville where they landed at stone wharves to discharge their treasures at the Tower of Gold and the Tower of Silver, after which there was a grand procession up the street to the great cathedral to give thanks for a safe return.

In the roughly 250 years this flota system operated, some 5 billion pesos in treasure was transferred from the New World to Spain. In today's money that would be equivalent to about $50 billion. One fifth of the treasure belonged to the king, who took more if he was hard-pressed; this substantial tax, and the possibility that the king might suddenly increase his share, encouraged smuggling. Silver bars were made in a standard size, weighing about 80 pounds, each one stamped with assayers' seals, customs markings, and owner's identification. Smaller bars were contraband, and stories abound of captains being sentenced to the galleys when they were caught smuggling silver, sometimes disguised as anchors or cannon balls. About 5 percent of the treasure shipped was gold; the rest was silver.

The flota system was shot through with theft, corruption, bribery, and inefficiency of all sorts. The ships were often in poor condition, and the fleets moved at an average speed of about 3 knots; their position on the ocean was determined by a vote of the navigators. Sometimes they encountered hurricanes or enemy fleets, but mainly they got through; over 98 percent of the voyages were completed. Even so, because there were so many ships involved for so many years, the flotas left a trail of wrecks around the Caribbean, along the coast of Colombia and Mexico, as well as on the shores of Cuba, Florida, and the Bahamas.

On New Year's Day, 1656, a Royal Armada of treasure galleons under the Marquis de Montealegre sailed from Havana, Cuba. The Marquis was aboard the lead ship or *capitana;* sailing close at hand was the second-in-command ship or *almiranta,* the *Nuestra Señora de la Maravillas;* both were laden with treasure. Upon reaching the Gulf Stream, the fleet moved with it up the low coast of Florida, which was visible to the west. On the evening of January 4, they "reached a place known as the *mimbres* (probably meaning a shoal with eelgrass) which is at latitude 27 degrees and a quarter." A paragraph in a letter from the Marquis to his majesty explaining the loss of the ship speaks vaguely of a shipwreck at midnight resulting from a collision.

Apparently the *almiranta,* in the lead, suddenly noticed it was in shoal water, put the helm over hard, and was rammed and sunk by another ship, which then sailed off into the darkness. Next morning when the sun came up, the *Maravillas* was resting on the bottom in 4 fathoms of water with forty-five men clinging to its projecting masts. The rest of the fleet had sailed on except for one other galleon captained by Juan de Hoya, which was at anchor nearby making repairs to its rudder. It sent a launch to pick up the survivors (only 10 percent of those who had been aboard), apparently requiring them to bring enough treasure to pay their passage home.

After encountering several storms that stripped away sails and rigging, De Hoya's galleon was blown down the eastern side of the Bahamas, through the Mona Passage, and back to Cartagena, arriving there in March, a bedraggled hulk. Thus notified of the shipwreck, salvage ships, using the pearlers of Marguerita Island as divers, were sent out to salvage the treasure. Although a list of the treasure originally shipped survives, which describes the markings on every bar, no list of the amounts of treasure originally salvaged was available.

Seafinders was able to obtain a salvage boat that had previously been equipped to handle sand. The *Grifon* was a converted shrimp boat whose interior held cabins instead of shrimp tanks, and it had a large propeller whose wash could be directed straight downward by a huge pipe elbow. Thus, when anchored over the proper spot, this "prop-wash diverter" could blow or "blast" away the sand cover and expose the hard coral bottom below.

ON THE LITTLE BAHAMA BANK

With the *Grifon* refurbished and ready for another round of treasure hunting, the Seafinders party set out from Fort Pierce, Florida, for the Little Bahama Bank some 80 miles to the east. Before us there was 1,000 square miles of shallow coral and sand between Grand Bahama Island and a few keys along the bank's northern extremity. The captain was an ex-commercial fisherman, very dubious about the idea of looking for an old wreck under the sand when we could be fishing. His negative feelings were offset by those of Marx, who was eager to try again to find the wreck he was sure was out there somewhere.

I was skeptical about the possibility we would find anything in the most likely area (about 50 square miles), because it had been carefully searched several times before. However, I believed firmly in the scientific method. If the wreck was there, we could find it even if it produced only a very small magnetic signal. With that in mind I personally ran the magnetometer and took all position fixes during the search. We would either find the *Maravillas* or be able to say definitely it was not in the sand-covered region on the edge of the bank near 27 degrees. Rhoda went along as expedition cook, in which capacity she provided a good-humored outlook as well as excellent meals for a generally dour crew, which was made of up four young fellows who helped Marx with the diving and anchoring work.

After nearly a month of practice, looking at the suspect area from low-flying aircraft and towing the magnetometer to get used to its operation, we began the formal search on July 12. Every day's work was carefully recorded in my notebook, and on a master search chart I made for plotting Loran-A fixes.

Our point of beginning was a framework of iron that looked like a wrecked tower but was known as the pipe wreck. It was several miles from the edge of the bank and reasonably close to the specified wreck site. The wind was light, the sky blue with a sprinkling of white clouds, and the water warm and clear. We attached a distinctive buoy to the metal frame to give us a visible marker in the otherwise green sea and set to work.

The plan was as follows: The small fiberglass boat we used for diving would lay out a series of bright red balloon buoys, about 100 yards apart, along a north-south line about a mile long. Then, a hundred yards to its east, a parallel line of buoys would be installed, thus creating a well-marked channel for the magnetometer survey.

Using the avenue of buoys for guidance, the *Grifon* would tow the magnetometer sensing head up and down the channel, moving eastward about a hundred feet with each pass. Our magnetometer continually measured changes in the strength of the earth's natural magnetic field and recorded it on a moving chart paper in the cabin. Other metals have no effect, but any iron object, depending on its size and orientation, will modify the earth's field by causing the magnetic lines of force to bunch as they pass through the iron. The distance between the sensing head and the iron object is important, because the signal dies out rapidly as the lines of force spread to resume their normal position. Therefore we overlapped search pathways enough that no spot on the bottom would be further than about 60 feet from the magnetic sensing head.

When all the space between the lines of buoys had been searched, the first line of buoys would be leapfrogged to make a third line, and the next channel would be swept. Thus, in 100-foot steps, *Grifon* moved slowly across the white sandy bottom.

At the north and south ends of each run, I would get a position fix with Loran-A. Some technique was required to match up the positions of the dancing green figures on the scope, but after a little practice it became easy, and the result was that we knew the ship's position within a couple of hundred feet every moment all that summer. Without the Loran fix, every place on the Bank looked like every other.

While the captain steered and Rhoda cooked, I would watch the magnetometer needle trace a smooth steady curve on the chart paper. About three times a day, the needle would suddenly swing wide as the magnetic sensor penetrated bunched lines of magnetic force. Then I would leap up, dash out on the stern, and heave over a concrete block attached to a small fisherman's buoy to mark that anomaly.

Marx and his helpers in the small boat would immediately check it, diving down to see what caused the signal. Sometimes they would find a length of pipe, a small anchor, or part of an engine. If they could not find anything, we would later moor the *Grifon* over the spot and blast the sand away with the wash of its propeller. A few feet down we would find a piece of iron junk, often fittings left from a wooden sailing ship wreck or an anchor.

On the first day, my notes show we completed fourteen lines and found a V-shaped iron anchor at L 1556 / H 1405. It was a small beginning, but it gave us confidence in our search system. The next day, after covering thirty-nine lines each 1.2 miles long, we located another old anchor. The third day we found the first shipwreck under the sand; it turned out to be a ship of the 1800s.

We had some doubts about how well the magnetometer was working, and on the next trip home for supplies sought out Fay Fields, a local expert who had spent years in "magging" for treasure wrecks. He had once looked for the *Maravillas* from a helicopter moving at 50 knots, towing the magnetometer head just above the water. But he did not think much of that method, which was based on the dubious assumptions that the ship was in one piece and that it carried over twenty iron cannon and three anchors, which would produce a magnetic signal of several hundred gammas. He advised us to keep doing what we were already doing and look at every magnetic anomaly, however small. That was good advice.

The following day there was a storm at sea, so I flew to Washington for a meeting of the Naval Research Advisory Committee. This group of high-powered scientists and engineers was somewhat surprised to find one of their members was on a treasure hunt, but they were much interested and agreed that the scientific method would find a ship if it was there.

A few days later the *Grifon* was out on the bank again, and soon we were covering fifty-one lines (more than a square mile) in a single long day. One of the anomalies from which we blasted the sand turned out to be an old

Spanish anchor of about the right period with a waterlogged buoy carved from a log still attached. Perhaps it was that of a Spanish salvage ship. That completed the first tier, a band a mile wide across the heart of the most promising area. We had covered about 8 square miles, and I estimated that at least four more tiers or 32 square miles remained to be searched. *Grifon* returned to Fort Pierce for rest and resupply.

On August 3, we finished thirty-seven lanes and found three anomalies: 100 gammas, 90 gammas, and 60 gammas. Then a news broadcast mentioned a hurricane off Puerto Rico, and Marx wanted to head for home, but finally he agreed to remain if we took the small boat aboard. The hurricane never showed up, and we got thirty-five lines the next day, finding anomalies we numbered 19 to 23. This group of targets required the *Grifon* to spend the following day blasting holes to see what was buried in the sand.

The blaster was the invention that made this kind of searching possible. When the big pipe elbow was lowered and secured, the propeller was engaged. As its diverted wash hit bottom, there was a sudden billowing sand storm in the clear water. Clouds of sandy water began to move outward in all directions from a spot under *Grifon*'s stern. Divers directly below the blaster could dig into the bottom to hold position against the current it created, and the visibility was good enough so they could see wisps of sand tear away from the bottom as it did its work. One could burrow with bare hands and loosen the sand which would be swept clear, uncovering artifacts. Often we found iron deck fittings or old tools badly rusted and covered with coral sheaths. As soon as enough was found to date the wreck as post-Spanish, we would move on to the next buoy and blast there.

At one of these sites the sand was deep, and we had a chance to see how far into it the blaster could dig. In water 20 feet deep, the prop-wash created a conical hole whose bottom was 32 feet. It had 30 degree side slopes, rimmed with a mound of sand that came to within 16 feet of the surface. This raised the question of how we would handle deeper sand if that were necessary, and I began sketching caissons that we could install using OSE's vibrating corer system. Luckily they were not needed.

In this deep sand area we uncovered what seemed to be parts of two wrecks, one above the other, both dated about 1850. We knew these had been found by previous treasure hunters, because they had left their cinder-block anchors with buoy lines still attached.

Blasting anomaly 20 exposed a piece of pipe 30 feet long with an iron crosspiece near one end. This was rumored among the treasure hunters of the Florida Coast to have been put there by an unnamed con man to promote money from possible investors or, alternatively, to throw other treasure hunters off the track by masking a wreck beneath. In any case, it made Marx suspicious, so we dragged the pipe away and remagged the place, but found nothing.

Every day after work the divers would catch fresh fish or lobster for dinner; then we would sleep on deck under velvety skies. The sea was marvelously quiet once the generator engine was shut off for the night; we could hear the family of porpoises that had befriended us softly blowing as they passed alongside. It is hard to complain about a life like that, but after a month the daily grind of magnetometer towing, moving buoys, and checking anomalies begins to drag.

Now Marx had new doubts about the accuracy of the position given by the Spanish, and he restudied the "Phips' charts" made by the King's cartographer that showed the "plate wreck" position to be at 17 degrees 20 minutes. This was well north of where we were working, so he thought we should move up there next trip, after we had done another two days' searching on the west end of each tier. He also fretted about some of the Loran fixes, "because we're in the Bermuda triangle" (which allegedly affects electronic devices) and wanted to choose arbitrarily which ones to trust. I reminded him that our electronics had worked very well, and there was no need to worry about the supernatural.

Marx became increasingly morose and belligerent; he thought he should be in command of the expedition, although he had no suggestions for doing anything better or more efficiently than we were already doing. I learned later that the divers had become suspicious of the daily notes I made in my diary from which this account is taken. On our next trip to Fort Pierce, I arranged for Rhoda to stay ashore and fired all the crew. Then, after further discussions with Marx, I allowed him to keep two of them and replace the others.

On Friday August 25, we were back on the banks again in the southern tier, rechecking a very small magnetic anomaly of only 35 gammas that had been found the week before but had not been blasted. When we started the blaster, one of the anchors did not hold; Marx, in a vile mood, swam out to reset it and found potsherds lying on the bottom. When we started the blaster, almost at once it uncovered several old jar necks, a jade axhead, and a small silver tray with a pair of navigational dividers stuck to it.

The material was just becoming really interesting when Bert Webber, another treasure hunter, hailed us from a green boat a few hundred feet away. Bert, wearing bathing trunks and a revolver strapped to his bare leg, gave us his opinion that the wreck of the *Maravillas* that we were both searching for was further north. We did not argue when he claimed to have the exploration rights and waved good-bye as he headed north.

On the following day, the blaster removed enough sand to uncover a pile of disturbed ballast stones and two bronze cannon with the coat of arms of Philip IV. Near these were lead harquebus balls, clay smoking pipes, olive jars, and a few loose silver coins. Suddenly the work had become very exciting. The adrenalin flowed as all of us were on the bottom digging and moving ballast. Soon we found what had been bags of silver coins with the contents stuck together and a few of the large silver bars. Marx used twelve bottles of air in nine hours of diving that day; the rest of us used four to six bottles. "This," I recorded, "is the turning point of Seafinders."

Was this the wreck of the *Maravillas?* Bob cleaned the silver sulfide off a couple of coins, using vinegar and aluminum foil, and a date, 1655, became visible. We had the right wreck!

The next day Bert Webber was back, this time wanting to come aboard the *Grifon.* I said to Bob, "Let's go over and see him first before he comes here. We don't want him to know we've found anything." So we rowed over and chatted about the weather and the boredom of searching for treasure. Before Bert went on his way, he left a marker buoy of his own about a thousand feet away; by nightfall we had moved it a mile south.

On Monday August 28 I called the office of the Bahamian Ministry of Transport on the *Grifon*'s radio. Because Dr. Doris Johnson, the minister, was not in, her deputy, Hartis Thompson, took the call; I told him we had found the wreck of the *Maravillas* in one of the pinpoint leases and had a dated coin to prove it. Luckily, I made a tape recording of our conversation, because later he denied that I had reported the finding as the lease required.

Then treasure began to come up thick and fast: more bags of silver coins weighing 20 to 30 pounds apiece, a smashed pot, two silver plates, a third bronze cannon, small tools, and a pile of 288 silver coins. The captain's eyes bugged out with astonishment as these came aboard and piled up on deck. Driven by the fun of finding something new every few minutes, we all worked frantically. Very little wood was found; after 300 years underwater, a wreck consists mainly of stones, cannons, and silver. When I came up that afternoon, I was sore all over from lifting and prying; my hands were so knicked it hurt even to write notes.

The following day we started slowly and stiffly, but once we got into it, the treasure flowed again: 655 loose silver coins, one gold one, several bags of stuck coins, a sword handle, and five 60-pound silver ingots. That night we returned to Fort Pierce about midnight, loaded the treasure into cars, and headed for home. I awakened Rhoda who kept saying excitedly, "I don't believe it," as she helped drag boxes and bars of silver into our motel room. Then we collapsed into a deep sleep.

The next morning I called Bill Weaver in New York to tell him the news and notified the U.S. customs agent in Fort Pierce, who asked only if the wreck was in any country's territorial waters. I told the agent that it might not be, but we had leased the property from the Bahamas to avoid a legal battle over who owned this piece of submerged property so far from any land.

When Marx and I met later that day to discuss the next moves, he told me that he had found a sack of silver coins stashed amid dirty laundry in the back of the captain's car. There was no need to fire him; he did not intend to return. Before the day was out, at least three persons called to say that our ex-captain had offered to sell them the Loran position of the wreck. Disgusted with that perfidy, they had turned him down, but there was the possibility that he would find someone who would take his offer.

Treasure does that to a man. This fellow, who had often grumbled about how foolish we were to hunt for treasure and had tried to talk us out of looking for it, now decided he was entitled to a large percentage of whatever we found.

By that night every treasure hunter in Florida knew we had found the famous plate wreck, the lost *almiranta* of the *mimbres*. The next day we learned that one of the divers had skipped with a bag of silver as evidence of the find and was trying to persuade someone in the Bahamas to back him in a competing expedition. Then Marx told me that some imitation gold coins he carried along to use as photo props for magazine articles had been stolen from his cabin. This was not a crew to be trusted.

In the afternoon we rented a safe deposit cabinet at the First National Bank in Satellite Beach and moved the bars, bags, and coins over there. The bank official I talked to was reluctant to take the stuff because it "increased our chance of being robbed." This was the first of many such arrangements with banks.

Once the wreck had been found, I sent Marx and the *Grifon* with a new captain back out to continue the salvage. From here on, it was merely necessary to dive in clear warm water less than 30 feet deep and pick up whatever could be found as the blaster moved the sand off the treasure that lay on the hard coral substrata. Marx laid out a pattern the blaster would follow to uncover systematically an area of bottom. It was hard work moving the ballast rocks out of the way and being underwater for many hours at a time, but once the really difficult problem of finding the wreck was solved, the salvage itself presented no special problems. Among the artifacts we brought up was a length of modern diving hose with some fittings attached, obviously discarded by a previous finder, perhaps the one who found the silver bars Marx saw in the barrel of flour. We were not the first to find the *Maravillas.*

I stayed in Fort Pierce to organize the next steps. These included settling legal details with Real Eight Company and the Bahamians. There were custom forms to be filled out and people calling frequently who swore they had prior rights, or who wanted to worm into the deal, join the company, shake us down, or threaten us in various ways. It brought home to me the huge difference between being a treasure hunter and a treasure finder.

With hindsight Bob Marx came to the conclusion that he and he alone had found the wreck of the *Maravillas.* This ignored the facts that the search system we had been following for six weeks had eliminated almost all the rest of the sandy area where the wreck was likely to be and that the reason the *Grifon* was at that specific location was to blast on the small magnetic anomaly we had found the week before. The only iron on the wreck turned out to be an old Spanish small-boat anchor.

I would not have found the wreck without Marx, and he would not have found it without me and the Seafinders organization—certainly not by the methods he was using before I came along. But most treasure hunters have superegos which need a lot of reinforcement, and I was not surprised by his unwarranted claims. My policy was to keep the find as quiet as possible for as long as possible; public credit could wait.

TREASURE OR JUNK?

The most immediate problem was to bring the treasure being picked up each day back to Florida and get it in a bank to reduce the very real threat of hijacking. Once it was known we had found the famous wreck, nearly every day a light aircraft would circle above the *Grifon* or a boat would pass nearby; my movements ashore were watched, and radio conversations with the *Grifon* were monitored by various groups. So I decided to get a fast boat to move the treasure back to Fort Pierce under cover of darkness. For this Seafinders bought *Seabird,* a twin-engined planing boat that would skim across the usually rough Gulf Stream at 25 knots.

Seabird was just right for clandestine operations, and I began running out to the *Grifon* after dark so as to attract minimal attention, taking a mechanic with me, just in case. If all went well, I could be back in Fort Pierce by 2 A.M.—but sometimes all did not go well, and a couple of those trips nearly ended in disaster. With only a compass for navigation and no radio aboard to call for help, there were risks in running some 80 miles each way across the Gulf Stream at night. The boat was new, and although it would move very swiftly when both engines were running, on the first crossing, one engine overheated and had to be shut down. Then it no longer planed on the surface but slogged along at about 6 knots. The *Grifon* and the treasure wreck were almost due east of Fort Pierce, but the fast-flowing Gulf Stream carried the slow-moving boat far north of its intended destination.

Finally I made out the light of a flashing buoy ahead and ran in close alongside it so as to read the number. It was buoy number 41, the northernmost tip of the Bahamas bank; the next land was the other side of the Atlantic. Once the *Seabird* was on the bank, out of the main current, it plodded south for some 15 miles until the faint masthead light of the *Grifon* pricked the black night ahead. Next day we transhipped the treasure that had been recovered, and I headed back, slowly. Boatyard mechanics removed an obstruction in one of the new engine heads that had caused the overheating.

Two days later Marx called in by radio with the message, "I've caught a big grouper." That was our code phrase meaning there's a lot more treasure ready to be moved ashore. Again I took the *Seabird* across and again had almost the same trouble as before; this time the boat picked up a floating sheet of plastic in one of the cooling-water intakes. Again I found buoy number 41 and missed a second opportunity to make a trans-Atlantic drift voyage.

After that, the crossings became routine, and on each return trip *Seabird* carried plastic trays full of silver coins, bits of Chinese porcelain, 19 small silver bars (contraband), seven 80-pound silver bars, a handful of small emeralds and amethysts and three golden suns (also contraband). The suns were rough segments of spheres about 4 inches across, apparently made by pouring molten gold into a depression made in the sand by a cannon ball. The last and largest item was an elephant tusk. How it came to be part of the cargo of a galleon headed east, one can only guess.

Every day a complete list of the material recovered was updated, and each trip back added to the pile of stuff in the bank vault. We were completely frank and honest with the Bahamians, so they would have no reason to become suspicious or alter the terms of our concession. Dr. Doris Johnson, the minister of transport, decided to fly to Fort Pierce to visit the treasure in the vault. The pile of old silver, rough and blackened with a sulfide covering, plus a small amount of gold, could not have been very impressive, but she thought the Bahamian government could set up a nice tourist display with their one-fourth share. She seemed pleased with what we had done and decreed that none of the stuff could be withdrawn without her representative, Rick Penn, and me both signing. In the meantime she said, he would be on our payroll along with an "inspector" she would assign to live aboard the *Grifon* and keep an eye on what we were doing.

On my next trip to Nassau, Mr. Penn suggested we should give him $60,000 for insuring that our lease would continue, hinting that this would go to unnamed higher-ups. We managed to maneuver past that one, although we never paid any bribes and none were directly requested, rumors of such goings-on spread; an election was coming up, and this was an attempt by rival politicians to embarrass the government.

Almost from the day we first found the wreck, we had asked the Bahamian government to provide help by keeping pirates away, specifically for a boat to guard the site when the *Grifon* returned to Florida for fuel, supplies, and to rest the divers. But when they eventually sent a guard ship, the *Lady Moore,* its duty was to watch us, not to protect the site from others. It would leave when we did, and we worried about leaving the site unguarded. We were not sure the Bahamians could even find the site if we weren't there, and we doubted that its crew could recognize treasure when they saw it. Once when *Grifon* raised a flag, they reported to the minister that we had been "signaling to confederates."

After questions were raised in the Bahamian Parliament, I was summoned to see Prime Minister Pindling, who said, "You must bring all the things back to our country. That is the law." I hesitated for a moment, reconsidering the advice from our lawyers not to do so and wondering how the shareholders in New York would react. Then I tried to persuade the PM what I firmly believed, namely that "passersby would not bother to steal the stuff if it were dumped on the courthouse steps. It's junk."

Like others with an inflated opinion of the value of old Spanish silver, the PM said, "Mr. Bascom, you have been careful not to be rude but I understand you very well. You must make me a report giving details of location, depth, what has been found, etc." Of course I agreed, marveling at how few layers of bureaucracy separated the bottom and the top in this small country.

On October 6 at 9:30 P.M. the *Grifon* was just pulling into its usual berth in Fort Pierce. I was sitting on a log under a street lamp at the pier making out paychecks for the crew when someone came up behind me and said, "I want to talk to you."

I answered, "I'll be with you in a minute," and kept writing, not knowing the speaker was holding a pistol to my head. Then a disturbance at the *Grifon* caught my eye, and I looked up to see armed men swarming over its seaward rail from a boat that had come up along its outer side.

My first thought was that these were hijackers, but actually it was a raid by special U.S. Treasury agents and the U.S. Coast Guard, both of whom had lots of questions. It seemed they were the only ones in Florida who did not believe our story of Spanish treasure. They were astonished to find the Bahamian inspector aboard and to learn I had spoken with the customs chief on the telephone only that afternoon. To be sure, they searched the boat thoroughly, even breaking open some of the frozen fish in the freezer to make sure nothing was hidden inside. Apparently one of our friends had tipped them off that the treasure story was a cover for smuggling dope.

On October 11 we made what was supposed to be a firm agreement with the Bahamian Transport Ministry to divide the treasure recovered to date during the first week in November. That was OK with Seafinders, but the Bahamians insisted that all the material would have to be moved to the Royal Bank of Canada in Nassau. We were not enthusiastic about that, but it seemed to be the best chance to resolve the matter. Somehow the information leaked that treasure would be coming out of the bank, and the Fort Pierce sheriff's office got a tip that a gang of men was preparing to hijack it during the transfer from the bank to the West Palm Beach airport.

As we moved the sacks and bars on hand trucks out the door of the bank and piled them into a rented truck, the special police officers I had hired for the day spotted some of the would-be hijackers parked across the square, watching our moves. They were outgunned and gave up but we were nervous, especially when halfway to the airport, the truck broke down. We transferred the heavy silver to the trunks of the two police cars; with bodies sagging down on the overburdened springs, they made it to the airport, where the stuff was loaded into a twin-engined Beechcraft. When the plane landed in Nassau, we were met by an armored car, but it too broke down, and we transferred the silver again, reaching the Royal Bank of Canada just before its vaults automatically locked for the night.

As we arrived, a group of gorgeous dark women in bright dresses (who were in town for a convention of models) blocked the bank's front steps for a few moments in what could have been a scene out of a James Bond movie.

I rented three large deposit boxes and, with the help of bank clerks, began transferring the load of silver for the sixth time. Just then all the lights in Nassau went out. Did this presage a bank heist? No. It was only a power failure, but the timing made the bank officials nervous. By the light of lamps and candles, receipts were typed out that said only Mr. Penn and Mr. Bascom together could remove the material. Exhausted from manhandling silver all day, we retired to a bar.

It soon became apparent that the division of treasure would not happen anytime soon, a breach of faith by the Bahamians they preferred to describe as "a misunderstanding." Although Seafinders assets were still considerable, if you counted the treasure we didn't really have, the company was out of cash for operating expenses. Its original shareholders were unwilling to risk more money if we couldn't get our hands on the share of the silver we'd already found.

By this time the adventure and glamour of Spanish galleon treasure had faded considerably, and I began to think anew about the actual value of our finds. Assuming these would be fairly divided eventually, 25 percent would go to the Bahamas. Then 10 percent would go to Real Eight Company for the lease, 15 percent to someone to clean the coins properly and make them salable in an auction, and the auctioneer would get another 20 percent. Until that sale the material had to be stored, transported, and guarded; that would take 5 percent more. (That's 75 percent so far.) Moreover, the special dates on the coins that had once made them rare would now be plentiful (we had found an estimated 12,000 pieces of eight by then), and their value to collectors dropped accordingly.

The best estimate of treasure value I had been able to make was about $400,000 in 1972 retail prices. That seemed like a pretty good return for a few month's work, but with 75 percent of the value committed, before the expense of search and recovery was deducted, there was no way that Seafinders, Inc., would break even. Operating expenses continued. Rhoda and I also had lost much of our faith in human nature as many of the people we had been associated with had lied, stolen bags of silver, and tried to sell the wreck location. In spite of our attempts to be scrupulously honest, the Bahamians had gone back on their promises.

I had tried to avoid publicity, but Marx was bursting to get public credit for "his" find, and in December he got carried away while talking to an Associated Press reporter. He was directly quoted as saying the "Bahamians had cheated us out of $2 million in treasure." It was "our wreck and we would guard it with machine guns." Then he talked to a CBS reporter on the phone, this time being a little stronger with his language, and a recording was broadcast in the Bahamas in which he said, "those black . . . bleeps . . . are not going to steal *my* treasure." Our Bahamian lawyer cabled to advise me to "put a muzzle on Marx," but by then the damage was done. I was in a poor negotiating position when I talked with the minister.

The worst problem was Marx's exaggeration of the worth of the material brought up. He had agreed that $400,000 was a pretty good guess as to its value; now that he had publicly raised that to $2 million, the Bahamians wanted to know what we'd done with the other $1.6 million worth.

After months of watching a pile of junk silver inspire greed, pettiness, and deceit, I decided the fun was over and resigned as the unpaid president of Seafinders, Inc., and walked away from my share of the treasure. We had loved our life on the Bahamas Bank, and Rhoda felt it was one of the most exciting summers we ever had.

I had often said that every red-blooded boy should experience at least one good treasure hunt in his life ("good" meaning with fun and adventure). The *Maravillas* was mine.

XII

Ocean Pollution?

SCCWRP

Most of the world's environmental problems have arisen because of the unrestricted growth of the human species and its technology. This has caused pollution of various kinds on land (including lakes, rivers and underground water), in the air, and in mostly land-locked tidal estuaries such as Chesapeake and San Francisco bays. This leads to the important question of whether the ocean, which covers a large part of the earth, deserves to be called polluted.

The ocean is not without substantial natural defenses against any acts of man or nature. Its vast size and constant motions mean that anything added is quickly diluted to very low concentrations; its great depths and the constant rain of particles to the bottom means that sedimentation will soon cover any pollutants that reach the ocean floor; its chemical nature prevents it becoming either too acidic or basic; too oxydizing or reducing. Now that international laws prevent the hunting of rare mammals, none of the hundreds of thousands of species of oceanic animals and plants seem likely to be extincted by man (as is happening on land).

Ocean plants and animals are accustomed to dealing with natural circumstances that would be called pollution if man were involved. These include the addition of hot mineralized volcanic water, natural leaks from oil reservoirs, the presence of billions of tons of metals and of salt, the existence of many toxic organic chemicals (manufactured by algae to protect themselves against bacteria), and the fact that ocean creatures necessarily live in a thin soup of their own fecal material that might be described as natural sewage.

Just because the ocean has these natural defenses does not mean that man should deliberately test them. We must be careful about materials that are released to the sea in substantial quantities and high concentrations because it is possible to pollute a small area for a short time. For example, plastic and paper objects should not be disposed of where they will drift on beaches. Oil should not be deliberately released into the sea (although the detrimental effects of crude oil on the ocean environment have been greatly exaggerated, there is no need to push our luck). High concentrations or large quantities of some synthetic organic chemicals should not be disposed of in the sea (however, the release of most chemicals at low concentrations will not cause environmental problems). Let me repeat that I am speaking about the open ocean and its exposed coastal waters, *not* mostly land-locked bays and estuaries such as Puget Sound and New York or Boston harbors.

During a 12-year long study of the chemical and biological effects of waste water discharged into the ocean via long pipes off southern California I learned some things that are contrary to widely accepted beliefs. These suggest that some of the actions now being taken to prevent pollution in exposed coastal waters are counterproductive and may actually increase our environmental problems on land, in the air and with energy use.

My intense involvement in environmental matters began in 1973 when John Isaacs suggested I take over the direction of the Southern California Coastal Water Research Project (SCCWRP), where he was chairman of the scientific advisory board.

SCCWRP, he explained, is an environmental laboratory that was set up to get some scientifically reliable data about the effects of municipal wastewater on sea life in coastal waters. The cities of the Los Angeles-San Diego region discharge over a billion gallons a day of primary-treated waste water offshore. Some environmentalists think that the effluent should receive additional (secondary or bacterial) treatment. The SCCWRP scientific board believes that is not a sensible thing to do because it will create worse environmental problems on land.

When public meetings are held to discuss the matter, these often end in shouting matches between the dischargers and the environmentalists, because both sides have very strong opinions about the effects of discharging waste water into the sea but neither has any facts to support its position. SCCWRP's job, Isaacs continued, is to remain neutral while obtaining scientific data that will settle those arguments. It has a budget of half a million dollars a year, a small staff of scientists, and a commission of publicly elected officials who are supposed to assure that there will be no interference by dischargers or politicians no matter what the findings are. Other members of the scientific Consulting Board contributed other pieces of the story.

Professor Erman Pearson, head of the sanitary engineering department at the University of California at Berkeley and an international consultant on ocean waste disposal, explained the results of treatment. The sludge (solids) removed by primary treatment is a natural substance, easily disposed of and widely used as plant fertilizer. But secondary treatment creates twice as much sludge of a very different kind that is not beneficial for crops and has a water content too high to permit it to be put in most landfills. It would have to be trucked long distances to acceptable disposal sites or, if it were incinerated, it would require a lot of energy. Either way the result would most likely be either air or groundwater contamination.

Professor Pearson continued, the southern California public works agencies who are responsible for the discharge are proud of their outfalls which are widely copied in other parts of the world and are generally believed to be a success. (Recent examples are in Sydney, Australia and Rio de Janiero, Brazil). These are concrete pipes up to 12 feet in diameter that extend from 1 to 5 miles offshore into water about 180 feet deep, where they end in diffuser sections, often over a half-mile long.

Diffusers are large pipes with small holes in the sides every hundred feet, through which jets of grayish effluent spurt into the ocean. Upon release, the discharged fresh water, which is relatively light and warm, mixes with the seawater while rising until the mixture is the same density as the surrounding sea water (about halfway to the surface). In a matter of minutes the effluent has been diluted at least 150 to one. Except when winter storm waves temporarily destroy water stratification (and double the dilution) none of the waste water reaches the surface.

This initial dilution results in all waste constituents meeting the state's ocean water quality standards, and the dilution continues to increase as the mixture drifts away with the ocean currents. Dilution is indeed a solution to pollution.

Professor Don Pritchard, a physical oceanographer from the Chesapeake Bay Institute, explained how the dilution was calculated; Dr. Richard Lee of the Hawaiian Public Health Department spoke of the absence of any cases of disease transmitted by bacteria in the sea; and Dr. John Ryther, an experimental biologist from Woods Hole Oceanographic Institution, minimized the importance of shifts in invertebrate populations which are a very local indication of changes in food supply.

Those five scientists were pretty much of one mind about the condition in the sea off southern California, having discussed the subject repeatedly for three years. They agreed that (1) The existing wastewater disposal practices would not have any substantial adverse or irreversible effects on the environmental quality of the Southern California Bight if persistent pesticides such as DDT were controlled at their source. (2) The secondary treatment of sewage would be of little value for discharges into open coastal waters because these do not need the oxygen that process adds and the additional reduction of solids is of minor value, (3) More scientific study was needed to understand the effects of various waste constituents on sea creatures.

They knew that the large but unquantified discharge of residues from the manufacture of DDT into the sea over a period of years had caused substantial harm to sea animals, but they noted that the principal source of it had been shut off when the Montrose Chemical Plant had stopped discharging through the Los Angeles County outfall in 1972. They were of the opinion that the DDT levels in the environment would steadily drop and in twenty years or so would be gone.

I accepted the job and with the help of the staff began studying the coastal waters of Southern California in a new way. We set up a chemical laboratory and began taking our own samples and making our own measurements of metals and synthetic organic chemicals (mainly DDTs and PCBs) so we would have independent data on their extent and influence. My attitude was one of suspicion towards all sides in the argument, including dischargers, regulating agencies, and environmentalists, all of whom seemed more interested in reinforcing their past positions than understanding the actual effects of each waste component on ocean life.

Scientific studies confirmed that two kinds of problems had been caused by outfall discharges. One was that about 10 percent of the solid particles released by the outfalls tended to stick together and settle to the sea bottom within a few miles of the discharge point. This had the effect of altering the composition of the bottom muds and their invertebrate populations in elongated oval patterns down-current from the discharge. Off the Palos Verdes peninsula, the extra solids in the water had cut off the light to deeper kelp plants for several years, temporarily destroying a kelp forest that later recovered.

The other was that the damaging effects of the DDT residues would be around for a long time. Local seabird populations, mainly brown pelicans and cormorants, had been greatly reduced; sea mammal pups were sometimes stillborn; many species of fish (especially those with fatty tissues) that lived near the Los Angeles County outfall carried high levels of DDT in their bodies that probably caused fin erosion disease and tumors. But from the beginning our measurements showed that DDT levels in animals were steadily dropping; animal and plant health was improving. By 1980 these populations had regained their former numbers.

SCCWRP set up a research plan whose object was to answer questions: How much of each possible pollutant is discharged by each outfall each year? What are the patterns of pollutants in the sediments and how do these change the corresponding animal populations? How much DDT, PCB, metals and other chemicals are present in various organs of sea animals? and What is the rate of decay of pollutants in the ocean? What range of conditions can be considered normal and how much do the bottom and the animals that live close to outfalls vary from that norm?

Our objective was to understand; routine monitoring measurements were made by others. The need for new kinds of data required the SCCWRP group to pioneer a series of techniques for navigating, sampling the bottom and its animals, reporting waste characteristics, quantifying invertebrate communities, studying internal changes in fish, and attaining statistical validity. Before long, other laboratories and monitoring groups copied our methods, and many of these became standard in environmental work. We made the first detailed biological and chemical charts of the bottom off southern California and reported our scientific findings in detail in annual and biennial reports. These were written and illustrated in a style intended to make them readable by non-scientists, but they were eagerly sought by scientific laboratories in other parts of the world who used them for guidance.

BASIC PRINCIPLES

Collecting data on animals and measuring changes in bottom chemistry is essential, but it alone does not bring understanding of the effects of pollutants. That requires a good deal of thinking and making comparisons with other places in the world. One must ask what the new information means and how it relates the health and abundance of the sea life around an outfall to that of a similar area where there is no discharge, and to the ocean as a whole. But first let us agree on a set of definitions; in SCCWRP reports the following apply:

The *ocean* is the big blue area on a globe that covers 72 percent of the earth and has a volume of 1.37 billion cubic kilometers. It does *not* include rivers, lakes, or shallow, mostly landlocked bays and estuaries whose volume is insignificant by comparison. It does include *open coastal waters* that are exposed to the winds, waves and currents of the ocean. *Pollutants* or *contaminants* are substances that can cause toxic effects if they are present at high enough concentrations; their mere presence at a very low concentration does not mean there is pollution.

Pollution, briefly, is a "damaging excess." Its international (UNESCO) definition is: "The introduction of substances into the marine environment resulting in harm to living resources, hazards to human health, hindrance to marine activities, including fishing and impairment of seawater quality." The *Marine Pollution Bulletin,* a distinguished British journal, noted that this definition "has the great advantage of establishing the concept of harm as a necessary element in pollution. If damage is to be established it must be shown to occur on a scale that can be related to the changes produced by other, natural, environmental variations." The US Congress stated the subject more succinctly: "Our waters should be swimmable and fishable."

Although tiny amounts of pollutants can be detected practically anywhere in air, water and land (with modern chemical apparatus it is possible to identify many chemicals that are present at less than one part per billion), it would be extremely rare for any of these to reach damaging concentrations even in a very small area of ocean.

The generally accepted method of deciding if environmental damage has been done at a specific location is to compare the suspected area with another similar area not affected by man and look for differences. The pristine area selected because it has similar depth, bottom conditions, wave exposure, etc., is known as a "control". The assumption is that plant and animal life in the control area, away from man's activities and pollutants, is the way nature intended that kind of an area to be. Any differences between the suspect area and the control are assumed to be the effects of pollution.

That concept sounds simple but in practice it is hard to find a similar pristine area; when one does the animals living there sometimes have more health problems than those living around outfalls or places alleged to be polluted. For example, the livers of fish living off San Simeon, California on the unpopulated open coast are in much worse condition than those that live close to outfalls.

No one knows why these biological anomalies exist, but part of the reason seems to be that the natural variability in sea life from place to place and from year to year is far greater than was estimated before good measurements were made; this variability requires scientists to take replicate samples of the same animals of the same size and sex at the same location to work out the range of normalcy. SCCWRP studies required from 3 to 25 replicates, depending on the funds available and the need for accuracy. We found that the variation between animals at control sites is often much greater than in areas where pollutants are present.

Another consideration is that the process of evolution has given all animals defenses for withstanding natural toxicants in their food supply; they are likely to consume greater quantities of these than of man-made toxicants. The natural toxicants are produced by plants intent on protecting themselves against bacteria in the sea, and scientists such as Dr. William Fenical at the Scripps Institution of Oceanography have spent much of their careers identifying complex chemicals in algae that are at least as toxic as DDT.

Media accounts often refer to "the delicate balance of marine ecosystems" and how easily these can be disrupted by man but professionals are much more likely to use the word "resilient" to describe the way these ecosystems have quickly restored themselves after being disrupted. Referring to the public concept of life in the sea as a trophic pyramid with food pathways like those on land, John Isaacs wrote:

In this imaginary textbook picture, plants initiate the cycle and various herbivores, primary and secondary carnivores, are each born into their alloted niche in the pyramid to constitute a vast, delicately balanced structure. A house of cards, ready to collapse into lifeless chaos with the slightest jiggling by the profane hand of man.

Regardless of the possible veracity of this picture on land, this idealization of the living systems of the open sea has misled us into a hysterical, "the ocean is dying" syndrome and irrational proposals and actions.

Actually almost all of the creatures of the open sea are born into a highly variable system in which they have little assurance of the nature of their prey, predators, associates, or competitors during their span of existence. They find repast in anything grossly accessible to their mode of feeding and are eaten by anything to which they are available. Individually and collectively they are opportunistic, unspecialized, and resilient.

Partly as a result of testimony before the House Subcommittee on Water Resources of the Committee on Public Works by Isaacs, Pearson, myself, and others, the 301h waiver regulation was added to the Clean Water Act. It permitted dischargers of wastewater to the ocean to avoid secondary treatment if their discharge met certain standards. This took into account the important differences in geography between, say, Colorado and Florida or Illinois and Hawaii. It is not sensible for inland states and island or coastal states to take exactly the same action. We thought we had made some progress, but this was an illusory triumph.

Far from following the will of Congress, the Environmental Protection Agency fought it by creating unreasonable requirements that could not be met and rarely waived secondary treatment for ocean dischargers as was intended.

That agency showed little concern that its policy would create worse long-term effects on air, land, groundwater and energy use. Nor does it seem to be bothered by the fact that drinkable water on land is put at risk by its attempt to protect open coastal waters that do not need further protection.

SOME ENVIRONMENTAL MYTHS

Thousands of kinds of chemicals in small amounts reach the ocean every day by various routes; it makes sense to narrow the problem by asking which ones we should be most concerned about. Let me answer that by pointing out some widespread misconceptions about "toxic" materials that are discharged through pipelines into moderately deep water well offshore in southern California. The three types are sewage, metals, and synthetic organic chemicals.

Myth 1 is that ordinary human sewage is harmful to animals in open coastal waters. As director of the Institute of Marine Resources, John Isaacs was incensed by regulations that attempted to control the discharge of human wastes into the ocean. His opinion was that "The return of organic waste and plant nutrients resulting from the most natural of acts to the sea is most probably beneficial. The benefits of putting the same material on land is clear to any farmer but the advantages to the sea are not so easily appreciated. The sea is *starved* for basic plant nutrients and it is a mystery to me why anyone should be concerned with their introduction into coastal seas in any quantity we can generate in the foreseeable future.".

John liked to point out that the anchovies living in southern California waters, then estimated at 3.5 million tons, produced five times as much fecal material as the 10 million humans living along southern California coast, and they were only one of thousands of species of animals living there. "Why should the human product be worse?" he would ask dramatically. "Don't you know that most sea animals live in a soup of fecal material and feed on it directly?" He scoffed at homeowners who happily import concentrated sewage sludge from Milwaukee (Milorganite) to fertilize their gardens but who claim to be disturbed by the idea that diluted Los Angeles sludge is discharged 7 miles offshore into moving water. Isaacs also favored the disposal of primary sewage effluent into some parts of the Mediterranean to help replace the loss of the Nile's nutrients after the construction of the high dam at Aswan. Of course, the release of sewage into the sea must be done thoughtfully, miles offshore, with great dilution, as it is done off southern California.

As most fishermen know, fish are more plentiful around sewage discharges. Nowhere is this more evident than around the Santa Monica Bay sludge outfall where, over a period of years, SCCWRP's bottom trawls consistently brought up more fish, more species of fish, and larger individual fish than any place else we trawled on the coast. It seemed clear that sewage particles (about one-third of which are uneaten food scraps) are beneficial, especially after Dr. Alan Mearns and Jack Word analyzed the data and published the fact that California's ocean outfalls increase the amount of sea life in direct proportion to the amount of solids released. They showed that the releases into Santa Monica Bay are responsible for an additional 5,600 tons of fish and other animals.

The sludge outfall produces an environmental effect much like that caused by the volcanic vents on the deep sea floor discovered by the research submarine *Alvin*. The black smoke that issues from those natural "chimneys" is made of metallic sulfides; the nearby bottom is a reducing environment; and the nutrients released support large numbers of specialized clams and worms.

Hundreds of high quality color photos of the fish and the bottom of Santa Monica Bay show its clarity and scenic beauty to be equal or better than that of exhibition aquaria at marine parks; one picture of the sewage being discharged from a diffuser pipe covered with a garden of *metridia* (sea anemones) and surrounded by rock fish was used for the cover of *Sea Technology* magazine.

Fecal material from all mammals contains large numbers of harmless coliform bacteria; each of us carry billions of them in our guts. They are present nearly everywhere in coastal waters but if their number rises above 1000 per 100 milliliter of seawater, that is taken as an indication of the possibility that pathogenic (disease causing) bacteria could also be present. Of course, if there are no epidemic diseases in the population ashore (usually there are not) few pathogens are likely to reach the ocean.

When I first looked into the question of pathogens in the ocean I asked about the health records of those who work in the Los Angeles sewage treatment plant and are exposed every day to an atmosphere made misty by churning sewage. Their health record was at least as good as other city employees.

Because 90 percent of the coliforms discharged die off in the first hour, it is virtually impossible for them to reach the beach in quantity after being released miles offshore. Even when the coliform count rises above the accepted standard, the chance of a swimmer getting a disease is essentially zero. In studies made of the Point Loma kelp bed about a mile offshore where excess coliforms were occasionally reported, Dr. John Conway of San Diego State University interviewed divers (the most exposed group of ocean users) after 1,371 dives. These people, mostly male, dived an average of five times a month to depths of about 50 feet. Ninety percent of them took seafood for eating, and a quarter of them ate it raw, sometimes while underwater. Their rate of gastrointestinal symptoms was 8 per 1,000 dives, which is less than half the number considered by to to be unsafe on bathing beaches. Dr. Conway concluded that the health risks are low.

Myth number 2 is that "heavy" metals added to the ocean are damaging to sea life or humans. These are often spoken of as "toxic" metals because under some special circumstances they can be toxic; however it would be extremely rare to find any indication of metal toxicity in the ocean. The ocean is the natural repository of fantastic amounts of metals and it seems strange that anyone would think that tiny additional amounts added by man (especially of the metals that humans are in intimate contact with every day) would cause a problem.

There are already about 12.6 billion metric tons of arsenic in the ocean, 3.7 billion tons of zinc, 2.9 billion tons of copper and 1.8 billion tons of chromium (based on world ocean average concentrations given by Professor Ken Bruland of the University of California at Santa Cruz). Somehow the plants and animals of the sea live very well in this metallic solution. But at one hearing on an offshore oil drilling platform an environmentalist alleged that this operation would cause pollution by discharging an additional 40 pounds of arsenic a year. Actually, at the site of that platform, the sea water flowing past the platform every day contains over 40 pounds of natural arsenic.

Nearly every home has copper or zinc-lined water pipes and chromium-plated bath and kitchen fixtures; we eat the contents of tin cans with silver tableware. This means practically everyone is constantly exposed to very low concentrations of these so-called toxic metals with no known effect. But we are told that sea creatures will be harmed if equally low concentrations of these same metals are discharged into the sea or are present on the sea bottom as essentially insoluble sulfides. In the latter case the metals must be extracted with strong acid before they can even be measured, an indication of their unavailability to sea animals.

Something is wrong with this picture! Present environmental regulations reflect a great misunderstanding of metals toxicity in the sea. In order to follow those regulations that are not only useless but sometimes counterproductive, our country spends billions of dollars to prevent trivial additions of metals from reaching the ocean.

Why do regulators and environmentalists think metals are toxic in the sea? First, there is the oft-repeated horror story of the mercury disaster at Minimata, Japan. In the late 1950s methyl mercury chloride, a rare organic form of mercury which is thousands of times more toxic than ordinary metallic mercury and is also soluble in body fats, was discharged directly into Minimata Bay from a plastics plant. There it was passed up the food web from plankton to copepods to fish to people, doing terrible damage to animal nervous systems as it went. But most mercury in use today is in the virtually non-toxic liquid metal form used in thermometers, not in the organic form. Now that the world has been warned it is hard to see how a Minimatalike accident could happen again.

A careful study made for the National Science Foundation by Dr. David Young and his assistants at SCCWRP showed that metals (except the organic form of mercury which is natural in ocean fish at low levels) do *not* move upward through the marine food web and do *not* bioaccumulate (remain in the body of the predator). Fish living around outfalls that discharge metals have no more metals in their tissues than fish caught in mid-Pacific.

Many experiments run by marine laboratories in small tanks of *filtered* seawater demonstrate that, if free ions of metals are present at high enough concentrations, they can be toxic. But these tests do not represent reality in the sea because free ions of metals are rare in coastal waters. If introduced, they quickly attach themselves to the huge numbers of particles there and are no longer easily biologically available.

In some EPA experiments clear water was contaminated by metal ions, and at high concentrations, these did indeed cause toxic effects on test animals—*except* in tanks where fine particles were added to the water. There the metal ions immediately attached to the particles and were no longer toxic.

All animals from the tiniest worm to man automatically regulate the amounts of metal in their bodies. Some metals, such as zinc and copper, are needed for the production of enzymes, and any unneeded excess of these metals is stored on a protein named metallothionein made by several vital organs. Other metals, including those that can be toxic in other situations and are not required for body functions, (with the possible exception of lead which is more difficult to measure), also become attached to the metallothionein. This natural protection effectively detoxifies any metals that are ingested by preventing them from reaching tissues where they might cause damage.

Metals discharged from the reducing environment of a primary treatment plant are usually in the form of virtually insoluble sulfides and are essentially inert. But secondary sewage treatment can convert these compounds to soluble oxides. To quote Isaacs' Congressional testimony on how secondary treatment makes the situation worse: "The toxicity of metals is closely related to their valence state, and it makes no sense at all to reduce the metal content of a discharge by 60 percent only to increase the toxicity of the remaining metals by factors of thousands. Chromium and mercury are cases in point where the usual low valence states are essentially nontoxic, but their more oxidized states are violent poisons."

Myth 3 is that the tens of thousands of synthetic organic chemicals that are in use and have been labled toxic can be harmful to sea animals at the concentrations that can be reached in the ocean. It is obvious that no one should release substantial amounts of these materials into the sea. The question is: Are small amounts, inadvertantly released or released at very low concentrations, a matter to be concerned about?

The toxicity of various chemicals, determined by extensive tests on white rats, is published in the *Registry of Toxic Effects of Chemical Substances.* The *Registry* lists the amount of each chemical in parts per million required to produce a median lethal dose (MLD). Some chemicals require concentrations of 10,000 parts per million to kill half the rats tested; others are so lethal that only a few parts per million are needed. It would be very rare for an animal to get a dose as high as a part per million in the ocean where concentrations are rarely over one-thousandth that level. However, if the chemical is also soluble in an animal's body fats (lipids), it can bioaccumulate.

So one must ask about the relative solubility of each chemical in lipids and water. If the chemical is readily soluble in water, it has very little chance of hurting an animal because even if some is taken into its body, it will simply pass on through. But if the chemical is soluble in body fats, it can be retained in the animal and, over a period of time, can accumulate in vital organs until a high concentration is reached. This bioaccumulation of synthetic organic chemicals such as DDTs and PCBs can make them dangerous. Eventually, when the animal is eaten, the fat-soluble chemical will be retained in its predator. In time, the chemical may reach a concentration that can cause toxic effects, especially in the animals at the top of the food web such as birds, swordfish, porpoises, and people.

By combining the toxicity data with the lipid solubility data one can discover which chemicals are not only very toxic but readily bioaccumulate. This is done by dividing a number that represents the solubility of each chemical in lipids by its MLD toxicity to obtain a figure called a Toxic Bioaccumulation Factor. The resulting number has no significance in itself. Its value is that it helps set priorities on effective toxicity, the largest numbers being the most dangerous.

The final step was to find out how much of the chemicals with the higher effective toxicities are actually discharged into the ocean. When we did that we learned that only seven chemicals had any reasonable chance of being a problem in southern California coastal waters: DDT, DDE, DDD, Arochlor 1242, Arochlor 1254 (the previous two are PCB's), hexachlorobenzine, and pentachlorophenol. These happened to be ones we had been tracking for some time and knew were steadily decreasing. Thus my midnight fears that dangerous unknown chemicals were being discharged that might surprise us some day were groundless. All of these compounds are broken down by metabolism, attach to particles and are buried in the bottom, and are steadily diluted to lower concentrations. This is how the ocean cleans itself.

Myth 4 is that there is a substantial risk of getting cancer from eating fish that contain trivial amounts of carcinogenic chemicals. I once used the generally accepted formula to calculate that if a person ate the national average (9 grams) of fish a day containing the average amounts of DDT and PCB found in Santa Monica Bay fish (and continued this for 70 years), his chance of getting cancer was one in 13,888. Again, that number has no special significance; the reason for making such a calculation is to find out if there's a large risk or a small one. In this case it was easy to see that the chance of getting cancer from eating fish was very low by comparison with other risks, including that of getting skin cancer from being exposed to the sun while catching the fish (about 1 in 5).

The statistical chances that any individual American will die of any of the following causes during a 70-year lifetime are useful for comparative purposes. There is about 1 chance in 70 that a person will be killed in a motor vehicle accident, one in 200 that he or she will be murdered, one in about 300 that any individual will die in a fall, or of drowning, or in a fire, or of lung cancer (For heavy smokers living in cities it's 10 times that great or about 1 in 30). Relative to those figures there is essentially no risk in eating the fish in question. Unfortunately the news stories did not explain how low the risks really are and the bad publicity greatly damaged the local fishing industry.

Dr. Bruce Ames, chairman of the biochemistry department at the University of California at Berkeley, winner of the Tyler Ecology Prize, and a member of the governor's scientific advisory committee on toxic chemicals, agreed. He said, "Very little cancer is caused by man-made chemicals and pollution; most cancer probably results from naturally occurring mutations. Eating a fish contaminated with 100 times the average level of DDT or PCB would contribute a possible hazard that is comparable to eating a peanut butter sandwich and is very small compared to other common minimal risks, such as drinking a glass of beer". Elsewhere he commented that "a far more serious risk comes from having a glass of wine with dinner because it contains a known carcinogen, alcohol, at a concentration of 50 million parts per billion."

POLITICS AND WASTE DISPOSAL

All the evidence SCCWRP gathered over a dozen years of intensive study showed that the condition of California coastal waters and the resident animals steadily improved after the DDT discharge was stopped in 1972. By 1984 the amount of DDT going into Santa Monica Bay was barely measurable (0.03 part per billion), and the levels of DDT in fish and mussels was only one tenth of what it had been a decade earlier. Of course the metals discharged over the previous twenty years or more were still on the bottom (I had personally mapped their concentration and location), but now we were sure they did not harm. In any case, both the DDT and the metals were steadily being buried by natural sedimentation. The kelp and the pelicans were back and the rigorous discharge standards set by the State of California's Ocean Plan were being met. In public presentations I said repeatedly, "Our coastal waters are not perfect but they're pretty good and steadily getting better", a demonstrably true statement but one that irritated environmentalists who imagined rampant pollution in Santa Monica Bay.

Suddenly, in 1985, the pollution of coastal waters became a political issue; unsupported allegations, some of which were over 20 years out of date, caught the attention of the news media and were treated as exposés. These made better headlines than the dull, complicated truth and they stirred controversy.

At one public hearing, to make sure my ideas were properly reported, I passed out twenty-eight pages of graphs and data that documented the steady reduction of pollutants in the discharge and the improvements in the animals in the bay. These were ignored. Politicians and the media preferred to make a scare story out of the very remote possibility of getting cancer from eating the fish caught in the bay even though these met the EPA and the Department of Agriculture toxicity standards.

Anyone who wishes to check the scientific data will find that the concentrations of toxic bioaccumulating chemicals are very low and still falling in southern California waters, that animals in open coastal waters are not intoxicated by metals, that bacteria have rarely, if ever, caused a serious disease to ocean swimmers, and that sewage is acceptable food for detritus-feeding sea animals.

These are examples of the myriad of facts that the public and its officials should consider before risking the land and air environments (and spending billions of dollars) to make what may not be improvements in coastal waters. Other things could be done with that amount of money and effort that would better improve the environment of our coastal region. But, as Albert Einstein pointed out, "A confusion of priorities characterizes our age."

Annoyed by the sustained ineptitude and political motivations of the regulatory agencies, John Isaacs testified before a Congressional subcommittee that those agencies had "adopted the pose of the medieval churches, with regard not for what is true or right, but rather for what defends their established policies and for what supports their own delusions of power, omniscience, and infallibility. The beleaguered scientist who has evidence of the fallibility of these agencies or the underlying faults in their regulations, can only recant his findings (if he wants any more research support) and content himself with the muttered aside, "Nevertheless it moves", as did Galileo, following his confrontation with the awesome forces of the hierarchy of his time."

In July 1986 an all day symposium on the future of the world ocean was held to celebrate the twenty-fifth anniversary of the University of Rhode Island's Graduate School of Oceanography. It was sponsored by the Providence *Journal* and moderated by Edwin Newman, formerly of NBC news. The panelists were Dr. John Guilland, of Imperial College, London, a world expert on fisheries; Dr. Kenji Okamura, special assistant fo the Japanese minister for science; Cecil Olmstead, an international attorney, expert on the law of the sea; Dr. Dixy Lee Ray, a zoologist, formerly governor of Washington and Assistant Secretary of State for the US Bureau of Oceans; Admiral Stansfield Turner, ex-president of the Naval War College and ex-head of CIA; Dr. John Knauss, dean of the graduate school of oceanography and former chairman of the National Advisory Committee on Oceans and Atmosphere; and me.

To the surprise of the 300 people present, plus a television audience and the sponsors, this group was of the unanimous opinion that ocean pollution was pretty much a myth. Dr. Guilland noted that "in the Mediterranean, fish are far more abundant in allegedly polluted harbors than in clean open waters." Dr. Ray said that "coastal waters are now far cleaner than they were a few decades ago," and Dean Knauss pointed out that "the environmental movement, in order to make its point, has overstated its case. People in that movement should take their responsibilities more seriously." I noted that many people confuse the ocean with small bodies of inland waters such as Narragansett and Chesapeake Bays that do require careful management. The enormous volume of the ocean and its self-cleansing properties make it very difficult to pollute more than a very small area for a short time.

A way must be found to replace environmental decisions based on imagined problems and alleged damage with ones based on scientific findings and common sense. The overall environmental problems of ocean, air, and land must be addressed simultaneously so that so-called solutions do not merely shift the problem from one medium to another. Whatever problems of pollution exist in inland bays and estuaries should certainly be fixed, but confusing these with the virtually unpolluted ocean is not of benefit to anyone. Our country is entitled to more thoughtful, long-range environmental plans that do not play on irrational fears, or squander energy and land, or spend huge sums on low-priority problems.

NUCLEAR WASTES

One special category of waste that might be disposed of beneath the bottom of the ocean is the high-level nuclear wastes created by power plants. At the moment this material is being accumulated at the nearly 100 plant sites spread all over the country, in some cases close to cities or drinking water supplies. Over a long period of time it surely is more of a hazard to keep these wastes where they are generated than to move them to centralized storage locations.

Already some 10,000 tons of spent nuclear fuel has been accumulated in this country, which needs to be permanently buried somewhere. In 20 years that will have risen to 100,000 tons—and three times that much for the world as a whole. Action will be required soon.

There are several technically acceptable locations on land for these wastes in deep, dry, abandoned mines in Missouri and New Mexico, whose surrounding rocks seem not to have moved or changed in 250 million years, or at the nuclear test sites in Hanford, Washington, or Nevada. Two problems have so far prevented them being used. The first is the question of exactly how the radioactive waste will be transported from the power plant that created it to the burial location with minimal risk. This leads to the second difficulty, namely that there are public objections to hauling it across many states or burying it there. Every state governor stands ready to fight any proposal for either of those acts, apparently without regard for the national welfare. The London Dumping Convention bans the disposal of radioactive wastes in the sea unless it has been "rendered harmless by containment".

The ocean has several inherent advantages as a nuclear waste repository: The depths of the sea (several miles of seawater) offer excellent shielding from nuclear emanations. These wastes generate heat, and there could be no better dissipating medium than the cold water of the ocean bottom. If there were to be a leak, maximum dilution is available and of course, there are no people for hundreds of miles in any direction. Experience with the rapid dissipation of radioactivity from nuclear weapons tests suggests that there is very little chance of damage to marine resources in the remote depths of the ocean.

Because a large number of nuclear power plants are located near the ocean's rim, the risks of a transportation accident are greatly diminished. Finally, there is no governor who can immediately veto the idea of using the ocean.

For many inland nuclear plants, it is doubtless best to use mines to bury nuclear wastes, in accordance with careful studies on which considerable effort has been spent. But the ocean bottom should be considered an acceptable alternative that reduces some of the risks. The rule should be that at any site, there must be no evidence of volcanic or seismic activity, and the sedimentation has been continuous for a million years. The waste must meet the requirement of the London Dumping Convention that it "has been rendered harmless by containment."

The United States should continually reconsider the option of ocean bottom burial for some of its nuclear wastes. If it is used thoughtfully, neither the ocean nor mankind will be damaged.

...ceanographers gathered at ...öteborg, Sweden, to plan for the ...ternational Geophysical Year. First ...o rows, left to right: Anton Brunn, ...enmark; Roger Revelle, U.S.; J. M. ...arruthers, U.K.; me; Le Clerese, ...ance; Columbus Iselin, U.S.; Georg ...ust, West Germany; Hakon ...osby, Norway; K. Hidake, Japan; ...me. Trotskaya, U.S.S.R.; Carl ...ossby, Sweden. Last row, center, ...hn Lyman and Maurice Ewing, ...th U.S.; the tallest man in that row ...Jacques Piccard, Switzerland. ...ordstrom)

The house at Pirae, Tahiti, a wonderful combination of split bamboo and coconut frond, where we lived during most of 1957. (WB)

...nitra, Rhoda, and me dining ...Bora-Bora. All restaurant ...ployees and all customers on ...e island are included in this ...cture. (WB)

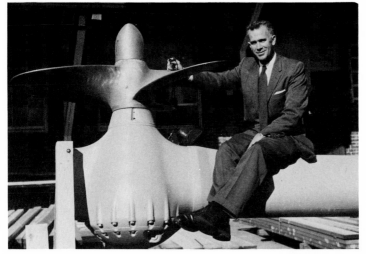

One of the outboard moto
used in the first dynamic-
positioning system. (NAS)

Surveying across the
Namib Desert's sea of
sand with a theodolite
and a tellurometer to
establish navigational
beacons for diamond
prospecting ships. (OSE)

April Fool!
Cuss it
J Steinbeck

John Steinbeck, who shared a
cabin with me aboard *CUSS 1,*
wrote "April Fool! Cuss it"
on this Polaroid photo I took
of him on that historic day in
1961.

The console on the bridge of *CUSS I* with a central joy stick that simultaneously controlled the four outboard motors. (NAS)

Rockeater drilling for diamonds off the Skeleton Coast. (OSE)

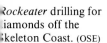

Colpontoon, a diamond mining barge, was driven ashore in Chameis Bay by a great storm. In trying to prevent that beaching, the *Collinstar*, a large tug, was lost with all hands. (OSE)

OSE's senior engineers. Left to right, back row: Larry Brundred, Bob Snyder, Skip Johns, Ed Horton, Francois Lampietti, Jack McLelland. Front row, Pete Johnson, me, Al Dimock. (OSE)

One of the WEBS buoys (built to protect delicate instruments against the shock waves of a nuclear explosion) about to be launched from an LSD. (OSE)

A 15-foot-long sea monster photographed at a depth of 1,490 meters in the eastern Mediterranean near Cyprus by John Isaacs monster camera. (Scripps)

Isaacs and me fishing for a sea monster off Baja, California. (OSE)

The *Tam Huu* in the backwaters of the Saigon River with its original crew aboard. A month later it had been converted to surveying offices and living quarters in Cam Ranh Bay. (WB-OSE)

An OSE survey party was working on the bank of the Mekong River when American troops arrived. In a few hours the machine gun at right was set up and a new bar and brothel established. (OSE)

Makeshift hydrographic survey boat on Cam Ranh Bay, Vietnam. (OSE)

Sounding a Vietnamese River required us to be constantly vigilant. (OSE)

Artist's conception of our cable-stayed-girder bridge for the Strait of Gibraltar built on tension leg platforms, 1970. (OSE)

Mock-up of the Deep-Oil-Technology, Inc. wellheads and submersible work chamber that was later installed underwater. Prospective customers could operate it themselves. (OSE)

The Jacobson's *J-Star*, used to find and raise the wreckage of a 727 from Santa Monica Bay. (OSE)

OSE's underwater dredge on the beach at Fort Pierce, Florida. (OSE)

The *Alcoa Seaprobe*, the first deep-water search-and-recovery ship. It pioneered the pipe-handling system that was used by the CIA. (ALCOA)

The prop-wash diverter used in the treasure hunt. When the ship is at anchor and this big elbow is swung down into place, the moving water from the propeller "blasts" away the sand beneath and can excavate a sizable hole. (WB)

A morning's find of treasure on the deck of the *Grifon*. Pieces of eight are in piles of 10 and 50. The six large silver bars weight about 80 pounds each. (WB)

The U.S. Maritime Administration's concept of a wind-driven oceanographic research ship named the *Willard Bascom*, operating in Antarctic waters. (Nixon Galloway)

Synthetic-Aperture-Radar image made by Seasat of the Columbia River entrance an Willalpa Bay showing internal waves on the continental shelf.
(Jet Propulsion Laboratory)

XIII

From the Past to the Future

WIND SHIPS

Toward the end of 1979, the world experienced an acute fuel shortage and a steep rise in oil prices. Marine diesel fuel had increased from $3 a barrel in 1972 to $34 and was still climbing. One effect of this was that the U.S. Oceanographic fleet had a $6 million overrun in fuel costs. This resulted in a couple of American scientific ships being held up in Chile and Peru while their sponsors scrambled for funds to buy enough fuel to bring them home. When the matter was discussed in a meeting of the academy's Ocean Science Board, I suggested that it might be worth while to reconsider using wind-powered ships in the tradition of the *Atlantis,* the Swedish *Albatross,* and the *E. W. Scripps.*

The board regarded this as a novel idea and asked me to form an ad hoc panel of experts to determine whether it was feasible and sensible to build a sailing ship for deep sea oceanographic research that would perform all the usual scientific functions while using one fourth the amount of fuel of the present ships and without extending the time at sea of the scientists and crew.

Six experts on modern sailing ships were chosen as panelists, including Lloyd Bergeson of Wind Ships Development Corporation, Corey Cramer, master of the *Westward* and head of Sea Education Association, Captain R. T. Dinsmore, superintendant of the Woods Hole fleet, Dr. Gustaf Arrhenius of Scripps who had sailed around the world on the Swedish *Albatross,* and Frank MacLear, specialist in automated sail handling.

Using data on the eleven largest existing oceanographic ships in the American fleet (156 to 245 feet long), our panel estimated their average annual fuel cost at $424,000. We felt that worthwhile savings could be achieved if ships of this size could operate on wind power without increasing other operating costs.

More important, the panel found that, in the roughly 80 years since the age of large commercial sailing ships came to a close, there have been very substantial technological advances in materials, sailing technology, navigational methods, and the understanding of hydrodynamic drag. Most of these advances could be used to create an entirely new kind of vessel that would perform all the needed oceanographic functions. Its job would be to measure, sample, core, and trawl in really deep water while providing comfortable living quarters and convenient laboratory space for months of ocean living.

The new materials included such things as aluminum for deck houses and masts, polyester fabrics for sails, stainless steel or Kelvar fiber for lines, and improved paints to reduce fouling. The new sailing technology included a boomless and sparless fore-and-aft rig, combined engine and wind propulsion, powered sail-furling, and tension-limited sheets that would prevent the ship from being blown over. Power distribution could be accomplished by SCR rectifiers for speed control of electric motors, adjustable pitch propellers, and hydraulic motors on deck winches. Automatic gyro pilot steering, possibly guided by a computer using satellite information, would be used to lay a course hundreds of miles in advance through the most favorable winds. And, of course, engines to meet any requirement were available, because it was clear from the beginning that some combination of sails and engines would be the best answer. All together these items would give a tremendous advantage to a modern sailing ship over those of the past.

The mental picture many persons have of men aloft on the yard arms of a square rigger rounding Cape Horn in a storm, trying to reef sails barehanded a hundred feet above the deck, was not what we had in mind. The ship we proposed would have no yards or men aloft; sails would be handled by automatic winches controlled from the bridge. The hull would be strengthened to withstand encounters with ice, because it would probably operate at high latitudes where encounters with bergs were likely.

The invention that made automation possible was the use of triangular sails that could be wound up like a window shade around a head stay. When we began, "luff-roller-furled-sails" were widely used as jibs for small sailboats and several larger vessels, especially the 125-foot-long *Audelia,* an English yacht which we thought made a very good model for our design.

In order for the new ship to maintain a schedule in low or adverse winds, it would need to be able to move under power at about 12 knots like the other ships of the oceanographic fleet. Its advantages were that it saved fuel when the winds were favorable, and it had a higher speed (15 to 20 knots) when the wind was strong and the course along a good point of sailing.

A few words about sailing may be helpful to those who do not quite understand how a sailboat can go in any direction its skipper chooses in spite of the wind direction. It is evident that a sailing ship can go in the same direction the wind is blowing, but sailing "into the wind" seems mysterious to some. Thousands of years ago the lateen sail, the keel, and the rudder were invented to make that possible. A cloth sail, whose leading edge is attached to a nearly vertical spar, forms an airfoil section similar to the one that causes an airplane to fly. This "lifts" the ship, but in a horizontal direction. The ship sails forward for the same reason that a glider flies when moving air flows over the curved surface; basically, there is more air pressure on one side of the sail than the other.

Although sail boats or ships cannot move directly into the wind, some can sail to within 45 degrees of its direction by setting the sails to obtain lift in that direction; then, by tacking, alternately right and left, they work their way upwind. The question for each hull and sail plan is, How close to the wind?

One advantage of a motor-sailer is that with the ship partly under power, it can sail closer to the wind, perhaps 30 degrees in special conditions. The ship we had in mind would be supplied with satellite information on the winds for hundreds of miles in all directions, and it will know its position precisely. If this data is fed into a shipboard computer that steers the ship along calculated courses and automatically sets the sails, wind power can be a very efficient means of travel.

Many of the arguments favoring sailing that are given here apply mainly to oceanographic ships under 4,000 tons that linger in remote high-latitude seas where there is ample wind and no need to make port for months at a time. Hauling cargo under sail in a larger ship is quite another matter and does not have many of these advantages.

The panel agreed on a set of ship "performance specifications" that met the original requirements of saving "50 to 80 percent of the fuel of a comparable-sized motor ship without requiring more crew" on some voyages. We duly reported these to the board, but it bothered me that our generalized plan and specifications were hard to discuss without a specific ship plan in mind to look at, so I sketched a ship to illustrate our proposal.

The Ocean Science Board then published our work in a small blue booklet that featured my sketch and suggested that appropriate government agencies sponsor a detailed design.

The booklet was distributed widely in Washington, and about a year later I received an invitation from Constantine Foltis and Tom Hooper, naval architects at the U.S. Maritime Administration in Washington, to stop around and talk about a sailing ship they had designed based on our report. There, I was surprised to find they had located a set of "lines" (the shape) of a sailing ship hull that had been "towed" by David Taylor Model Basin and the drag characteristics determined. On this hull they had made great progress in the design of living and laboratory space, the sail plan and machinery arrangements. MarAd's ship would be 315 feet overall, 50 foot beam, 2,900 tons loaded, carry 24,750 square feet of sail on four masts, and support forty-seven people for sixty days, half of whom would be scientists.

After several sessions in which we discussed design details and oceanographic operations with such a ship, they unfolded the final version of their plan. It showed a beautiful new kind of ship, and they had generously printed "Willard Bascom" across its bow.

At this writing the price of marine diesel has dropped back to about $18 a barrel, but the time will come again when fuel costs are important, and new oceanographic ships will be needed. Let us hope someone remembers this design and makes use of it. Already there are cruise ships such as *Wind Song* that look remarkably like the final design that MarAd named after me.

SATELLITES AND OCEANOGRAPHY

"Oceanography from a satellite? To a generation of scientists accustomed to Nansen bottles and reversing thermometers, the idea may seem absurd." So wrote Dr. Gifford Ewing, long a proponent of oceanography from aircraft, after the first conference on satellite oceanography at Woods Hole in 1964. Columbus Iselin agreed. The next year Arthur Alexiou of the Naval Oceanographic office suggested that satellites could "provide global perspective which could possibly lead to new understanding of oceanic processes" and the "capability to survey areas normally inaccessible on a large scale basis (Arctic, Antarctic, and South Pacific)."

In 1968, at the request of NASA, the National Academy of Sciences conducted another more thorough study of the subject; again the findings were recorded by Giff Ewing, aided by a group of "Nestors" (after the wise man who advised the Greeks during the siege of Troy). "The satellite," they said, "is the only feasible vehicle from which the whole surface of the sea can be viewed. Satellite oceanography is inherently directed toward observing the upper layers of the sea, the part that is stirred by the wind and lit by the sun. Fortunately the layer of ocean exposed to that overview is by far the most significant because it directly touches the lives of most mankind."

With great prescience, that report identified the kinds of measurements that could be made and specified in general terms all the instruments that are available today. It pointed out the problems ahead of acquiring "sea truth" (making independent measurements at the sea surface to confirm or calibrate the satellite's measurements), and of persuading oceanographers to use the results. "The great strength of the satellite observatory," it said, "is its ability to look at the world ocean on a time scale that is small compared with that of many important dynamic processes."

Developing and testing instruments to make the needed measurements, finding appropriate vehicles for each one, and obtaining financial support consumed another decade. Finally, the national program came together in Seasat, a satellite launched in June 1978 that was "dedicated to establishing the use of microwave sensors for remote sensing of the earth's oceans."

Seasat A was placed in a nearly circular orbit at an altitude of 800 kilometers that covered the earth between the latitudes of 72 north and 72 south. It completed 14.3 revolutions of the earth every day and covered 95 percent of the global ocean every thirty-six hours with two of its wide-swath sensors. Data from four of its five sensors were recorded on board for later transmission to a Space Tracking Data Network receiving station. These gave continual global coverage of surface winds, sea surface temperatures, wave heights, atmospheric water, and ocean topography. The fifth instrument, an SAR (explained presently), acquired data at a rate that precluded storage and could only be turned on when the spacecraft was in view of one of the five tracking stations.

Seasat A was a triumph, except for two things. It failed because of a massive electrical short circuit after only ninety-one days of operation. Second, although it produced marvelous new information about the ocean, the deluge of data frightened many scientists. Like the sorcerer's apprentice, they had unleashed a flood. Years later some were still struggling to analyze the information obtained by the three-month-long experiment.

At that time I was a member of the Ocean Science Board of the National Academy of Science, enthusiastic about this new ocean capability but with no special knowledge of it. In 1980 I was appointed chairman of a subgroup called the Panel on Satellite Oceanography, charged with writing a report for NASA on the role of satellites in oceanography. Then began over a year of intently studying what had been accomplished so far and what the future possibilities were. Here are some of the highlights.

Images from spacecraft had already done much to change simplistic concepts of the oceans. Great currents were observed to take unexpected directions; open holes appeared in the polar ice in winter; huge masses of plankton, previously thought to be a coastal phenomenon, were seen in midocean; topographical features 2,000 meters deep affected the motions of surface water; and there were large depressions in the ocean surface caused by variations in gravity. Let us look into these remarkable findings in more detail because they illustrate the new possibilities.

The movements of the Gulf Stream in the Atlantic and the Kuroshio across the North Pacific have been known for a long time, or so we thought, until satellite findings changed our views. Generally the Gulf Stream moves north and east just off our Atlantic coastal shelf, shedding warm rings of water a hundred kilometers across as it goes. With the help of the infrared radiometers carried on the Tiros-N and NOAA-6 satellites, the tracks of a dozen of these warm-core eddys were followed for several years. They were found to move in the opposite direction to that of the Gulf Stream, and endure from about 1 to 9 months. Eddy number 79-B for example, moved nearly 600 miles west and 300 miles south in 6 months, a totally unexpected result.

In the Pacific a satellite was used to track four drifting buoys towed by parachute drogues 100 meters below the surface. These were launched amid the high speed core of the Kuroshio just off Japan, and for a while all four followed a well-known meander of that current. It was expected that they would continue on along a roughly great circle route across the open Pacific toward Vancouver Island, Canada. Instead, one went north into the subarctic gyre, one went south to the Philippines, and two went eastward and circled Jingu Seamount whose top is 900 meters deep. Surprise!

Although the California current is generally believed to flow southward offshore of the California coast and outside the Channel Islands, satellites taught us some new things about it. West of Cape Mendocino a rocky escarpment, whose crest is 2,000 meters deep, extends due west for 50 kilometers or more. One would expect that to be too deep to influence surface currents, but an infrared satellite scan showed that this deep ridge was forcing cold water to the surface.

In another experiment two buoys deployed in the California current off San Francisco were tracked by a satellite. One went due north to the Columbia River entrance and then turned out to sea; the other went south of Point Conception before turning westward at a place where it had been expected to turn eastward. Surprise again! Obviously a lot of complexities in ocean currents remain to be unraveled.

The first overall view of the Arctic and Antarctic ice sheets was produced by the scanning microwave radiometer on a satellite named Nimbus-5. This device made some unanticipated discoveries through thick clouds and the polar night, as it investigated the question of how much open water exists amid the polar ice packs at various times of the year. In late summer 1974 a succession of these open areas, called polynyas, one as large as England, moved across the Arctic Basin. In the Weddell Sea (Antarctic) a huge area of open water has existed for a number of years. Curiously, it seems to grow in seasons when one might expect it to freeze over. Because the heat flowing from open water is over a hundred times greater than that flowing from ice-covered water, these events are significant to scientists studying the earth's heat balance. The why of polynyas is not yet known.

In cloudless daylight, an instrument known as the color scanner measures the wave lengths of light radiating from the water surface to determine the amount of chlorophyll (a measure of the phytoplankton population) present. This device was intended for coastal waters, but when it was turned on over the Central Pacific in deep water, great masses of phytoplankton were observed. These data have already resulted in an upward revision of the estimate of ocean productivity.

The scatterometer measures radar reflectivity of the sea surface in all weather. It permits the velocity and direction of all the winds of all the world to be determined. Imagine how valuable this can be to fishing fleets, cargo or passenger ships, towboats or sailing ships (like that described) that are looking either for favorable winds along their routes, or trying to avoid storms.

Seasat carried a radar altimeter that could measure the distance from the satellite to the water surface with a precision of about 7 centimeters (3 inches) from 800 kilometers in space. Because the satellite flew in a smooth orbit and the relative elevation of the geoid (the surface of the sea at mean sea level) was known, variations from normal in the height of the sea surface could be mapped. One of Seasat's most impressive discoveries was a depression in the surface of the ocean over the Puerto Rico Trench that is 15 meters (about 46 feet) deep. A hole in the top of the sea? It is caused by an increase in the earth's gravity where the dense rock of the mantle comes near the surface.

Seasat's ultimate instrument, the one that overwhelmed even its enthusiasts with data, is the Synthetic Aperture Radar. As the satellite moves in its orbital path, the SAR's antenna scans a 100-kilometer-wide strip of earth or ocean below and measures the backscatter reflections. In order to resolve objects on the sea surface at such a long range with a resolution of, say, 10 meters, a very long linear antenna is needed that cannot be carried on a satellite. So the SAR uses a short antenna moving rapidly along a sequence of positions (the flight path) to form a long one synthetically. When the "phase history" of the data from those positions is processed later, a high-resolution image of the sea surface is produced. Every object larger than 10 meters (waves, ice floes, ships, and rocks) shows on what looks like a high quality photograph. Someday, when a capability is developed to "crunch" data (process it conveniently) at 10,000 times the present rate, the SAR may be the principal instrument of oceanography.

All the above findings show the importance of satellite measurements to oceanographers. There are also other benefits of satellites to seagoers, including precise navigation with NAVSTAR, radio communications via satellite links, and a search-and-rescue system that detects emergency signals and relays them to shore stations where the position of the ship in distress is computed.

Our panel listed a dozen objectives for NASA's oceanography effort, but we put specific emphasis on the development of a capability to "measure sea level on a routine basis to an accuracy of plus or minus two centimeters, relative to the geoid." When that is available, it will be possible to track all the world's currents by measuring the slope of the sea surface. Even in the ocean, water flows downhill.

I thank fate and Dr. Warren Wooster, then chairman of the Ocean Science Board, for giving me the opportunity to learn a little about satellite oceanography, but in recent years my interest has waned again, probably because this is not a very adventurous form of oceanography. Although the scientific findings can be intellectually exciting, it takes a special sort of person to sit at a computer all day sorting the data from a big dish antenna pointed at the sky or to spend years figuring out algorithms for converting the millions of data bits into useful information about what the ocean is doing.

One of the problems with satellite oceanography is that not many scientists have been attracted to it; at the time of the study we guessed the total number might be forty. This means that not enough people are talking about its findings and hopes, or pressuring NASA and Congress to contribute more funds for new instruments, satellites, and graduate students. The result has been slow growth of what is undoubtedly the most promising new aspect of oceanography.

HOME FROM THE SEA

Many other stories could be told about scientific adventures in the ocean during the last few decades, but those presented here are sufficient to convey the scope and technical complexity of oceanography. During my watch it expanded from a small club of curiosity-driven but poorly equipped scientists into what could be called a scientific industry today. Guided mainly by intuition and a few simplistic measurements, these fellows laid the foundations of ocean science.

The sea was no rougher then than it is now, but with smaller ships, not designed to do scientific work at sea, our predecessors bounced around more, got wetter and colder, and took more risks to get data. When Sir Hubert Wilkins' oceanographic submarine *Nautilus* had serious mechanical problems in the Arctic and seemed to be in danger of sinking, its crew got out on the ice. When they noticed that Dr. Harald Sverdrup was not among them, a search found him in the forward compartment of the submarine with its chamber open to the ocean below, still measuring water temperature and salinity. Curiosity about the nature of the Arctic Ocean had priority over personal safety.

My generation of oceanographers, which began work in the 1940s, had the good fortune to deal with what was, from a scientific and technological point of view, a nearly virginal ocean. Driven by scientific curiosity, we explored the same seas as the pre-nineteenth century adventurers who used their opportunities to chart previously unknown lands. Our objectives were to learn the nature and mechanics of the ocean and the rocks beneath it; in a different way, our findings were as important as theirs. I hope that the spirit of adventure and the thirst to be first will continue in the generations that follow us.

On the maiden cruises of the modified *CUSS I,* the *Alcoa Seaprobe,* and the *Gulfrex* I spoke to the scientists and crews of those ships in romantic words to involve them psychologically in the objectives of those expeditions: "We sail in the tradition of Columbus and Magellan, looking for new knowledge that will benefit mankind." Like Columbus and Magellan we had naysayers. These pessimists warned us about modern forms of sea monsters and the waterfall at the rim of the earth that represented the technically impossible. But, as our predecessors did, we found that those stories were invented to frighten the fainthearted. The impossible turns out not to be so difficult once someone boldly takes the risks and shows how it can be done.

Then the same people who once claimed the project was impossible decide that it was really very easy.

My ambition to be in the front line of ocean science and technology was fulfilled by being present on the first major geophysical exploration of the Pacific basin with modern instruments, by observing the effects of the largest nuclear explosions, by drilling for science in oceanic depths, by exploring for diamonds and metals off several countries, and by pioneering new kinds of ships, searching methods, deep water archaeology, and environmental studies.

Over the years I became a member of a number of scientific societies and government/military science advisory committees. There were also three clubs with a scientific bent: The Cosmos Club, the Adventurers' Club and the Explorers Club. It had been a boyhood ambition to belong to the Explorer's Club and rub elbows with the heroic fellows who had led the way to the earth's poles, examined its jungles and deserts, climbed its highest mountains, and visited space. Each year since 1906 the club has awarded its highest honor, the Medal of Exploration, to such famous men as Robert Perry, Roald Ammundsen, Auguste Piccard, Sir Hubert Wilkins, General Jimmy Doolittle, Neil Armstrong and Edmund Hillary. At the 1980 annual banquet in the Waldorf Astoria in New York I was honored with the medal for my part in the exploration of the ocean.

One problem that scientist-explorers face is how to communicate what they have done and learned to a public, which has traveled little and does not have much understanding of geography or the nature of the earth. For example, few persons who have not personally crossed the ocean on a small, slow ship appreciate its size. The *Queen Mary* is tied up in Long Beach, California, and guided tours end up on her afterdeck from which Catalina Island, twenty-five miles away, is easily visible. Occasionally visitors ask the tour guide, "Is that Japan?"

I hope that these stories will (1) help give a new perspective to the scale of the ocean and the perception of what oceanographers do and (2) attract new talent to ocean sciences and ocean engineering. If some young readers join the next generation of ocean researchers and doers, this effort will be well rewarded, for there is much more to be done in each of the dozen or so areas of ocean work that have been mentioned.

It was a lot of fun being one of the small group who rode the crest of the wave through the golden years of oceanography. My principal regret is that many of those associates are no longer around to reminisce about those exciting early days and the way our ideas developed as we met with triumph and disaster. Some of those who did survive have been helpful in jogging my memory and straightening out the stories reported here. My recollections were also refreshed by many boxes of papers, photos, notebooks, and diaries that accumulated over the years.

Probably there will be less personal risk in future ocean studies, even as my generation risked less than those who sailed unknown seas in caravels. The new ships and submarines and aircraft are much better designed and built; communication and navigation are easy and precise; equipment is much more reliable; technicians are better trained. This means the chances of accomplishing something on schedule and returning from expeditions with large quantities of good data are greatly improved.

My intention from the first had been to learn everything I could about ocean work rather than confine my studies to any one scientific specialty or technology. This strategy was something of a disadvantage in the world of academic specialists, but that was more than offset by the value of seeing the ocean sciences in broader perspective and being able to apply the concepts and scientific tools of other specialties to the problems at hand.

As John Isaacs, my generalist friend, once said in an address to the Pacific Science Congress, "The scientific hierarchy demands deeper penetration of nature, not broader comprehension. Yet it is the development of increasing breadth and comprehension as well as penetration that we must espouse if we are to understand the complexity of nature. Our crucial task is to learn how the pieces fit together."

No one can foresee the future very well; usually prognostications fall far short of what actually comes to pass, but I will suggest a few technologies that might be used to make man's harvest of the ocean more efficient. Perhaps a new material can be devised (now jokingly known as nonobtainium) that will be light, workable, corrosion-proof, and with several times the strength of present materials. Possibly fish or other marine animals can be genetically changed to improve their size and edibility as much as turkeys have been improved. Maybe duterium can be extracted from the sea and be used as a fuel or that the heat of undersea volcanoes can become a practical source of energy. Perhaps some atoll lagoons will be made into huge fish farms and special structures something like oil platforms will be built in coastal waters as fish havens. Possibly sea barriers will be constructed on a grand scale, capable of holding back rising sea level at coastal cities, or creating new and larger harbors, or making perimeters around airports in shallow waters. Perhaps it will be possible to obtain panoramic pictures of large undersea features so we can directly see what trenches, faults, and canyons look like. I fully expect that deep water oil operators and bridge foundation builders will make use of tension leg platforms and that archaeologists will find complete ancient wrecks in deep reducing environments using the techniques I have suggested. That is the system: dream about what could be useful; then work to convert dreams into reality.

The next generation would do well to follow the suggestion of Thomas Carlyle: Go as far as you can see; then you will be able to see further. When you do, may you have as much fun as I have had.